NANO-SIZED MULTIFUNCTIONAL MATERIALS

SYNTHESIS, PROPERTIES AND APPLICATIONS

Edited by

NGUYEN HOA HONG

ELSEVIER

Elsevier
Radarweg 29, PO Box 211, 1000 AE Amsterdam, Netherlands
The Boulevard, Langford Lane, Kidlington, Oxford OX5 1GB, United Kingdom
50 Hampshire Street, 5th Floor, Cambridge, MA 02139, United States

Library of Congress Cataloging-in-Publication Data

A catalog record for this book is available from the Library of Congress

British Library Cataloguing-in-Publication Data

A catalogue record for this book is available from the British Library

ISBN: 978-0-12-813934-9

For information on all Elsevier publications visit our website at https://www.elsevier.com/books-and-journals

Publisher: Matthew Deans
Acquisition Editor: Simon Holt
Editorial Project Manager: Leticia M. Lima
Production Project Manager: Joy Christel Neumarin Honest Thangiah
Cover Designer: Greg Harris

Typeset by SPi Global, India

CONTENTS

CONTRIBUTORS

Timur Sh. Atabaev
Department of Chemistry, School of Science and Technology, Nazarbayev University, Astana, Kazakhstan; Department of Physics and Astronomy, Seoul National University, Seoul, Republic of Korea

Orion Ciftja
Department of Physics, Prairie View A&M University, Prairie View, TX, United States

Xun Geng
School of Materials Science and Engineering, UNSW, Sydney, NSW, Australia

Souraya Goumri-Said
College of Science, Physics Department, Alfaisal University, Riyadh, Saudi Arabia

Nguyen Hoa Hong
Department of Physics and Astronomy, Seoul National University, Seoul, Republic of Korea

Nguyen Que Huong
Department of Physics, Marshall University, Huntington, WV, United States

Mohammed Benali Kanoun
Physics Department, College of Science, King Faisal University, Saudi Arabia

Varsha Khare
Institute of Advanced Machinery and Design, Seoul National University, Seoul, Republic of Korea

A.T. Raghavender
Department of Physics, Electronics, and Communication Engineering, Nishitha College of Engineering and Technology; Abhigyaan Labs Private Limited; Aksharaarka Enterprise Private Limited, Hyderabad, India

Sanjiv Sonkaria
Institute of Advanced Machinery and Design, Seoul National University, Seoul, Republic of Korea

Jiabao Yi
Global Innovative Centre for Advanced Nanomaterials, School of Engineering, The University of Newcastle, Callaghan, NSW, Australia

PREFACE

Matters behave very differently at the nanoscale. Our book will give an overall view on how materials can be downscaled to nanometer size in order to tailor and control their properties. These advanced, low-dimensional materials including quantum dots, nanoparticles, and ultra-thin films, which are certainly targeted to have multifunctional properties. After demonstrating how to downscale those materials to nanosize in order to make them become multifunctional, our chapters will then show the readers how new technological methods can be used in electronics, medicine, energy, and the environment.

The authors of this book, who have been working at many universities all over the world, will bring to the readers a fresh approach in materials research. The authors expect this book to be a valuable reference resource for materials scientists, materials engineers, chemists, physicists, and bioengineers who want to learn more about the remarkable properties of nanosized materials.

ABOUT THE EDITOR

Dr. Nguyen Hoa Hong is professor and head of the Nanomagnetism Lab at Seoul National University, South Korea. She previously worked for CNRS at the University of Tours, France, and the University of Leipzig, Germany. Her research interests focus on nanoscaled magnetic materials, and this is her seventh book as author and second book as editor.

CHAPTER 1

Introduction to Nanomaterials: Basic Properties, Synthesis, and Characterization

Nguyen Hoa Hong
Department of Physics and Astronomy, Seoul National University, Seoul, Republic of Korea

1 INTRODUCTION

Nanometer-scaled materials could be defined as specimens that have at least one dimension of less than 100 nm. Nanomaterials include ultrathin films, quantum wire, and nanoparticles-quantum dots, in general.

For more than a decade, nanomaterials have been very attractive to science and technology due to their particular properties, which are not usually seen in bulks of the same compounds.

Those remarkable features may include special optical, magnetic, and electrical properties, among many others. These unique properties could potentially have great impacts on electronics, spintronics, medical applications, and other devices.

Some nanomaterials may be found in nature, but in most cases, they should have been fabricated in laboratories by some engineering methods so that they can contain demanding properties that are targeted for industrial purposes. Nanomaterials have been intentionally designed for industrial applications. They can be applied in cell phones, CD-ROMs, magnetic recordings, cosmetics, drug delivery in medicine, imaging, water splitting, etc. Engineered nanomaterials are synthesized in special ways in order to take full advantage of their tiny size and their particular properties, which are generally not observed in their conventional bulk counterparts.

There are two main reasons why nanomaterials have properties that are different from corresponding bulks: (1) a large surface area and (2) confinement (quantum) effects. Due to their structures that contain many small grains, nanomaterials have a much greater surface area to volume ratio than their conventional counterparts, leading to a possibility of having greater

Nano-sized Multifunctional Materials
https://doi.org/10.1016/B978-0-12-813934-9.00001-3

chemical reactivity; this may affect pressure inside the lattice. Moreover, at the nanometer scale, confinement effects should play a key role in shaping the properties and characteristics of materials. As a result, new optical, electrical, and magnetic behaviors may occur.

Even though the concept of nanostructured materials is "fashionable" now, such materials have existed and been widely used for centuries. One famous example is the Lycurgus Cup of the fifth century that is currently on display at the British Museum. The cup has gold particles that were dispersed in the soda glass. Therefore, it has a green appearance but reflects purple due to transmitted light, which is the consequence of the interaction of gold-silver and copper nanoparticles with the light [1]. Many similar applications of nanostructures were used in pottery for centuries. More popular uses of nanostructured materials include image photography, which exploits metal nanoparticles that have large surface area, and painting, which uses pigments based on metallic or semiconducting nanoparticles.

Over the last few decades, strong technological development has enabled manipulation of the synthesis of nanostructured materials with shaped-up properties to meet the needs of applications. One can actually tailor up the properties of low-dimensional materials by the methods of physical chemistry. There is always the matter of how to control the size, shape, crystal structure, and surface formation to bring the materials to meet the requirement of any targeted application. Controlling as to targeting is never an easy task for any research team.

In Fig. 1, you can find some images of nanostructured materials as examples.

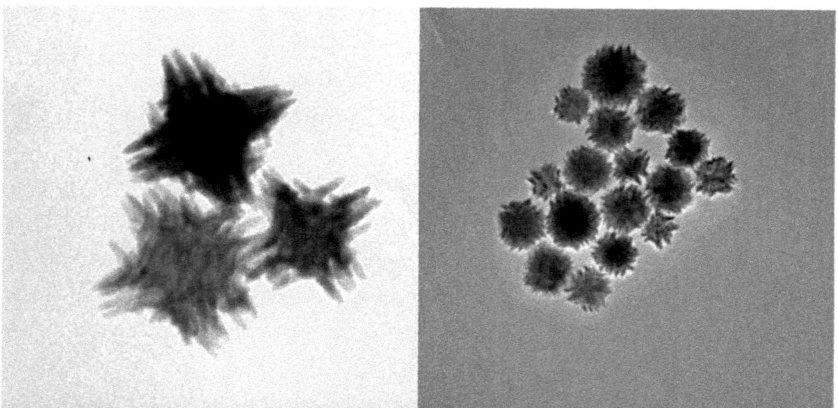

Fig. 1 TEM images of urchin-shaped MnO nanoparticles.

As mentioned above, one might be called "nanomaterial" if it has at least one dimension in the range of 1–100 nm. Nanomaterials can be three dimensions (nanoparticles); two dimensions (nanometer-thick films), one dimension (nanowires) or zero dimensions (e.g., quantum dots). They can exist in single, fused, aggregated, or agglomerated forms with spherical, tubular, and irregular shapes. The most popular types of nanomaterials are nanotubes, dendrimers, quantum dots, and fullerenes.

Some typical examples of nanomaterials are summarized below:

(i) Nanostructured semiconductors are known to show various non-linear optical properties. They may also show quantum confinement effects leading to special properties, such as the luminescence in silicon powders and silicon germanium quantum dots as infrared optoelectronic devices. Nanostructured semiconductors are used as window layers in solar cells.

(ii) Nanometallic powders have been used for coatings.

(iii) Single nanomagnetic particles are monodomains. One also expects that, in magnetic nanophase materials, the grains correspond with domains while boundaries correspond to disordered walls. They could be used in magnetic recording. Recently, magnetic nanoparticles have also been used widely for imaging and drug delivery.

(iv) Nanostructured metal-oxide thin films are widely used for the realization of gas sensors due to their enhanced sensitivity and selectivity.

(v) Nanocrystalline silicon films are known for highly transparent contacts in thin film solar cell and nanostructured TiO_2 films are known for their high transmission and significant surface area enhancement leading to strong absorption in dye sensitized solar cells.

(vi) Ultrathin films are used to coat smart phones with superinduction and optical properties.

(vii) Polymer-based composites with a high content of inorganic particles leading to a high dielectric constant have potential for photonic band gap structure.

2 REMARKABLE PROPERTIES OF NANOSTRUCTURED MATERIALS

2.1 Properties of Quantum Dots

Quantum dots (QDs) are nanocrystals of a semiconductor material whose size is between single molecules and bulk crystalline solids. A typical picture of quantum dots can be seen in Fig. 2 [2]. QDs are remarkable because

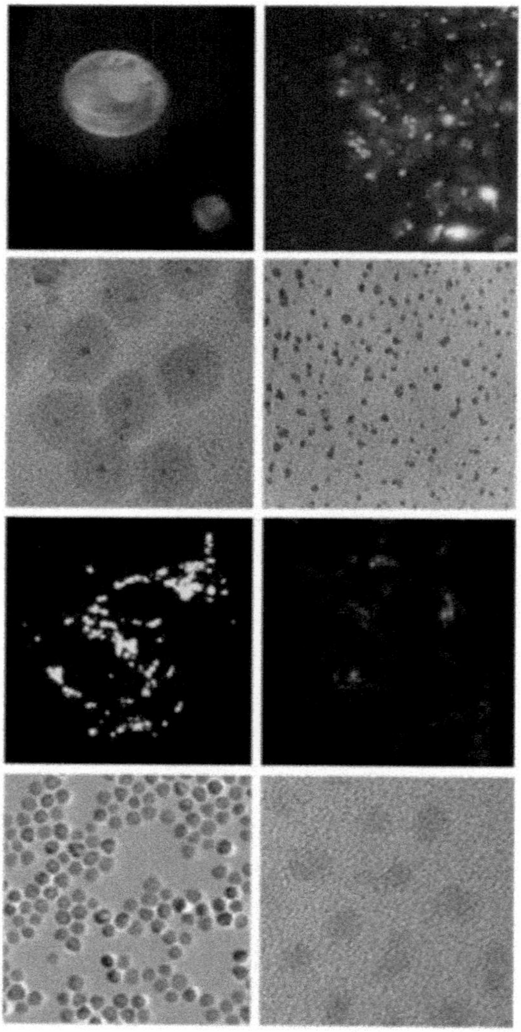

Fig. 2 Differently luminescent nanocrystals for in vitro cell imaging *(Source: Internet-reference: D.S. Kumar, http://www.drsakthikumar.com/quantumdots.php.)*

their tiny size produces physically confined electrons, which are also called "quantum confinements." As a result of their spatial properties, they are subject to many particular optical, electronic, and chemical properties that are not found in other materials. The size dependence shows that those properties can be easily controlled by tuning the size of QDs. When more methods to reliably synthesize variable sizes of monodisperse nanocrystals can be developed, QDs would become low-cost. These unique qualities have given QDs potential for a lot of scientific and industrial applications.

The unique properties of QDs are formed basically due to their size. As semiconducting materials, they have an intrinsic band gap through which electrons can be bridged by excitation of incident light. However, unlike bulk semiconducting materials, QDs are too sparse to create the continuous valence and conduction bands as other macroscopic semiconductors. Differently, QDs produce an electronic structure that is more reminiscent of the discrete electronic states found in single atoms; the larger the QD, the smaller its band gap can be. As a consequence, the larger a QD becomes, the more continuous its electronic structure is and its band gap becomes closer to that of the bulks (see Fig. 3 and Ref. [3]). Unlike a single atom, however, this size dependence allows the band gap energy to be modulated by varying the size of the QD [3].

The optical properties of quantum dots occur partially due to a unique effect known as quantum confinements. When a photon of sufficient energy (equal to or greater than the band gap) strikes a QD, it may excite an electron from the valence band to the conduction band, leaving a positive hole in its place. This leads to an electron-hole pair (also known as an "exciton"), which is a common phenomenon in semiconducting materials. In a QD, the average exciton size (the exciton Bohr radius) is smaller than the size of the QD, producing a confinement energy as the exciton is squeezed into the material.

The total energy of a fluorescing photon can be modeled as the sum of the band gap of the QD, the quantum confinement energy, and the bound exciton energy. In the size regime of 10–100 Å, quantum dots have highly tuneable absorption and emission spectra. This high degree of control makes QDs able to be optimized for many applications. The most fashionable application of QDs is for drug delivery. Fluorescent-carrying drug dots are able to be tracked, making them very helpful for imaging in cancer diagnostics, for example.

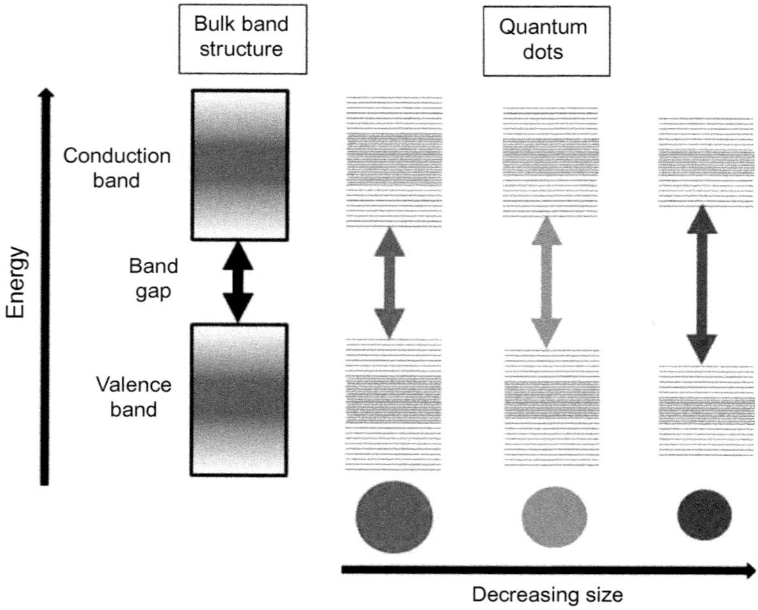

Fig. 3 Splitting of energy levels in quantum dots due to the quantum confinement effect; semiconductor band gap increases with decrease in size of the nanocrystal. *(Source: Internet; reference: https://www.sigmaaldrich.com/technical-documents/articles/materials-science/nanomaterials/quantum-dots.html.)*

2.2 Properties of Ultrathin Films

We call films with thickness between 1 and 50 nm "ultrathin films." Once one dimension is reduced to be very tiny, the films are considered two-dimensional materials. They have confinement effects in one dimension. These confinement effects can influence their properties very much. In fact, depending on each purpose one can tune structural, optical, or mechanical– or the magnetic properties of ultrathin films, by modifying the size, or in other words, varying the thickness. Another important factor besides the confinement effect is the substrate effect, which is equally important because the constraints of the substrate leading to the interface effect can modify the properties of films enormously.

Many groups have discovered how the material properties (electrical, thermal, optical, magnetic, mechanical, chemical) of thin films may deviate from those of the bulks [4]. Such differences in observed parameters can be due to different sizes, shapes, and orientations of crystal grains, the presence of interfaces, quantum confinement, constraints, and/or morphological effects related to the film growth. Nanometer-scale surface morphology can, for example, strongly modify the dielectric response. This is particularly important in the case of metallic films in the optical regime [5]. Furthermore, even in the case of structurally perfect films, the (local) dielectric function is a classical concept that, in the nanometer limit, must be reevaluated to account for the spreading of the electron wave functions and the finite-size effects [6, 7].

Some group reported that 3–4 nm-thick gold films grown on silica substrates could have an optical transmission much lower than that of bulks. This decreased transmission of continuous ultrathin films can be accounted for, however, in a conventional transfer-matrix calculation by assuming "diffuse interfaces" on each side of the gold film. For thinner films, transmission is additionally affected by broad bandwidth plasmonic absorption associated with geometrical features. The observed reduction of the optical transmission is particularly important in the context of using ultrathin gold films as transparent electrodes where maximizing optical transparency is crucial [8].

Magnetically, it has been observed that by making films very thin, one could make some material that is originally paramagnetic in bulks become ferromagnetic, or make the magnetic moment much greater in comparison with that of thicker films (see Fig. 4 and Ref. [9]). As for more details, please read Chapter 7 in this volume. This can happen due to defects and oxygen vacancies that may be more favored at the surface of the film, or at the

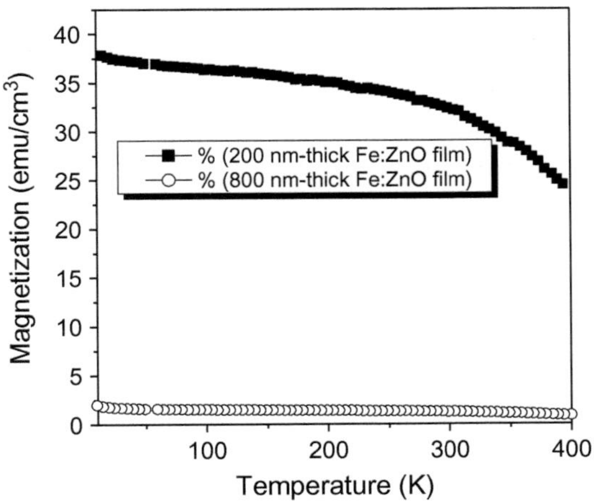

Fig. 4 Magnetic moment versus temperature taken at 0.2 T for Fe-doped ZnO films with different thicknesses.

interface between the film and the substrate. On the other hand, one cannot rule out the confinement effects that exist in thin films that are known to be two-dimensional materials.

2.3 Properties of Nanoparticles

The principal parameters of nanoparticles are their shape (including aspect ratios where appropriate), size, and the morphological substructure of the substance. In the presence of chemical agents (surfactants), the surface and interfacial properties may be modified. Indirectly, such agents can stabilize against aggregation by conserving particle charge and by modifying the outermost layer of the particle. Depending on the growth route and the lifetime of a nanoparticle, very complex compositions should be expected. In the typical route of fabricating nanoparticles, for example, many different agents tend to condensate on the particle while cooling down, and the particle indeed is exposed to various ambient atmospheres. Therefore, complex surface chemical processes are expected and have been identified only for a small number of particulate model systems. At the nanoparticle-liquid interface, polyelectrolytes have been used to modify surface properties and the interactions between particles and their environment. They have been exploited in industries, including adhesion, lubrication, stabilization, and controled flocculation of colloidal dispersions [10].

Both physical and chemical properties of nanoparticles are based on atomic and molecular origins in a complex way. The electronic and optical properties and the chemical reactivity of small clusters are completely different from the better-known property of each component in the bulk or at extended surfaces. Complex quantum mechanical models are required in order to be able to predict the evolution of such properties with particle size, and typically very well-defined conditions are needed to compare experiments and theoretical predictions.

At the nanometer scale, interactions between particles can be dominated by weak Van der Waals forces, stronger polar and electrostatic interactions, or covalent interactions. Depending on the viscosity and polarization of the fluid, particle aggregation can be determined by the interaction between nanoparticles. By surface modification, the tendency of a colloid to coagulate can be enhanced or degraded. The details of nanoparticle-nanoparticle interaction forces and nanoparticle-fluid interactions are very crucial for physical and chemical processes. It is difficult to characterize properties due to the small amount of molecules involved in the surface active layer. Surface energy, charge, and solvation are relevant parameters to be considered. The interaction forces, either attractive or repulsive, would importantly govern individual and collective nanoparticles. This interaction between nanoparticles will result in aggregates and/or agglomerates that may influence their behaviors.

3 POPULAR FABRICATION METHODS OF NANOMATERIALS

3.1 Quantum Dots

3.1.1 Aerosol Synthesis

The typical synthesis of GaAs quantum dots includes the nucleation and growth of the semiconductor particles, passivation of the particle surface with a higher band gap semiconductor, cluster classification, deposition, and incorporation into a semiconductor device. A critical step in an aerosol process is to induce a homogeneous nucleation of particles from the vapor phase. Several processes have been developed for vapor phase epitaxial growth of GaAs. The most popular methods with possible applications are vapor phase epitaxi (VPE) and organometallic chemical vapor deposition (MOCVD). For the synthesis of GaAs clusters, MOCVD seems to be better because an introduction of the two components of the semiconductor can be decoupled, and the particle stoichiometry through reaction kinetics can be controled. Additionally, MOCVD is better suited for hetero-epitaxi of AlGaAs as a passivating agent. An initial high-temperature stage should be designed to promote a homogeneous nucleation of GaAs clusters. Provision is made for a differential mobility classification of the clusters prior to passivation with AlGaAs [11].

3.1.2 Cluster Production Using Nonthermal Vapor Sources

GaAs particles can be first produced using spark sources. However, the resulting particles can be nonstoichiometric due to preferential vaporization of volatile arsenic from a gallium arsenide source. To produce stoichiometric GaAs clusters by direct vaporization, a method that vaporizes both gallium and arsenic with equal efficiency is needed. Single crystal silicon particles have been generated using the spark source. There are two methods for vapor-phase nucleation from nonthermal sources: (a) laser ablation and (b) the tile exploding wire technique. Both methods can lead to the formation of stoichiometric nanometer-scale clusters by vapor phase nucleation, therefore overcoming a key difficulty in the achievement of stoichiometry in the vapor phase growth of III–V clusters that resulted from the large difference in equilibrium anion and cation vapor pressures [11].

3.2 Thin Films Made by Pulsed Laser Deposition Technique

The pulsed laser deposition (PLD) technique has been widely used for the deposition of thin films. A ceramic target and a substrate on which the film should be deposited are fixed in a chamber. The chamber is kept in an ultra-high vacuum and oxygen gas flows from the tube inside the chamber. Pulses are intentionally shot directly at the target, creating a plasma plume. This plasma plume touches the substrate so that the film can be grown on it. See schema in Fig. 5.

3.3 Thin Films Made by the Chemical Vapor Deposition (CVD) Technique

This method is usually used in semiconductor industries to produce thin films for electronic devices.

In this technique, the substrate should be exposed to one or more volatile precursors, which would react and decompose on the surface of the substrate in order to produce the targeted composition. In general, residues are also produced as side products, which should be removed by designed gas flow through the reaction chamber.

People use the CVD technique in microfabrications to fabricate materials such as semiconducting films (silicon, nitrides, etc.), graphenes, various high-k dielectric materials, and many others.

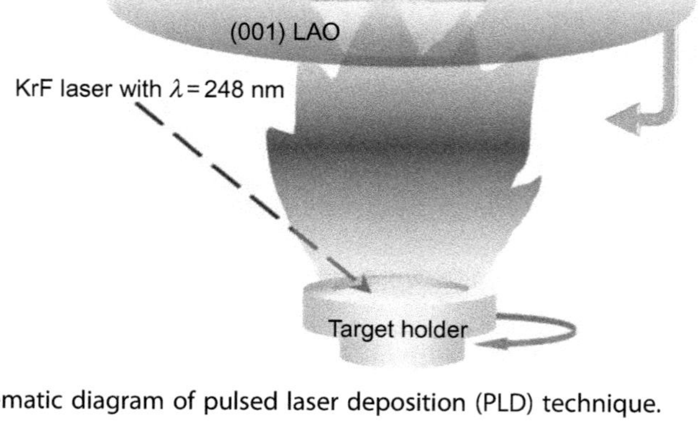

Fig. 5 Schematic diagram of pulsed laser deposition (PLD) technique.

3.4 Nanoparticles

Particles can be prepared using the coprecipitation method. For example, the aqueous precursor solution can be obtained by dissolving metal nitrates with appropriate molar concentrations. Then, the excess of NH_4OH is added to the precursor solution under a vigorous stirring to initiate the precipitation process, thereby obtaining mixed metal hydroxides. The precipitates were collected by centrifuge, washed several times, and then dried in an oven at the appropriate temperature for many hours. The dried precipitates are then calcined at high temperatures in the air and ground several times. The size of particles can be controlled by varying solvents and applying different thermal treatments [12].

4 CHARACTERIZATIONS OF NANOMATERIALS: POINTS TO PAY ATTENTION TO

4.1 Measurements for Quantum Dots

Usually, morphology, composition, crystal structure, band gap energy, optical properties, and the nature of the surface ligand are important to be characterized in semiconducting QDs. Transmission emission microscopy (TEM) imaging, therefore, is a useful tool to determine the size, size contribution, and shape of QDs. On the other hand, energy-dispersive X-ray diffraction spectroscopy (EDS) can tell us about the composition of the samples while high-resolution TEM (HRTEM) can be used to give information on crystalline structure. Depending on the specific type of dots, the appropriate solution should be used.

4.2 Measurements for Thin Films

4.2.1 Transport

4.2.1.1 Make Contacts

One should check first to see if the sample has a resistance that is smaller than the internal impedance of the voltmeter (be careful with insulating samples because you may not expect how large the resistance can be when the temperature gets lower). If the resistance is too large, the measurement may not be completed due to an overloading (OL). For low-resistance samples, the contacts may be simply made by silver paste and gold wires. For thin films, it is also good if you use indium to first fix four points of contacts, and then put silver paste on top of them. The advantage to using indium is that we can define the shape of the contacts easily, and after the R-T measurement, we can remove them thoroughly to use the sample for other purposes. When using indium, use a clean toothpick to press it onto four positions where you want your contacts to be. Be sure to use silver paste that is well mixed and has an appropriate wetness (not too wet but not too dry). Fix the sample into the sample holder by using low temperature/vacuum greases.

In fact, the current always tends to choose an "easier way" to go, and therefore, in some homogenous samples, we may obtain a negative resistance. To avoid this, it is better to create terminals by using Pt-Pr sputtering first, and then apply indium and silver paste on the top (if you want to make contacts this way, be sure that you measure all other measurements before because, after creating terminals, the sample is not supposed to be used for magnetic measurements or X-rays, etc.).

If the resistance of the sample is extremely small or extremely large, a microfabrication using lithography in clean rooms should be used to create paths. Based on the formula of $R=\rho l/s$, one can see that if you want a sample to be measurable, if its resistance is too large, you may reduce the "measured" resistance by decreasing l or increasing s. If it is too small, you can increase the "measured" resistance by increasing l and reducing s. See an example of creating necessary paths in Fig. 6.

If your films are bilayer and you want to measure the resistance over the "real sample," you should do etching to make the bottom layer displayed so that you can put two current contacts on the parts belonging to the bottom layer, and two other contacts for voltage can be fixed to the top layer. An easier way to avoid doing etching is that, when you fabricate your bilayer sample, use a mask to cover the outer quarter of the bottom layer of the sample before depositing the top layer so that you can fix two contacts on these areas.

Because the gold wires used for an R-T measurement are often very fine, be sure to have tweezers that have very small but not very sharp tips to fix the wires while making contacts in order to avoid cutting your wires or removing your contacts.

4.2.1.2 Run a Measurement

If you want to measure magnetoresistance (MR), basically the sample should be fixed into the sample holder in two configurations: parallel to the applied magnetic field or perpendicular to it. When you fix/cover the sample, be sure that the wires do not touch one another because, if they get shorted during the measurement, they may give mistaken data.

The applied current must be small to avoid the OL (because if the current is large then the measured voltage will be large), but it must be large enough

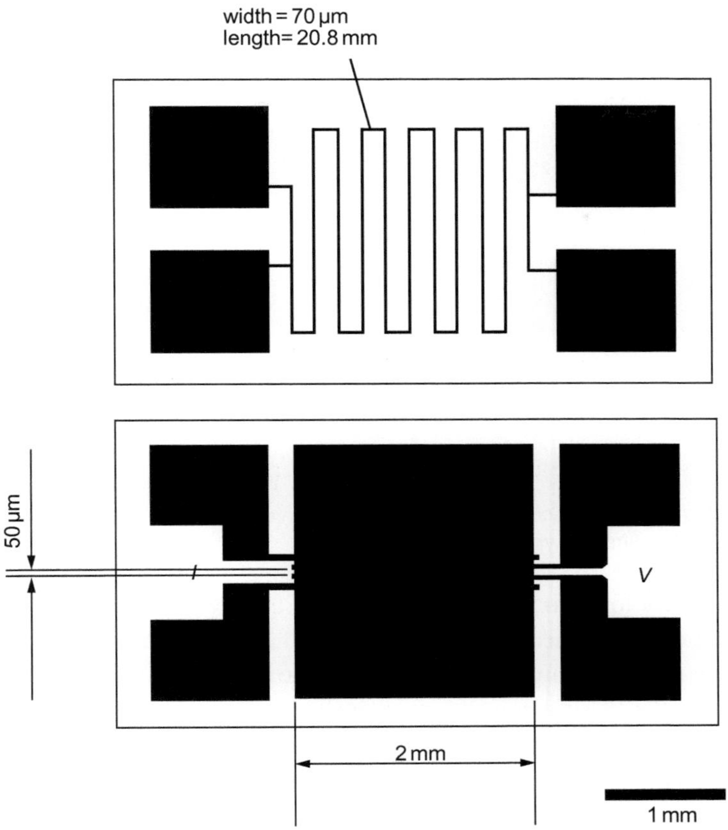

Fig. 6 Paths created by lithography for magnetoresistance measurements.

to avoid the case that the signals of the noises and of the sample are in the same order. Each voltmeter/program has been fixed with some cut-off voltage. When you see the measured voltage approaching the limit, you have to reduce the applied current. However, it is better to avoid doing this many times to avoid the electric field effect that may occur on the sample (in some manganite thin films, when we change the applied current/voltage, the feature is not recoverable, see Fig. 7) [13]). Therefore, it is perfect if you can choose an appropriate current that can stand for the whole measurement.

In principle, it is always safer to apply magnetic fields when the system is at low temperatures. In general, a standard MR measurement process will be carried out in this way: you measure R-T under zero field upon cooling, then apply a magnetic field at 5–10 K, then measure R-T under that field upon warming, then upon cooling at the same field, then again apply a

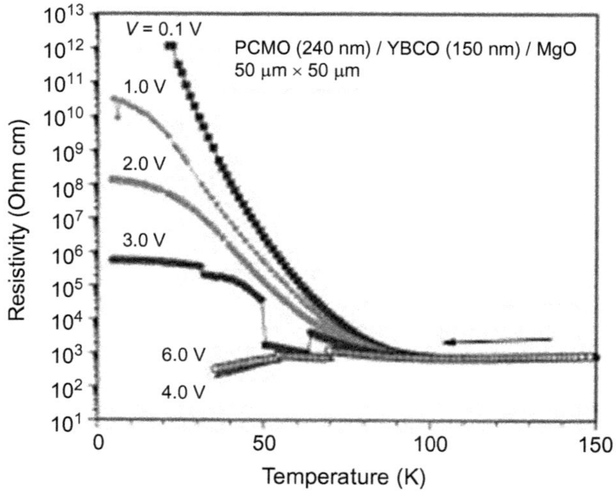

Fig. 7 Magneto-resistance versus temperature curves for bilayers of Pr-Ca-Mn-O and Y-Ba-Cu-O thin film grown on MgO.

higher field at low temperature. Finally, you quench the field to zero also at low temperatures when you want to finish all MR measurements, and measure under zero field upon warming up to room temperature before picking up the sample. Be sure your sample is at room temperature when you take it out in order to avoid any damage due to thermal shock.

In some cases (e.g., manganese oxide films), a "history effect" may be observed. For example, the curve measured at zero field upon warming has exactly the same feature as the curve at 5 T upon cooling measured just before, but not the same as the curve measured at zero field upon cooling that was done at the beginning [14, 15]. This feature needs to be noticed when you interpret some MR data.

Keep in mind that because there are many possibilities for an oxide film to be inhomogeneous, after removing the old contact and making the new contacts onto different positions, you rarely obtain the same magnitude of resistivity, even if the tendency of the R–T curve remains the same. On the other hand, due to the "aging effect" that is very often found in oxide samples after a few months, the sample cannot maintain the same quality as before.

When you measure a thin film sample, the thickness of the film should be taken into account when interpreting data. Because in some cases, due to strain effects, a film sample can be insulating or metallic depending on the thickness as well as the type of substrate [16–18].

4.2.2 Magnetization

The maximum size of the film sample using for magnetization measurements should be about 5 mm × 5 mm in order to load the sample in without any trouble. The film should not be too thin (to have big enough signals). Measurements in two configurations can be performed. In one configuration, the film plane is parallel to the magnetic field (vertical in the loading) and in another configuration, it is perpendicular to the magnetic field (it also means that the small piece of straw containing the film is placed normal to the parent straw). If you can move the sample skillfully (by using some bar to support it while putting in), then you can place it perpendicular without cutting the parent straw because the small piece of straw can fix itself firmly inside due to its four corners leaning to the inner part of the parent straw. If you are not sure that you can place it nicely, then you had better split the parent straw at the position you want to put the sample, place it in, and finally use nonmagnetic tape to wrap around for sealing and protecting. Be sure to have an angle of 90 degrees between the film plane and the straw (see Fig. 8).

To be sure to obtain the real magnetic signal of the sample, fixing by non-magnetic tape is required. To have an easy subtraction, and to avoid errors, it is better to use some pieces of tape with the same size to fix the sample and to fix the bare substrate when you measure the substrate alone. For the same reason, use the same straw for measurements of the film sample and the bare substrate that you use for the subtraction (for samples with strong ferromagnetism and with FM particles on the surface, as in the case of magnetic semiconducting films, the piece of the straw may be magnetized also).

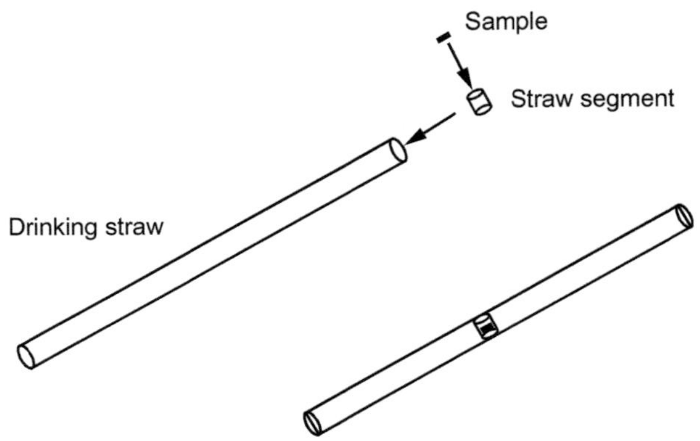

Fig. 8 Example for fixing samples for magnetization (versus temperature or versus magnetic field) measurements.

During loading, if you insert the sample too fast, or you do not purge very well, a solid oxygen lump may be formed on the sample. Therefore, when you measure *M–T*, you may see a "wrong" peak around 50–60 K and it may be mistaken as some intrinsic property of the film while, in fact, it is due to solid oxygen. In this case, it is recommended to heat the sample to room temperature to melt such a solid oxygen lump. After doing this, one can restart the sequence. Due to this reason, of course in principle you can perform the measurements in both zero field cool (ZFC, i.e., upon warming) and field cooling (FC) modes, according to your own purpose of experiments. However, in the case where you just need a simple *M–T* curve, to have T_C, or to have an idea about the magnetic nature of the sample, it is simpler to measure upon cooling to avoid that kind of trouble.

The first signal you see while centering, along with your knowing about the magnetism base of your sample (such as being ferromagnetic or paramagnetic at that temperature), will basically guide you to select a correct position to be the center of the sample (in some cases, it is easy when the signal is strong, but in some other cases, it can be rather subtle and it would be helpful to use your own knowledge about your sample). By seeing the first signals, you also can decide what field should be suitable to apply to perform the centering. For a ferromagnetic sample, a field of 100 Oe may be enough but for an antiferromagnetic sample, it may require 2000 Oe or larger. As mentioned earlier, the field quenching must be done by the oscillating mode. If the field applied for centering is small, one had better apply a big field, then quench again to be very sure to completely get rid of the magnetic residue.

In general, measuring a ferromagnetic thin film is much easier than measuring an antiferromagnetic sample because in the latter case, the signals are very small. As for ferromagnetic samples, a magnetization versus temperature (*M-T*) measurement (in order to determine T_C) used to be performed at low fields of 100–2000 Oe (depending on samples). However, the applied field must be large enough to be sure that the magnetic moment can get saturated. In the case of paramagnetic/antiferromagnetic samples, even a field of about 3 T may be needed. The temperature range is usually 5–400 K, but it can be larger, depending on facilities and the need measurement. In fact, the magnetization versus field (*M-H*) curve usually gives direct evidence of the nature of a sample at a specific value of temperature to show that that sample is paramagnetic, ferromagnetic, or antiferromagnetic, etc. However, because the performance of measurements is very time consuming, it is suggested to measure the *M–T* curve first to have a guide to other *M–H* measurements (at which temperature you can expect to observe what, e.g.).

Keep in mind that when measuring a thin film sample, the obtained signals should be the signals of both the film and the substrate. Therefore, when performing a measurement for a thin film sample, it is necessary to perform a corresponding measurement with the same sequence for a bare substrate. For example, if one measures *M–T* at 0.2 T from 400 K to 5 K for a thin film sample on an $LaAlO_3$ substrate, then the *M–T* curve at the same field and in the same temperature range is also needed for a bare $LaAlO_3$ substrate. If the sizes of the sample and the bare substrate are the same, the subtraction should be very simple because one just needs to deduct the data directly. Otherwise, the proportional difference must be taken into account while calculating.

One should be careful with the subtraction for an antiferromagnetic thin film sample because the signal is quite small, and basically most substrates used for oxide thin films such as $LaAlO_3$, $SrTiO_3$, MgO, etc., are diamagnetic. To avoid getting the wrong values of the magnetic moment, the mass and size of the sample and the substrate used for subtraction must be noted carefully and precisely.

The unit of magnetization can be emu/g or emu/cm^3. If you want to have the unit as emu/cm^3, use the DEKTAK profiler to determine the thickness and then calculate the volume of the film (to get that, multiply the thickness by the area of the substrate in which the film is deposited).

4.2.3 X-Ray Measurements

As for films, one should pay attention to fix the film sample very flat on the sample holder because otherwise, it may give wrong or weak signals concerning oriented peaks. Signals of substrates are always very big so that a zoom in is necessary to clearly observe the peaks of films. Sweeping slowly is also preferred for film samples.

4.3 Measurements for Nanoparticles

When performing X-ray diffraction measurements for nanoparticles, one should prepare a sufficient amount, and nanopowders/nanoparticles should be fixed homogenously on a flat holder to be targeted correctly for the beam. Normally, peaks are of several different orientations so that indexing must be done with some special care.

As for optical measurements such as photoluminescence, nanoparticles should be contained in some transparent glass container. Even though this kind of measurement is not hard to perform, repeating the measurements several times can help to ensure that the data are taken correctly.

As for magnetic measurements, nanoparticles should be given in a larger amount to be sure they are giving strong enough signals. Nanopowders/nanoparticles must be contained in nonmagnetic capsules, and the data of the bare capsule must be measured with the same sequence as for the particles plus the capsule, in order to be able to subtract them from raw data. Magnetization should be normalized by taking into account the mass of the whole amount of nanoparticles/nanopowders.

5 CONCLUSION

Nanostructured materials have remarkable properties that are not usually observed in bulk materials of the same compositions. The reduction in the spatial dimension and confinement effects lead to modifications of physical properties of the systems. We have generally discussed properties that are typically expected in nanostructured materials. The most popular methods of synthesis as well as important issues for characterizing nanomaterials have been also discussed, with hope to give a general understanding about this research domain. Details for different types of nanomaterials can be found in the following chapters in this book.

ACKNOWLEDGMENTS

The chapter is supported partially by grant 2017060271 of the National Research Foundation of Korea (NRF).

REFERENCES

[1] http://www.britishmuseum.org/.
[2] D.S. Kumar, http://www.drsakthikumar.com/quantumdots.php.
[3] https://www.sigmaaldrich.com/technical-documents/articles/materials-science/nanomaterials/quantum-dots.html.
[4] A. Goswami, Thin Film Fundamentals, New Age International Publishers, New Delhi, India, 1996.
[5] H. Fujiwara, Spectroscopic Ellipsometry: Principles and Applications, John Wiley and Sons. Ltd., Chichester, West Sussex, England, 2007.
[6] F.J. García de Abajo, J. Phys. Chem. C 112 (2008) 17983.
[7] S. Laref, J. Cao, A. Asaduzzaman, K. Runge, P. Deymier, R. Ziolkowski, M. Miyawaki, K. Muralidharan, Opt. Express 21 (2013) 11827.
[8] A. Kossoy, V. Merk, D. Simakov, K. Leosson, S. Kéna-Cohen, S.A. Maier, Adv. Opt. Mater. 3 (2015) 71.
[9] N.H. Hong, J. Sakai, F.J. Gervais, Mag. Mag. Mater. 316 (2007) 214.
[10] M. Mazilu, V. Musat, P. Innocenzi, T. Kidchob, D. Marongiu, Particulate Sci. Technol. 30 (2012) 32–42 and references therein.
[11] R.C. Flagan, H.A. Atwater, K.J. Vahala, J. Aerosol Sci. 22 (Suppl. 1) (1991) 31.

[12] T.S. Atabaev, J.H. Lee, J.J. Lee, D.W. Han, Y.H. Hwang, H.K. Kim, N.H. Hong, Nanotechnology 24 (2013) 345603 and references therein.
[13] J. Sakai, A. Kitagawa, S. Imai, J. Appl. Phys. 90 (2001) 1410.
[14] N.H. Hong, S. Imai, J. Sakai, H. Iwasaki, J. Phys. Condens. Matter 15 (2003) 1921.
[15] N.H. Hong, JAIST, Japan, (PhD thesis), (2001).
[16] W. Prellier, C. Simon, A.M. Haghiri-Gosnet, B. Mercey, B. Raveau, Phys. Rev. B 62 (2000)R16337.
[17] J. Zhang, H. Tanaka, T. Kanki, J.H. Choi, T. Kawai, Phys. Rev. B 64 (2001) 184404.
[18] J. Klein, J.B. Philipp, G. Carbone, A. Vigliante, L. Alff, R. Gross, Phys. Rev. B 66 (2002) 52414.

CHAPTER 2

Properties of Quantum Dots and Their Biological Applications

Orion Ciftja
Department of Physics, Prairie View A&M University, Prairie View, TX, United States

1 INTRODUCTION

Many advances in the field of nanotechnology have made possible the fabrication of various structures at extremely small length scales, typically the nanoscale size of atoms and molecules [1–7]. This has enabled scientists to fabricate novel artificial materials and devices in which charge carriers, typically electrons, are confined in a precise and controlled way. Quantum dots are typical nanostructures where electrons are confined in all three spatial dimensions. Many times, confinement in a quantum dot arises from electrostatic potentials created by applying lithographically patterned gate electrodes, or by etching on a two-dimensional (2D) electron gas in a semiconductor heterostructure. This leads to systems of confined electrons in a host 2D semiconductor quantum dot region with a size that varies from nanometers to few microns [8–14].

The number of confined electrons in a 2D semiconductor quantum dot can be anything from a single electron to thousands of electrons. This number can be precisely controlled by experimental tools and, under normal conditions, an electron's quantum mechanical nature dominates. In atomic systems, electrons are confined by the Coulomb attraction of the nucleus. In a 2D semiconductor quantum dot, the electrons are instead confined by an artificially created electrostatic potential. Because of this analogy to real atoms, semiconductor quantum dots are sometimes referred to as "artificial" atoms. Energy levels of electrons become quantized in a 2D semiconductor quantum dot due to the confinement potential. Therefore, the behavior of confined electrons in a 2D semiconductor quantum dot shows many parallels with the behavior of naturally occurring confined electrons in the atomic systems. However, there are also important differences from real atoms because the structure associated with confined electrons in an "artificial" quantum dot atom reflects both the 2D nature of the system as well as

Nano-sized Multifunctional Materials
https://doi.org/10.1016/B978-0-12-813934-9.00002-5

the non-Coulomb form of the artificially created confining potential. Unlike real atoms, the artificially created confining potential in a 2D semiconductor quantum dot is easily tuned externally. Freedom to choose the confinement potential is extremely desirable because it implies high tunability of physical properties using standard experimental methods. Therefore, confined electrons in an "artificial" atom provide excellent tools to study atomic-like behavior in a fully controllable way [15].

The interplay between quantum confinement, electron-electron correlation, and other external parameters provides a very fertile ground for the observation of a wealth of interesting phenomena in 2D semiconductor quantum dot devices. From a theoretical and modeling perspective, one may assume that electrons in a typical semiconductor quantum dot are confined strictly in 2D by means of a suitable confining potential that is electrostatic in origin. The parabolic confining potential is a very popular choice in many studies, despite the fact that the precise form of confining potentials varies from one experimental setup to another. The parabolic form of the confining potential is convenient to elucidate various atomic-like properties and explain the shell structure of 2D semiconductor quantum dots at a low magnetic field [16–18]. When one applies a magnetic field perpendicular to the 2D quantum dot plane, there is a change on the structure of single-electron levels as the magnetic field increases. This results in a more pronounced role played by electron-electron correlations. In this regime, the interplay between quantum confinement, electron-electron correlations, and the magnetic field leads to many interesting phenomena, including the appearance of strongly correlated fractional quantum Hall effect (FQHE) phases if a high perpendicular magnetic field is applied [19–33]. Transport properties of confined systems of electrons in a 2D semiconductor quantum dot are also interesting because of possible applications of single-electron tunneling phenomena of electronic devices [34, 35].

Overall, studies of confined systems of electrons in a 2D semiconductor quantum dot are of great interest from the perspective of nanoscale device technologies. From this perspective, systems of confined electrons in a 2D semiconductor quantum dot are seen as one of the most promising avenues for meeting future challenges on device technologies. These systems are also very important from a fundamental theoretical perspective because they represent an ideal setting to investigate the interplay between many quantum effects that otherwise are not readily available [36–53]. A complete study of the properties of confined electrons in a 2D semiconductor quantum dot requires a quantitative understanding of new physical phenomena occurring at nanoscale lengths.

In this contribution, we present an overview of common theories and models that describe the properties of confined electrons in a 2D semiconductor quantum dot in the absence as well as in the presence of a perpendicular magnetic field. We also provide a brief overview of a few concepts and approaches that are needed to understand their unique electro-optical properties and explain possible biological applications.

2 2D SEMICONDUCTOR QUANTUM DOTS UNDER CONFINEMENT

The detailed form of the confining potential depends on many factors as well as the specific experimental setup of a 2D semiconductor quantum dot. One of the simplest yet best choices for a confining potential in a 2D semiconductor quantum dot assumes that electrons are trapped by an isotropic parabolic confinement potential [54–56] of the form:

$$V(r) = \frac{m}{2}\,\omega_0^2\, r^2, \tag{1}$$

where m is the electron's mass, ω_0 is the frequency of the parabolic confinement, and $r = \sqrt{x^2 + y^2}$ is the 2D distance of the electron from the center of the parabolic well. Shown in Fig. 1 is a plot of such a potential in

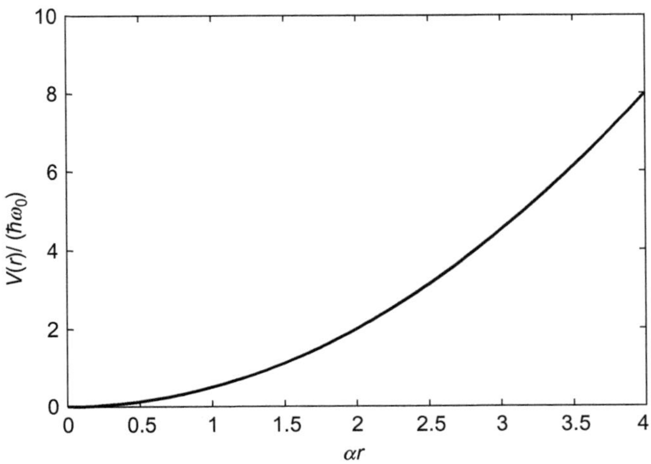

Fig. 1 Plot of $V(r)/(\hbar\,\omega_0)$ as a function of dimensionless distance, $\alpha\, r$ for a 2D parabolic confining potential.

dimensionless energy units, $V(r)/(\hbar\,\omega_0)$ as a function of the dimensionless distance, $\alpha\, r$ where

$$\alpha = \sqrt{\frac{m\,\omega_0}{\hbar}} \tag{2}$$

is the inverse harmonic oscillator length parameter ($\hbar$ is Planck's reduced constant). The parabolic model works well to describe major features of typical systems of confined electrons in a 2D semiconductor quantum dot [57–59]. In particular, this model is very well suited to describe atomic-like properties of 2D semiconductor quantum dots because the discrete energy levels of a 2D quantum oscillator are exactly known:

$$E_{nm_z} = \hbar\,\omega_0\,(2\,n + |m_z| + 1), \tag{3}$$

where $n = 0, 1, \ldots$ is the radial quantum number and $m_z = 0, \pm 1, \pm 2, \ldots$ is the z-angular momentum quantum number.

Circularly symmetric potentials such as the parabolic potential in Eq. (1) are suitable and easier to study. However, many times the region where the electrons are confined is not a perfectly circular domain. For this reason, anisotropic noncircular parabolic confinement potentials of the form:

$$V(x, y) = \frac{m}{2}\left(\omega_x^2\, x^2 + \omega_y^2\, y^2\right), \tag{4}$$

where ω_x and ω_y are confinement frequencies in respective x, y directions have also been investigated by several authors [60, 61]. Despite their wide use, both isotropic and anisotropic parabolic confinement models have limitations due to the implied infinite range and infinite height of the potential at large distances, an assumption that is unphysical for certain experimental regimes.

The parabolic confinement model is certainly appropriate to describe systems of few electrons with low energies where the majority of electrons are localized in the central region of the quantum dot domain. However, a parabolic confinement model may not be very suitable for large systems of electrons where a good number of them have energies close to the energy continuum threshold. Close to the energy continuum threshold, the shape of the confinement potential felt by electrons is no longer parabolic; therefore, an infinite range parabolic potential may no longer be justified.

The issue of infinite range and height of the parabolic potential can easily be addressed by considering simple 2D confinement potentials with a finite depth. Key features may even be captured by simple choices such as a finite 2D cylindrical well potential of the form:

$$V(r) = \begin{cases} -V_0; & 0 \le r \le R \\ 0; & R < r < \infty, \end{cases} \tag{5}$$

where R is the radius of the circular semiconductor quantum dot domain and $V_0 > 0$ is the depth of the potential well. If the confining region of the electrons resembles a rectangle, a similarly simple potential for consideration would be a finite 2D square well potential of the form:

$$V(x,y) = \begin{cases} -V_0; & -\frac{L_x}{2} \le x \le +\frac{L_x}{2}; -\frac{L_y}{2} \le y \le +\frac{L_y}{2} \\ 0; & \text{otherwise}, \end{cases} \tag{6}$$

where $L_x \times L_y$ is the area occupied by the confined electrons. The two potentials in Eqs. (5), (6) are favorite quantum mechanics textbook examples and their properties are well known. Within the context of this work, these two confining potentials represent a specific class of confinement potentials with sharp boundaries/edges in addition to having a finite depth.

Other potentials with finite depth but with smooth boundaries that do not become infinite at large distances have been proposed. One of them is a Gaussian confining potential [62] shown in Fig. 2, which has the form:

$$V(r) = -V_0 \exp\left(-\frac{r^2}{2R^2}\right), \tag{7}$$

where $V_0 > 0$ is the depth of the potential well and R is the range of confinement potential. Note that the Gaussian potential has no sharp edges and smoothly goes to zero in the $r \to \infty$ limit.

Another confining potential with smooth boundaries [63] that goes smoothly to zero in the $r \to \infty$ limit is written as:

$$V(r) = -\frac{V_0}{\left(1 + \frac{r^2}{R^2}\right)^2}, \tag{8}$$

where $V_0 > 0$ and R is an adjustable parameter. At small distances, the previous confining potentials are parabolic while at large distances they tend to a finite (zero) asymptotic value. While the confining potentials in Eqs. (7), (8) do not have some of the shortcomings of the infinite range and height

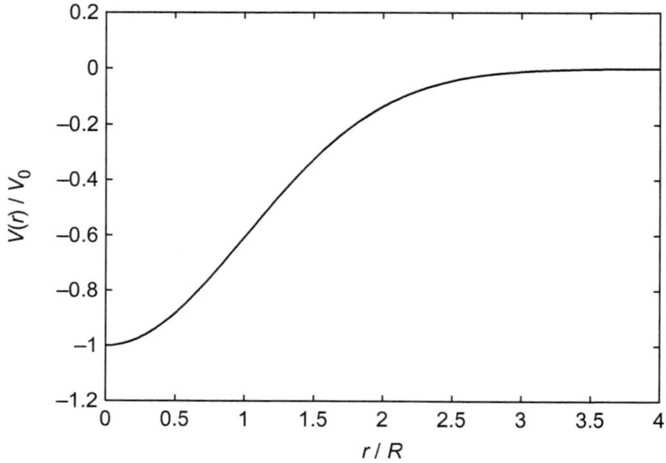

Fig. 2 Plot of $V(r)/V_0$ as a function or r/R for a Gaussian confining potential of the form: $V(r) = -V_0 \exp\left(-\frac{r^2}{2R^2}\right)$, where $V_0 > 0$ is the depth of the potential well and R is the range parameter.

parabolic potential, their physical origin and experimental justification are not entirely clear.

Even though the "ideal" parabolic confining potential has its limitations, such a choice is generally a very good choice for electrons confined around the central region of the quantum dot. The underlying physical reason has to do with the fact that even realistic confining potentials are approximated rather well by a parabolic form around the bottom of the quantum well. While this approximation works well for low-energy electrons (those around the bottom of the quantum well), it is not a very good approximation for electrons that happen to have energies close to the energy continuum threshold. Generally speaking, for all experimental situations, there is a "finiteness" of the domain in which electrons are confined. In particular, there are more complicated quantum dot systems such as laterally coupled 2D semiconductor quantum dots (for instance, double dots) where an infinite range and height parabolic confinement model cannot be used because it does not allow unbinding of states from the individual harmonic wells [64].

At this juncture, we point out that all confining potential mentioned so far is phenomenological. If one wants a more experimentally grounded approach, there are ways to introduce more realistic (yet still simple) confinement potentials that are better experimentally justified than the commonly used phenomenological potentials already mentioned. One possible way to view the problem of electrostatically confined 2D semiconductor quantum dots is to argue that a very typical experimental setup in such cases involves the application of gate potentials that cause electron depletion in the area near the gates. The area depleted from electrons acts as a positively charged region that most simply can be modeled as a uniformly charged 2D disk with a positive background charge. Even though the positive charge of the depleted region may not necessarily be equal to the total charge of the electrons confined in the quantum dot, for simplicity, one may assume that the overall system is charge neutral. This way, one relates to standard quantum models in condensed matter physics that require overall charge neutrality. In this setup, individual electrons feel a confinement potential originating from the uniformly positively charged 2D background disk. Differently from the infinitely high "ideal" parabolic confinement potential, the resulting 2D charged disk potential has a finite depth. The resulting 2D charged disk potential has a form that can be reasonably well approximated by a parabolic potential in the central region of the semiconductor quantum dot (for states with low energy) and as a Coulomb potential at large distances. More precisely, the potential has a parabolic form at the bottom of the quantum well (in the central region of the disk), but is no longer parabolic outside that region, where it behaves like a Coulomb potential that obviously has a finite (zero) asymptotic value at large distances.

An exact expression for the 2D charged disk electrostatic confining potential may be derived by considering the interaction potential between an electron with charge $-e(e>0)$ and a uniformly charged finite 2D disk (the depleted region) with total positive charge Q and radius R. Both parameters, Q and R, can be tuned experimentally. It has been calculated [65] that the resulting electrostatic confining potential energy between an electron and a 2D charged disk can be written as:

$$V(r)=-e\,V_0\,F(r,R), \tag{9}$$

where

$$V_0=\frac{2\,k\,Q}{R} \tag{10}$$

is the electrostatic potential created by the disk at its center ($r=0$) and

$$F(r,R)=\int_0^\infty \frac{du}{u} J_0\left(\frac{r}{R}u\right)J_1(u) \tag{11}$$

is a function that depends only on the ratio r/R (given in integral form), $J_n(z)$ are nth-order Bessel functions, $r=\sqrt{x^2+y^2}$ is the distance of the electron from the center of the disk, and $k = 1/(4\,\pi\,\epsilon_0)$ is Coulomb's electric constant. While the integral presentation of $F(r, R)$ is rather convenient, the quantity appearing in Eq. (11) can be calculated analytically, resulting in an expression involving special functions such as complete elliptic integrals of the first and second kind [66] and/or hypergeometric functions [67]. Two special values of this function are:

$$F(r=0,R)=1;\;\; F(r=R,R)=\frac{2}{\pi}. \tag{12}$$

A more in-depth overview of this topic is readily available [15].

However, considering the complexity of the nanomaterials world, there are situations in which the experimental setup in a 2D semiconductor quantum dot involves asymmetric application of several gate potentials with sharp geometric features. In all these instances, the area depleted of electrons, thus, the quantum dot region, is not circular. If, for simplicity, we consider the confinement region of the electrons as square in shape, then an accurate calculation of the properties of such a square-patterned quantum dot should be made by using the realistic confinement potential that originates from that particular configuration. Thus, if one considers that the confining region of the electrons is precisely a square domain, then one must calculate exactly the electrostatic confinement potential that results from such a geometry. To this effect, one starts by treating the area depleted of electrons as a uniformly positively charged square. The confinement potential felt by electrons with charge $-e(e>0)$ in the presence of a uniformly charged finite square (the depleted region) with length L and total positive charge Q can be written as:

$$V(x,y)=-e\,\Phi(x,y), \tag{13}$$

where $\Phi(x, y)$ is the electrostatic potential created by a uniformly charged square at the point $\vec{r}=(x,y)$ on the 2D plane. One can write the expression for such an electrostatic potential as:

$$\Phi(x,y)=k\,\rho_0\int_D d^2r'\;\frac{1}{|\vec{r}-\vec{r}'|}, \tag{14}$$

where k is Coulomb's electric constant, $\rho_0=Q/L^2$ is the uniform 2D charge density of the square, $\vec{r}'$ (and $\vec{r}$) are 2D position vectors, and D is the square domain uniformly filled with charge. We choose the origin of the system of coordinates at the center of the square, which implies that all the positive charge is found in the domain D: $-\frac{L}{2} \leq x' \leq +\frac{L}{2}$; $-\frac{L}{2} \leq y' \leq +\frac{L}{2}$. Note that the total charge of the domain Q and the length L can both be tuned experimentally. The calculation of the electrostatic potential $\Phi(x, y)$ is simplified by using the transformation [68, 69]:

$$\frac{1}{|\vec{r}-\vec{r}'|} = \frac{2}{\sqrt{\pi}} \int_0^\infty du\, e^{-u^2 (\vec{r}-\vec{r}')^2}. \tag{15}$$

The remaining steps are straightforward, with the final result for the electrostatic potential written in integral form as:

$$\Phi(X,Y) = \frac{k\,Q}{L} \frac{2}{\sqrt{\pi}} \int_0^\infty dt\, f(t,X) f(t,Y), \tag{16}$$

where $X = x/L$, $Y = y/L$, and $t = u\,L$ are dimensionless variables. The function $f(t, X)$ (as well as $f(t, Y)$) is given by the following expression:

$$f(t,X) = \frac{\sqrt{\pi}}{2\,t} \left\{ erf\left[t\left(\frac{1}{2} - X \right) \right] + erf\left[t\left(\frac{1}{2} + X \right) \right] \right\}, \tag{17}$$

where $erf(z) = 2/\sqrt{\pi} \int_0^z dx\, e^{-x^2}$ is an error function [70]. Though the integrals appearing in Eq. (16) are quite unyielding, an exact analytic expression has been recently obtained [71] for the Coulomb electrostatic potential created by a uniformly charged square at an arbitrary point on its plane:

$$\begin{aligned}
\Phi(X,Y) = \frac{k\,Q}{L}\frac{1}{2} \Bigg\{ & (1-2\,Y)\left[\frac{1-2\,X}{|1-2\,Y|} \sinh^{-1}\left(\left| \frac{1-2\,Y}{1-2\,X} \right| \right) + \frac{1-2\,X}{|1-2\,X|} \sinh^{-1}\left(\left| \frac{1-2\,X}{1-2\,Y} \right| \right) \right] \\
& + (1+2\,Y)\left[\frac{1-2\,X}{|1+2\,Y|} \sinh^{-1}\left(\left| \frac{1+2\,Y}{1-2\,X} \right| \right) + \frac{1-2\,X}{|1-2\,X|} \sinh^{-1}\left(\left| \frac{1-2\,X}{1+2\,Y} \right| \right) \right] \\
& + (1-2\,Y)\left[\frac{1+2\,X}{|1-2\,Y|} \sinh^{-1}\left(\left| \frac{1-2\,Y}{1+2\,X} \right| \right) + \frac{1+2\,X}{|1+2\,X|} \sinh^{-1}\left(\left| \frac{1+2\,X}{1-2\,Y} \right| \right) \right] \\
& + (1+2\,Y)\left[\frac{1+2\,X}{|1+2\,Y|} \sinh^{-1}\left(\left| \frac{1+2\,Y}{1+2\,X} \right| \right) + \frac{1+2\,X}{|1+2\,X|} \sinh^{-1}\left(\left| \frac{1+2\,X}{1+2\,Y} \right| \right) \right] \Bigg\},
\end{aligned} \tag{18}$$

where

$$\sinh^{-1}(x) = \ln\left(x + \sqrt{x^2+1} \right), \tag{19}$$

is the inverse hyperbolic sine function. Given the lack of circular symmetry and the dependence of the potential separately on x and y coordinates, one should not be surprised by the complicated analytic form of $\Phi(X, Y)$ in Eq. (18). The expression given in Eq. (18) is extremely useful because it allows us to calculate the electrostatic potential, thus the confining potential felt by an electron at any location on the 2D square semiconductor quantum dot region:

$$V(X,Y) = -e\,\Phi(X,Y). \tag{20}$$

As expected from physical arguments, the strongest confinement potential is felt by electrons localized at the center of the uniformly charged square. The complexity of the expression in Eq. (18) indicates that dealing with confining potentials, even under idealized conditions, is quite challenging. This task becomes even more complicated if one wants to calculate precisely the realistic confining potentials that arise in various experimental situations. In all such cases, rather than using model potentials, one must apply numerical ab initio methods. Ab initio calculations for semiconductor quantum dots that incorporate the detailed form of the confinement potential and many electron interactions are extremely challenging. They rely on a combination of Schrödinger-Poisson equations to obtain realistic ab initio confining potentials and then calculate numerically the electronic structure [72–74]. Overall, detailed studies of the properties of confinement potentials (number of bound states, discrete energies, effect of electronic interactions on the discrete energy spectrum, etc.) are important to shed light on various properties of 2D semiconductor quantum dots for regimes where parabilicity is not guaranteed.

3 2D SEMICONDUCTOR QUANTUM DOTS IN A MAGNETIC FIELD

Theoretical models for confined systems of electrons in a 2D semiconductor quantum dot involve several approximations where the most typical ones assume that: (i) the motion of electrons is exactly 2D; (ii) the confining potential is simplified to a manageable level; and (iii) the interaction potential between electrons is of the Coulomb form. If a magnetic field is applied perpendicular to the 2D quantum dot plane, such a magnetic field affects both the spin and orbital motion of electrons. As a result, spin effects are also pronounced and they do occur at ordinary magnetic fields. The main effect of the magnetic field on the 2D semiconductor quantum dot is to change the nature of single-electron energy levels from 2D harmonic oscillator states to Landau levels. In such a case, states with different angular momentum become degenerate and electron-electron correlations start to play an important role.

The properties of a system of N electrons parabolically confined in a 2D semiconductor quantum dot subject to a magnetic field perpendicular to the 2D quantum dot plane,

$$\vec{B} = (0, 0, B_z), \tag{21}$$

are generally calculated from the following Hamiltonian:

$$\hat{H} = \sum_{i=1}^{N} \hat{H}_0(\vec{r}_i) + \sum_{i>j}^{N} v(r_{ij}) + g\,\mu_B\, B_z\, S_z, \tag{22}$$

where the first term is a one-electron term, the second term represents the potential energy of interacting electrons, the last term is the Zeeman energy, g is the electron's g-factor,μ_B is Bohr's magneton, and S_z is the total spin of the electrons in the z-direction of the magnetic field. The interaction potential between electrons is taken as a Coulomb interaction potential:

$$v(r_{ij}) = \frac{k\,e^2}{r_{ij}}, \tag{23}$$

where k is Coulomb's electric constant and $r_{ij} = |\vec{r_i} - \vec{r_j}|$ is the separation distance between two arbitrary electrons, i and j, respectively, located at $\vec{r_i}$ and $\vec{r_j}$.

The single-particle Hamiltonian $\hat{H}_0(\vec{r})$ for an electron in a 2D parabolic confinement potential subject to a uniform perpendicular magnetic field is written as:

$$\hat{H}_0(\vec{r}) = \frac{1}{2m}\left[\hat{\vec{p}} + e\,\vec{A}(\vec{r})\right]^2 + \frac{m}{2}\,\omega_0^2\, r^2, \tag{24}$$

where m is the electron's mass, $\hat{\vec{p}}$ is the electron's 2D quantum linear momentum operator, $-e(e > 0)$ is the electron's charge, $\hbar\,\omega_0$ is the strength of the parabolic confinement potential, and $\vec{A}(\vec{r})$ is the magnetic vector potential that generates the magnetic field, $\vec{B} = \vec{\nabla} \times \vec{A}$. In a symmetric gauge, the magnetic vector potential is written as:

$$\vec{A}(\vec{r}) = \frac{B_z}{2}(-y, x, 0), \tag{25}$$

where $\vec{r} = (x, y)$ is a 2D position vector. The stationary Schrödinger's equation for the quantum problem in Eq. (24) can be solved exactly. As a result, the wave function corresponding to the Hamiltonian in Eq. (22) admits an exact analytic solution for the case of noninteracting electrons. This fact provides a strong justification for the wide use of the parabolic confinement potentials when studying systems of confined electrons in a 2D semiconductor quantum dot. The quantum problem in Eq. (24) was solved a long time ago by Fock [75] and Darwin [76] and the eigenstates are routinely called the Fock-Darwin(FD) states:

$$\Psi_{nm_z}(r,\varphi) = N_{nm_z} \exp\left(-\frac{r^2}{4\,l_\Omega^2}\right)\left(\frac{r}{l_\Omega}\right)^{|m_z|} L_n^{|m_z|}\left(\frac{r^2}{2\,l_\Omega^2}\right)\frac{\exp(-i\,m_z\,\varphi)}{\sqrt{2\,\pi}}, \tag{26}$$

where

$$l_\Omega = \sqrt{\frac{\hbar}{2\,m\,\Omega}}, \tag{27}$$

is an effective magnetic length, $n = 0, 1, \ldots$ is the radial quantum number, $m_z = 0, \pm 1, \ldots$ is the z-angular momentum quantum number, $N_{n\,m_z}$ is a normalization constant, and $L_n^{|m_z|}\left(\frac{r^2}{2\,l_\Omega^2}\right)$ are associated Laguerre's polynomials [77]. The frequency Ω appearing in the expression for the effective magnetic length l_Ω is:

$$\Omega^2 = \omega_0^2 + \frac{\omega_c^2}{4}, \tag{28}$$

where

$$\omega_c = \frac{e\,|B_z|}{m} \tag{29}$$

is the so-called cyclotron frequency. Even though the classical solution of the same problem exhibits a forbidden central region for the motion of charge [78], the quantum ground state wave function peaks at $r = 0$. The energies corresponding to $\hat{H}_0(\vec{r})$ (without the Zeeman term) are:

$$E_{n\,m_z} = \hbar\,\Omega\,(2\,n + 1 + |m_z|) - \frac{\hbar\,\omega_c}{2}\,m_z. \tag{30}$$

These energy values crucially depend on the magnetic field through the dependence on the parameter ω_c. A key characteristic energy length of the system is the dimensionless magnetic field parameter defined as:

$$\gamma = \frac{\omega_c}{\omega_0}. \tag{31}$$

The expression in Eq. (30) predicts that the energy levels with positive m_z shift downward and levels with negative m_z shift upward as the magnetic field increases. This implies that electrons undergo subtle transitions from one to another quantum number as a function of the magnetic field. When $\omega_0 = 0$ or in the limit of a very large magnetic field ($\omega_c \gg \omega_0$), the effective magnetic length l_Ω reduces the electronic magnetic length l_0:

$$l_\Omega \rightarrow \sqrt{\frac{\hbar}{m\,\omega_c}} = \sqrt{\frac{\hbar}{e\,|B_z|}} = l_0; \;\; \gamma = \frac{\omega_c}{\omega_0} \rightarrow \infty. \tag{32}$$

If we assume $\omega_0 \neq 0$, the effective magnetic length l_Ω can also be written as:

$$\frac{1}{l_\Omega^2} = 2\,\alpha^2 \sqrt{1 + \frac{1}{4}\left(\frac{\omega_c}{\omega_0}\right)^2}; \;\; \omega_0 \neq 0, \tag{33}$$

where the parameter α is the inverse oscillator length. By using the expression in Eq. (33), it is easy to recover the 2D harmonic oscillator states from the FD states in the limit of the zero magnetic field ($\omega_c/\omega_0 \rightarrow 0$).

One should note that (without the Zeeman term) the Hamiltonian of the system is spin-free, therefore the Zeeman energy can be taken into account afterward. As in any quantum problem, one must solve the stationary Schrödinger's equation to obtain the many electron energy spectrum and wave functions:

$$\hat{H}\,\Psi(\vec{r_1}, \ldots, \vec{r_N}) = E\,\Psi(\vec{r_1}, \ldots, \vec{r_N}), \tag{34}$$

for the Hamiltonian, $\hat{H}$ in Eq. (22). Clearly, this is a formidable task in the presence of interactions. With few exceptions [79–82], the general quantum problem cannot be solved exactly, even for small systems with $N = 2$ electrons [83]. As a result, one must resort to numerical treatments, for instance quantum Monte Carlo methods [84–98] to solve the general quantum problem in the presence of interactions. Thus, despite the apparent simplicity, few-electron 2D semiconductor quantum dots [99–101] are of particular interest because single-electron confinement energy, cyclotron energy for ordinary magnetic fields, and electron-electron correlation effects are all of the same order of magnitude. As a result, rich physics and many interesting quantum phenomena arise in such systems.

If the focus is on the electronic properties of the system, then the Hamiltonian $\hat{H}$ in Eq. (22) is an appropriate description for a system of parabolically confined electrons. However, for other confining potentials, typically those that are electrostatic in origin (such as the case of a charged 2D disk potential), one must amend the Hamiltonian by including all effects that originate from the presence of a charged background. In this manner one can also relate to other well-known models that apply to bulk systems of electrons. One such famous model in condensed matter physics is the *jellium*-model, also known as the uniform (or homogeneous) electron gas. The jellium model is essentially a quantum mechanical model of interacting electrons in a solid where the positive charges (i.e., ions) are assumed to be a background uniformly distributed in space. In the jellium model, the Coulomb self-energy of the positive background represents an indispensable term that enters the calculation of the total energy of any charge-neutral system. A typical jellium Hamiltonian can be written as:

$$\hat{H} = \hat{T} + \hat{V}_{ee} + \hat{V}_{eb} + \hat{V}_{bb}, \tag{35}$$

where in this case

$$\hat{T} = \sum_{i=1}^{N} \left[\hat{\vec{p}}_i + e\,\vec{A}\,(\vec{r_i}) \right]^2 \tag{36}$$

is the total kinetic energy of the electrons (in a magnetic field),

$$\hat{V}_{ee} = \sum_{i>j}^{N} v(r_{ij}) \tag{37}$$

is the electron-electron Coulomb repulsion potential,

$$\hat{V}_{eb} = \sum_{i=1}^{N} V(\vec{r_i}) \tag{38}$$

is the electron-background interaction term and

$$\hat{V}_{bb} = \frac{\rho_0^2}{2} \int_A d^2r_1 \int_A d^2r_2\, v(r_{12}) \tag{39}$$

is the background-background interaction energy of the positive background charge interacting with itself, namely, the Coulomb self-energy. In the jellium model, N electrons of charge $-e(e > 0)$ are considered embedded in a uniform neutralizing background with area A filled with positive charge $Q = N\,e$. The uniform density of the system is $\rho_0 = N/A$ and $v(r_{12}) = k\,e^2/r_{12}$. Note that the Zeeman term is not specifically written.

In spite of its simplicity, the jellium model is very instructive as it displays interesting many body effects of delocalized electrons in a metal, such as screening, plasmons, Wigner crystallization, and Friedel oscillations [102]. For a three-dimensional electronic system (3DES), the artificial and structureless background charge fills uniformly a cubic volume [103–105] or a spherical volume [106–109]. The 2D counterpart of a jellium model implies a background charge that fills uniformly a 2D disk or a square region [110]. In particular, the 2D disk geometry for jellium is commonly used in studies of two-dimensional electronic systems (2DES) in the FQHE regime [111–115]. The Coulomb self-energy of a uniformly charged 2D disk can be easily calculated [116]. A natural extension to a three-dimensional (3D) system would lead to a cylindrical jellium model [117] with the additional benefit of incorporating finite thickness effects in the third dimension [118].

4 DIVERSE REGIMES OF 2D SEMICONDUCTOR QUANTUM DOTS

Confined electron systems in a 2D semiconductor quantum dot are an ideal ground to study novel phenomena. In particular, the study of the strong magnetic field regime with all electrons fully spin polarized is relevant because this is the crossover regime between microscopic 2D semiconductor quantum dots and macroscopic 2DES of FQHE type [23–31]. The weak (or zero) magnetic field regime is also interesting [119–123] with the main focus on the Fermi liquid-Wigner solid crossover regime. The transitional regime of intermediate (weak) magnetic fields is expected to have prominent spin effects and requires further investigation. The richness and complexity of the phenomena associated with the behavior of 2D semiconductor quantum dot systems in a perpendicular magnetic field make them fascinating and at the same time challenging from a theoretical and modeling point of view. In the following we consider systems of parabolically confined electrons in a 2D semiconductor quantum dot and try to classify the various quantum regimes that influence the properties of confined electrons [15].

There are two main dimensionless parameters that determine the behavior of a 2D semiconductor quantum dot in a perpendicular magnetic field:

$$\lambda = \frac{k\,e^2\,\alpha}{\hbar\,\omega_0};\quad \gamma = \frac{\omega_c}{\omega_0}, \tag{40}$$

where the dimensionless Coulomb correlation parameter λ gauges the strength of the Coulomb correlation energy relative to the confinement energy while the dimensionless magnetic field parameter γ gauges the strength of the magnetic field relative to the confinement energy. The characteristic length scale of the system is: $l_{\Omega}=\sqrt{\hbar/(2\,m\,\Omega)}$ where $\Omega=\sqrt{\omega_0^2+\omega_c^2/4}$ is a characteristic frequency, $\omega_c=e\,|B_z|/m$ is the cyclotron frequency, and α is the harmonic (parabolic) oscillator inverse length. Another parameter related to λ that characterizes the interaction strength of the electrons is the Wigner-Seitz dimensionless density parameter r_s. In a homogeneous 2D electron system, such a parameter is defined by the radius of the circle (in units of the Bohr radius) that every electron effectively occupies so that the homogeneous density is:

$$n_0=\frac{1}{\pi\,(r_s\,a_{\mathrm{B}})^2}, \tag{41}$$

where

$$a_{\mathrm{B}}=\frac{\hbar^2}{k\,m\,e^2} \tag{42}$$

is Bohr's radius. In the low-density (strong interaction) limit $r_s\longrightarrow\infty$, classical considerations suggest the stabilization of a Wigner molecule [124] or a Wigner crystal-like phase [125, 126]. In contrast, at high densities (weak interaction) limit, $r_s\rightarrow 0$, a Fermi liquid-like description is expected to be valid.

The phase diagram of 2D semiconductor quantum dots in a strong perpendicular magnetic field is intricate. The overwhelming majority of studies to date have considered only the stability between competing liquid and Wigner crystal phases. It is likely that other, more exotic intermediate phases that have not been yet considered so far can stabilize in systems of confined electrons in a 2D semiconductor quantum dot. The most conventional of such quantum phases would be the equivalent of the charge density wave stripe state [127–129] seen in quantum Hall samples in a weak magnetic field. The possibility of other intermediate quantum phases with liquid crystalline order and broken rotational symmetry [130–138] cannot be excluded. Naive arguments may even suggest that, for a careful choice of parameters (temperature, magnetic field, density, etc.) because of confinement, some intermediate exotic liquid crystalline phases [139, 140] might be energetically more robust than their anisotropic quantum Hall counterparts [141].

In the low-density limit and at arbitrary magnetic fields, the system should resemble a Wigner molecule or a Wigner crystal state. In the high-density limit and at strong magnetic fields, the system should resemble a strongly correlated fully spin-polarizedLaughlin-likeFQHE-type liquid state. In the high-density limit and at zero magnetic field, the system should resemble a correlated spin-unpolarized Fermi liquid system In the high-density limit and for intermediate magnetic fields, the system should resemble a correlated partially spin-polarized Fermi liquid state with pronounced spin-dependent effects. In short, all these important physical constraints should be carefully incorporated in any description or microscopic theory dealing with confined electrons in a 2D semiconductor quantum dot in a perpendicular magnetic field. If one wants to start with a microscopic wave function, a trial wave function for N electrons should have, at the very least, the following form [15]:

$$\Psi_\gamma = \prod_{i<j}^{N} J(r_{ij}) \prod_{i<j}^{N} (z_i - z_j)^{n_p} D_\uparrow(\phi) D_\downarrow(\phi) \chi(S), \tag{43}$$

where $J(r_{ij})$ are Jastrow factors, z_j is the 2D position of electrons in complex notation, $r_{ij} = |\vec{r_i} - \vec{r_i}|$ is the distance between a pair of electrons, $D_{\uparrow\,\downarrow}(\phi)$ are Slater determinants of single-particle orbitals, $\phi(\vec{r})$ corresponding to $N_\uparrow$spin-up and $N_\downarrow = N - N_\uparrow$spin-down electrons, $\chi(S)$ is the spin function, $S = \{s_1, s_2, \ldots, s_N\}$ represents all spin coordinates, and $n_p = 0, 1, \ldots$ is an integer quantum number that determines the parity of the Laughlin polynomials, $\prod_{i<j}^{N}(z_i - z_j)^{n_p}$. The single-particle orbitals $\phi(\vec{r})$ that enter the Slater determinant functions may be chosen to be FD orbitals when describing a liquid state (at high density) or localized Gaussian orbitals [142, 143] when describing a Wigner molecule/crystal state (at low density). We expect the system to manifest its crystalline nature (Wigner molecule/crystal) in the low-density limit. It is expected that the presence of a perpendicular magnetic field will further stabilize the crystalline phases if a crystalline wave function is the starting premise. On the other hand, exotic liquid phases may emerge and may stabilize because of the presence of a perpendicular magnetic field in the high-density limit.

5 APPLICATIONS OF QUANTUM DOTS AND MAGNETIC NANOPARTICLES IN BIOLOGICAL SYSTEMS

Recently, magnetic nanoparticles and nanosized dots have become attractive tools for clinical applications. The nanometer size of the structure in conjunction with peculiar optoelectronic and magnetic properties [144–146] allow them to achieve easy accessibility to specific organisms for targeted means of function (e.g., for pathogens in infected tissues). Nanodots can be easily coated with biocompatible substances to increase their bio-persistence [147, 148]. Alternatively, their surfaces can be functionalized by various polymers. A variety of polymers can be selected based on their abilities to improve the solubility of particles in a particular fluid. Many biological applications have been proposed for magnetic nanodots over the past few decades. Some of these topics continue to be actively pursued by interdisciplinary research [149].

Two of the most common biomedical applications that involve magnetic nanoparticles are diagnostics and drug delivery. Compelling evidence has already shown that functionalized nanoparticles are capable of binding and targeting pathogens, thus enhancing their detection [150–153]. The magnetic property of these particles also facilitates their ability to locate tumor sites or areas where a therapeutic approach is needed. This makes them attractive tools for drug delivery. In the heart, the contraction of left ventricular (LV) myocytes (the actual working cells that produce the power of the heart) is essential to maintain the normal function of the heart. Reduced contraction is often associated with heart diseases (e.g., heart failure). Therefore, recent work aimed to test whether magnetic nanoparticles/dots can be applied directly to isolated cardiac myocytes and whether an increased magnetic field can improve the contractile function of LV myocytes [144]. The magnetic nanoparticles that we are interested in are spinel ferrites of MFe_2O_4($M = Mn$, Mg, Zn, Co, and other divalent cations or a mixture of these cations) [154]. Due to the high compositional flexibility, spinel ferrites offer great opportunities for designing and fine-tuning the magnetic properties of nanoparticles based on the magnetic interactions at the atomic level and at the crystal chemistry. Russell-Saunders couplings between the cations on the lattice sites of the metals make up the spinel structure to shape its magnetic properties. Among the family of ferrites, $CoFe_2O_4$ has been known to have a large magnetic moment. In addition to Co ferrite, we are also interested in making $ZnFe_2O_4$ due to the fact that Zn exists in the human body, so that $ZnFe_2O_4$ may be a safer ferrite to be used in human tissues or cells. The nanoparticles are coated with a polymer to increase their biocompatibility. This technique provides the advantage of diminishing the immune responses of the tissues to the foreign particles and/or avoiding cytotoxicity of human cells or tissues [155]. The polymer coating contains numerous free hydroxyls that willingly form hydrogen bonds in an aqueous solution. Thus, the many surface hydroxyls hold the particle and surface coat in suspension for an indefinite period of time.

The surfaces of the polymer can be functionalized with molecules (e.g., antibodies, peptides, etc.) that can direct them to specific sites in vivo, and can influence their distribution, reactivity, and retention at the target site [156, 157]. The coating can make the injection possible and safe while retaining the magnetic properties that are important to activate the response that we expect to see in the cells.

We investigated the effects of $CoFe_2O_4$ and $ZnFe_2O_4$ (coated with silicon) on the contractility of rat LV myocytes in a recent study [144]. The experimental evidence in these studies suggests that the delivery of magnetic quantum dots to the surface of a rat's heart considerably affects and improves its mechanical functions. Under normal conditions, LV myocytes contract in response to an electrical impulse elicited by the transport of charged ions through voltage-gated ion channels across the membrane (mainly Na^+, Ca^{2+}, and K^+). The amount of ion fluxes determines the homeostasis of the intracellular Ca^{2+} level, which is the critical signaling messenger for initiating and determining the amplitude of the myocyte contraction in each heartbeat. Therefore, the regulation of ion channels and electrical activity of the cardiac myocyte or intracellular Ca^{2+} will significantly affect its contractility. As expected, reduced contractions caused by defects of ion channels or intracellular Ca^{2+} release are the main mechanism for myocardial dysfunction in heart failure. Hence, these experiments [144] suggest that inducing a magnetic field by nanoparticles can improve LV myocyte contraction triggered by electrical stimulation. Whether the magnetic field increases the dynamics of ion fluxes or intracellular Ca^{2+} needs to be further investigated. Furthermore, a full understanding of the mechanism mediating magnetic application by nanoparticles to both healthy and diseased cardiac myocytes becomes important. Overall, clinical studies suggest that the delivery of magnetic quantum dots to the rat's heart considerably affects its mechanical functions. This is important because the current approach seems to address a major limitation of cell delivery systems for cardiac repair, which has been the ineffective localization and retention of physiologically relevant numbers of cells in the heart. At a basic level, the typical heart's beatings can be approximated as an oscillation system of ions, thus, charges. In order to maintain normal blood circulation, these oscillating systems of charges should beat in a regular pattern, thus with a well-defined (angular) frequency. Obviously, problems arise when the heart does not beat in a regular pattern at the given frequency. If this is the case, electrical currents applied by means of a defibrillator restore the regular heartbeat patterns. Thus, the functions are regained by inducing the heart to beat regularly again via electric pulses.

The present experiments suggest that a similar effect to the heart may be induced via magnetic fields rather than electric fields. This may occur due to the exposure of a rat's heart to the rather strong magnetization (thus the magnetic field) created by the magnetic quantum dots when they are in contact and interact with the heart cells. Obviously, a full understanding of the interactions and effects is way beyond all possible approaches. Nevertheless, one can still capture the basic physics by adopting a simple model that considers the heart beatings as an oscillating system of charges exposed to a relatively large magnetic field (magnetization).

Obviously, the oscillating systems are 3D. However, given that the oscillatory motion along the direction of the magnetic field will be unaffected by the magnetic field, we may focus our attention only on the oscillatory patterns arising on the 2D plane perpendicular to the magnetic field. For simplicity, we may consider the magnetic field to which the heart cells are exposed as uniform and locally pointing in the z-direction, $\vec{B}=(0,0,B)$. Thus, our concern is to understand the effect of such a magnetic field on the regularity of the pulsating patterns of a 2D (on the $x-y$ plane) oscillatory system of charged cells with positive charge $q>0$ and given mass m. We assume that, in the absence of the magnetic field, the original pulsating frequency of the 2D oscillatory motion is ω_0 (not necessarily the native one, though). Through this simple model, we want to describe how the magnetic field affects the beating patterns of the heart and whether the outcome of the effect is to induce transitions from irregular to regular heartbeat patterns (or vice versa). This outcome would be much desirable because it indicates the ability to control heartbeat patterns via magnetic fields rather than electric fields. For the simple model under consideration, one can write the Newtonian equations of the 2D motion (along the x and y directions) as:

$$\begin{cases} \dot{v}_x = +\omega_c v_y - \omega_0^2 x \\ \dot{v}_y = -\omega_c v_x - \omega_0^2 y, \end{cases} \tag{44}$$

where $\omega_c = qB/m > 0$ is the new frequency due to the magnetic field, v_x (and v_y) are velocities, and the "dot" symbol represents time derivatives. The solution of the above system of coupled differential equations of motion when extended to both velocities and coordinates is not trivial. Nevertheless, we were able to solve exactly this system of coupled differential equations [78].

It turns out that, under the most general conditions, the resulting motion and the time-dependent expressions for the coordinates $x(t)$ and $y(t)$ are rather complicated and depend on many factors. However, the key result is that, depending on various combinations of parameters, a magnetic field is able to induce dramatic transitions from irregular to regular patterns and vice versa (thus resetting the heart beatings much like a quantum dot "magnetic defibrillator"). We see that there are important effects and a strong influence of the magnetic field on the regularity of the heartbeat patterns of motion for various values of a magnetic field. These patterns depend on how the oscillatory system is behaving at the moment when it becomes exposed to the large magnetic field (various initial conditions, such as small or large initial velocity and so on). Clearly, magnetic quantum dots may cause important effects on the heartbeat patterns when interacting with the heart cells. For example, we were able to prove that even under the most general conditions, there are various combinations of the magnetic field that may stimulate the system to regain the regularity of the patterns of motion. For instance, we found out that the regularity of the patterns of motion is regained (namely, the cycles repeat themselves at times $t_n > 0$) at magnetic field values given by:

$$\Omega\, t_n = n\, 2\pi; \quad n = 1, 2, \ldots, \tag{45}$$

where $\Omega^2 = \omega_c^2 + 4\,\omega_0^2$ represents the influence of the magnetic field on the original oscillating patterns of the system. The overall conclusion is that the magnetization effects of the magnetic quantum dots under consideration have a considerable effect on the pulsating properties of the heart cells. Through these effects, the locally applied magnetic field is able to induce as well as turn on/off various regular beating patterns, thus resetting the heart's beatings and performing the same functions as an electric defibrillator.

To summarize, $CoFe_2O_4$ and $ZnFe_2O_4$ nanodot particles were synthesized for clinical studies. Both types of particles show interesting ferromagnetic behavior at room temperature. Experiments performed on rat heart cells show that heart cells survive normally with being injected by both Co and Zn ferrite particles in the incubation solution. In both cases, we observed a considerable improvement of rat heart cell functions. These results suggest that both Zn and Co ferrites can help improve the contraction of rat's heart cells. However, Zn ferrites may be advantageous over Co ferrites because they maintain the dynamics of myocyte relaxation. It can also be argued theoretically that the magnetic quantum dots with a certain magnetization may have a considerable effect on the pulsating properties of the heart cells. Through this effect, the locally applied magnetic field is able to induce as well as turn on/off various regular beating patterns, thus, resetting the heart beatings and performing the same functions as an electric defibrillator.

ACKNOWLEDGMENT

This research was supported in part by the National Science Foundation Grants No. DMR-1410350 and DMR-1705084.

REFERENCES

[1] L. Theodore, Nanotechnology: Basic Calculations for Engineers and Scientists, Wiley, Hoboken, NJ, 2005.
[2] M. Di Ventra, S. Evoy, J.R. Heflin Jr. (Eds.), Introduction to Nanoscale Science and Technology, in: Springer, New York, NY, 2004.
[3] L. Banyai, S.W. Koch, Semiconductor Quantum Dots, in: Series on Atomic, Molecular and Optical Physics, vol. 2, World Scientific, Singapore, 1993.
[4] L. Jacak, P. Hawrylak, A. Wojs, Quantum Dots, Nanoscale and Technology Series, Springer, New York, NY, 1998.
[5] Y. Masumoto, T. Takagahara (Eds.), Semiconductor Quantum Dots: Physics, Spectroscopy and Applications, in: Springer, New York, NY, 2002.
[6] Single Quantum Dots: Fundamentals, Applications and New Concepts, P. Michler (Ed.), In: Topics in Applied Physics, vol. 90, Springer, New York, NY, 2003.
[7] J.P. Bird, Electron Transport in Quantum Dots, Kluwer Academic Publishers, Dordrecht, 2003.
[8] G.W. Bryant, Phys. Rev. Lett. 59 (1987) 6.
[9] R.C. Ashoori, H.L. Stormer, J.S. Weiner, L.N. Pfeiffer, K.W. Baldwin, K.W. West, Phys. Rev. Lett. 68 (1992) 6.
[10] R.C. Ashoori, Nature 379 (1996) 6.
[11] L.P. Kouwenhoven, C.M. Marcus, Phys. World 11 (1998) 35.
[12] D. Heitmann, J.P. Kotthaus, Phys. Today 46 (1993) 56.
[13] M.A. Kastner, Phys. Today 46 (1) (1993) 24.
[14] S. Tarucha, D.G. Austing, T. Honda, R.J. van der Hage, L.P. Kouwenhoven, Phys. Rev. Lett. 77 (1996) 6.
[15] O. Ciftja, Phys. Scr. 88 (2013) 6.
[16] S. Tarucha, D.G. Austing, T. Honda, R.J. van der Hage, L.P. Kouwenhoven, Jpn. J. Appl. Phys. 36 (1997) 6.
[17] S. Sasaki, D.G. Austing, S. Tarucha, Physica B 256 (1997) 6.
[18] D.G. Austing, S. Sasaki, S. Tarucha, S.M. Reimann, M. Koskinen, M. Manninen, Phys. Rev. B 60 (1999) 6.
[19] S. Das Sarma, A. Pinczuk (Eds.), Perspectives in Quantum Hall Effects, in: Wiley, New York, NY, 1996.
[20] R.E. Prange, S.M. Girvin, The Quantum Hall Effect, Springer Verlag, New York, NY, 1990.
[21] D.C. Tsui, H.L. Stormer, A.C. Gossard, Phys. Rev. Lett. 48 (1982) 6.
[22] R.B. Laughlin, Phys. Rev. Lett. 50 (1983) 6.
[23] R.B. Laughlin, Phys. Rev. B 27 (1983) 6.
[24] F.D.M. Haldane, Phys. Rev. Lett. 55 (1985) 6.
[25] R. Morf, B.I. Halperin, Phys. Rev. B 33 (1986) 6.
[26] S.M. Girvin, A.H. MacDonald, P.M. Platzman, Phys. Rev. B 33 (1986) 6.
[27] A. Lopez, E. Fradkin, Phys. Rev. B 44 (1991) 6.
[28] B.I. Halperin, P.A. Lee, N. Read, Phys. Rev. B 47 (1993) 6.
[29] E. Rezayi, N. Read, Phys. Rev. Lett. 72 (1994) 6.
[30] O. Ciftja, S. Fantoni, Phys. Rev. B 56 (1997) 6.

[31] O. Ciftja, S. Fantoni, J.W. Kim, M.L. Ristig, J. Low. Temp. Phys. 108 (1997) 6.
[32] O. Ciftja, S. Fantoni, K. Gernoth, Phys. Rev. B 55 (1997) 6.
[33] O. Ciftja, Phys. Rev. B 59 (1999) 6.
[34] L.L. Sohn, L.P. Kouwenhoven, G. Schön (Eds.), Mesoscopic Electron Transport, of NATO ASI Series, in: Series E: Applied Sciences, vol. 345, Kluwer Academic Publishers, Dordrecht, 1997.
[35] D.P. DiVincenzo, Science 270 (1995) 255.
[36] U. Merkt, J. Huser, M. Wagner, Phys. Rev. B 43 (1991) 6.
[37] D. Pfannkuche, R.R. Gerhardts, P.A. Maksym, V. Gudmundsson, Physica B 189 (1993) 6.
[38] R.C. Ashoori, H.L. Stormer, J.S. Weiner, L.N. Pfeiffer, K.W. Baldwin, K.W. West, Phys. Rev. Lett. 71 (1993) 6.
[39] A. Kumar, S.E. Laux, F. Stern, Phys. Rev. B 42 (1990) 6.
[40] D.A. Broido, K. Kempa, P. Bakshi, Phys. Rev. B 42 (1990) 6.
[41] V. Gudmundsson, R.R. Gerhardts, Phys. Rev. B 43 (1991) 6.
[42] D. Pfannkuche, V. Gudmundsson, P.A. Maksym, Phys. Rev. B 47 (1993) 6.
[43] M. Fujito, A. Natori, H. Yasunaga, Phys. Rev. B 53 (1996) 6.
[44] H.M. Muller, S.E. Koonin, Phys. Rev. B 54 (1996) 6.
[45] B. Reusch, H. Grabert, Phys. Rev. B 68 (2003) 6.
[46] M. Koskinen, M. Manninen, S.M. Reimann, Phys. Rev. Lett. 79 (1997) 6.
[47] O. Steffens, U. Rössler, M. Suhrke, Europhys. Lett. 42 (1998) 6.
[48] M. Macucci, K. Hess, G.J. Iafrate, Phys. Rev. B 48 (1993) 6.
[49] M. Ferconi, G. Vignale, Phys. Rev. B 50 (1994) 6.
[50] P.A. Maksym, Phys. Rev. B 53 (1996) 6.
[51] F. Bolton, Phys. Rev. B 54 (1996) 6.
[52] J. Kainz, S.A. Mikhailov, A. Wensauer, U. Rössler, Phys. Rev. B 65 (2002) 6.
[53] A. Harju, S. Siljamäki, R.M. Nieminen, Phys. Rev. B 65 (2002) 6.
[54] P.A. Maksym, T. Chakraborty, Phys. Rev. Lett. 65 (1990) 6.
[55] A.H. MacDonald, M.D. Johnson, Phys. Rev. Lett. 70 (1993) 6.
[56] S.M. Reimann, M. Manninen, Rev. Mod. Phys. 74 (2002) 1283.
[57] O. Ciftja, Mod. Phys. Lett. B, 23 (2009) 6.
[58] O. Ciftja, Physica B 404 (2009) 6.
[59] O. Ciftja, J. Phys. Condens. Matter. 19 (2007) 6.
[60] K. Hirose, N.S. Wingreen, Phys. Rev. B 59 (1999) 6.
[61] P.S. Drouvelis, P. Schmelcher, F.K. Diakonos, Europhys. Lett. 64 (2003) 6.
[62] J. Adamowski, M. Sobkowicz, B. Szafran, S. Bednarek, Phys. Rev. B 62 (2000) 6.
[63] S. De Filippo, M. Salerno, Phys. Rev. B 62 (2000) 6.
[64] M. Helle, A. Harju, R.M. Nieminen, Phys. Rev. B 72 (2005) 6.
[65] O. Ciftja, J. Comput.-Aided Mater. Des. 14 (2007) 6.
[66] O. Ciftja, I. Hysi, Appl. Math. Lett. 24 (2011) 6.
[67] O. Ciftja, C. Wexler, Phys. Rev. B 67 (2003) 6.
[68] O. Ciftja, Phys. Lett. A 374 (2010) 6.
[69] O. Ciftja, Eur. J. Phys. 32 (2011) L55.
[70] M. Abramowitz, I.A. Stegun, Handbook of Mathematical Functions, Ninth Printing, Dover, New York, 1970.
[71] O. Ciftja, J. Electrostat. 71 (2013) 6.
[72] M. Stopa, Phys. Rev. B 54 (1996) 6.
[73] N.A. Bruce, P.A. Maksym, Phys. Rev. B 61 (2000) 6.
[74] R. Ravishankar, P. Matagne, J.P. Leburton, R.M. Martin, S. Tarucha, Phys. Rev. B 69 (2004) 6.
[75] V. Fock, Z. Phys. 47 (1928) 446.
[76] C.G. Darwin, Proc. Cambridge Philos. Soc. 27 (1930) 86.

[77] G.B. Arfken, H.J. Weber, Mathematical Methods for Physicists, fifth ed., Harcourt/Academic Press, New York, 2001.
[78] O. Ciftja, Cent. Eur. J. Phys. 11 (2013) 6.
[79] M. Taut, Phys. Rev. A 48 (1993) 6.
[80] M. Taut, J. Phys. A. Math. Gen. 27 (1994) 6.
[81] A. Turbiner, Phys. Rev. A 50 (1994) 6.
[82] M. Dineykhan, R.G. Nazmitdinov, Phys. Rev. B 55 (1997) 6.
[83] O. Ciftja, M.G. Faruk, J. Phys. Condens. Matter. 18 (2006) 6.
[84] W.L. McMillan, Phys. Rev 138 (1965) 6.
[85] R. Jastrow, Phys. Rev 98 (1955) 6.
[86] D. Ceperley, G.V. Chester, M.H. Carlos, Phys. Rev. B 16 (1977) 6.
[87] D. Ceperley, Phys. Rev. B 18 (1978) 6.
[88] K. Schmidt, M.H. Kalos, M.A. Lee, G.V. Chester, Phys. Rev. Lett. 45 (1980) 6.
[89] P.A. Whitlock, D.M. Ceperley, G.V. Chester, M.H. Kalos, Phys. Rev. B 19 (1979) 6.
[90] O. Ciftja, S. Moroni, S. Fantoni, J. Phys. Condens. Matter. 13 (2001) 6.
[91] O. Ciftja, S.A. Chin, F. Pederiva, J. Low. Temp. Phys. 122 (2001) 6.
[92] O. Ciftja, S.A. Chin, Phys. Rev. B 68 (2003) 6.
[93] J.B. Anderson, J. Chem. Phys. 63 (1975) 6.
[94] M.H. Kalos, D. Levesque, L. Verlet, Phys. Rev. A 9 (1974) 6.
[95] P.J. Reynolds, D.M. Ceperley, B.J. Alder, W.A. Lester Jr., J. Chem. Phys. 77 (1982) 6.
[96] D.M. Ceperley, B.J. Alder, J. Chem. Phys. 81 (1984) 6.
[97] H.A. Forbert, S.A. Chin, Phys. Rev. B 63 (2001) 6.
[98] G. Ortiz, D.M. Ceperley, R.M. Martin, Phys. Rev. Lett. 71 (1993) 6.
[99] F.M. Peeters, V.A. Schweigert, Phys. Rev. B 53 (1996) 6.
[100] M.B. Tavernier, E. Anisimovas, F.M. Peeters, Phys. Rev. B 70 (2004) 6.
[101] J.J. Palacios, L.M. Moreno, G. Chiappe, E. Louis, C. Tejedor, Phys. Rev. B 50 (1994) 6.
[102] C. Kittel, Introduction to Solid State Physics, seventh ed., Wiley, New York, 1996.
[103] G.D. Mahan, Many-Particle Physics, third ed., Plenum Press, New York, 1981.
[104] A.L. Fetter, J.D. Walecka, Quantum Theory of Many-Particle Systems, Dover Publications, 2003.
[105] O. Ciftja, Phys. Lett. A 375 (2011) 6.
[106] J. Martins, J. Bottet, Surf. Sci. 106 (1981) 6.
[107] W. Ekardt, Surf. Sci. 106 (1981) 6.
[108] M.Y. Chou, A. Cleland, M.L. Cohen, Solid State Comm. 52 (1984) 6.
[109] W.-J. Yin, X. Gu, X.-G. Gong, Solid State Comm. 147 (2008) 6.
[110] O. Ciftja, B. Sutton, A. Way, AIP Adv. 3 (2013) 6.
[111] O. Ciftja, J. Math. Phys. 52 (2011) 6.
[112] O. Ciftja, Physica B 404 (2009) 2244.
[113] O. Ciftja, Physica B 406 (2011) 2054.
[114] O. Ciftja, Physica B 404 (2009) 6.
[115] G. Dev, J.K. Jain, Phys. Rev. B 45 (1992) 6.
[116] O. Ciftja, Europhys. Lett. 74 (2006) 6.
[117] O. Ciftja, Physica B 407 (2012) 6.
[118] F.C. Zhang, S. Das Sarma, Phys. Rev. B 33 (1986) 6.
[119] C.E. Creffield, W. Häusler, J.H. Jefferson, S. Sarkar, Phys. Rev. B 59 (1999) 6.
[120] S.M. Reimann, M. Koskinen, M. Manninen, Phys. Rev. B 62 (2000) 6.
[121] R. Egger, W. Häusler, C.H. Mak, H. Grabert, Phys. Rev. Lett. 82 (1999) 6.
[122] J. Harting, O. Mülken, P. Borrmann, Phys. Rev. B 62 (2000) 6.
[123] A.V. Filinov, M. Bonitz, Y.E. Lozovik, Phys. Rev. Lett. 86 (2001) 6.
[124] C. Yannouleas, U. Landman, Phys. Rev. B 66 (2002) 6.
[125] H.M. Müller, S.E. Kronin, Phys. Rev. B 54 (1996) 14532.

[126] J.R. Trail, M.D. Towler, R.J. Needs, Phys. Rev. B 68 (2003) 6.
[127] A.A. Koulakov, M.M. Fogler, B.I. Shklovskii, Phys. Rev. Lett. 76 (1996) 6.
[128] M.M. Fogler, A.A. Koulakov, B.I. Shklovskii, Phys. Rev. B 54 (1996) 6.
[129] R. Moessner, J.T. Chalker, Phys. Rev. B 54 (1996) 6.
[130] E. Fradkin, S.A. Kivelson, Phys. Rev. B 59 (1999) 6.
[131] H. Yi, H.A. Fertig, R. Côté, Phys. Rev. Lett. 85 (2000) 6.
[132] C. Wexler, A.T. Dorsey, Phys. Rev. B 64 (2001) 6.
[133] L. Radzihovsky, A.T. Dorsey, Phys. Rev. Lett. 88 (2002) 6.
[134] H.Y. Kee, Phys. Rev. B 67 (2003) 6.
[135] K. Musaelian, R. Joynt, J. Phys. Condens. Matter. 8 (1996) 6.
[136] C. Wexler, O. Ciftja, Int. J. Mod. Phys. B 20 (2006) 6.
[137] O. Ciftja, C.M. Lapilli, C. Wexler, Phys. Rev. B 69 (2004) 6.
[138] C. Wexler, O. Ciftja, J. Phys. Condens. Matter. 14 (2002) 6.
[139] O. Ciftja, C. Wexler, Phys. Rev. B 65 (2002) 6.
[140] Q.M. Doan, E. Manousakis, Phys. Rev. B 75 (2007) 6.
[141] M.P. Lilly, K.B. Cooper, J.P. Eisenstein, L.N. Pfeiffer, K.W. West, Phys. Rev. Lett. 82 (1999) 6.
[142] L.H. Nosanow, Phys. Rev. Lett. 13 (1964) 6.
[143] L.H. Nosanow, Phys. Rev. 146 (1966) 6.
[144] N.H. Hong, A.T. Raghavender, O. Ciftja, M.-H. Phan, K. Stojak, H. Srikanth, Y. H. Zhang, Appl. Phys. A 112 (2013) 6.
[145] C.C. Berry, A.S.G. Curtis, J. Phys. D Appl. Phys. 36 (2003) 6.
[146] A. Ito, M. Shinkai, H. Honda, T.J. Kobayashi, Biosci. Bioeng. 100 (2005) 1.
[147] S. Dutza, W. Andraa, R. Hergta, R. Mullera, C. Oestreichc, C. Schmidta, J. Topferd, M. Zeisbergera, M. Bellemann, J. Mag. Mag. Mater. 311 (2007) 6.
[148] A. Gupta, M. Gupta, Biomaterials 26 (2005) 6.
[149] Q. Pankhurst, J. Connolly, S.K. Jones, J. Dobson, J. Phys. D Appl. Phys. 36 (2003) 6.
[150] D.G. Georganopoulou, L. Chang, J.M. Nam, C.S. Thaxton, E.J. Mufson, W.L. Klein, C.A. Mirkin, Proc. Natl. Acad. Sci. USA 102 (2005) 6.
[151] J.K. Herr, J.E. Smith, C.D. Medley, D. Shangguan, W. Tan, Anal. Chem. 78 (2006) 6.
[152] J.M. Nam, C.S. Thaxton, C.A. Mirkin, Science 301 (2003) 6.
[153] J.E. Smith, C.D. Medley, Z. Tang, D. Shangguan, C. Lofton, W. Tan, Anal. Chem. 79 (2007) 6.
[154] C. Liu, B. Zou, A.J. Rondinone, Z.L. Zhang, J. Am. Chem. S. 26 (2000) 6.
[155] L.M. Lacava, V.A.P. Garcia, S. Kuckelhaus, R.B. Azevedo, N. Sadeghiani, N. Buske, P.C. Morais, Z.G.M. Lacava, J. Mag. Mag. Mater. 272–276 (2004) 6.
[156] V. Lavhasetwar, C. Song, W. Humphrey, R. Shebuski, R.J. Levy, J. Pharm. Sci. 87 (1998) 6.
[157] M. Geiser, S. Schurch, P. Gehr, J. Appl. Physiol. 94 (2003) 6.

CHAPTER 3

Mn^{2+} Emission in Mn-Doped Quantum Dots

Nguyen Que Huong*
*Department of Physics, Marshall University, Huntington, WV, United States

1 INTRODUCTION

In this chapter, we will study the effect of 3D confinement of an Mn-doped quantum dot on its emission processes.

Over the past several decades, theoretical and experimental studies show that reduced dimensional semiconductor structures own unique properties that are completely different from the bulk semiconductors that have never been observed before. Electronic structures and the optical properties of the semiconductor mesoscopic structures depend sensitively on their sizes and start to show interesting behaviors of confined structures when the sizes of these structures are comparable to the electron-hole effective Bohr radius. Being confined in all three dimensions, quantum dots have the biggest quantum size effect and therefore, are very attractive objects for experimental and theoretical investigation, for both basic research and applications [1–8].

In bulk II–VI semiconductor compounds, the doping of transition metal or rare-earth elements is used to obtain efficient magnetic and optical properties. With recent developments in nanotechnology such as colloidal synthesis and molecular-beam epitaxy, magnetic ion-doped quantum dots have become of great interest and have been studied intensively. Successful incorporation of magnetic ions into single colloidal nanocrystals in different labs has been reported, opening the door for many applications using doping dots in spintronics and quantum information technology. Optical spectra from single quantum dots doped by magnetic elements such as Mn, Fe, Cr, and Co have been investigated [9–12]. Impurity states in a doped nanocrystal are reported to play an important role in the electronic structure, transition probabilities, and optical properties of nanocrystals.

Mn^{2+} is a good example of the magnetic impurities. Well known as an activator for photoluminescence (PL) and electroluminescence (EL), the ion Mn^{2+} has been among the first magnetic ions to be doped into a

Nano-sized Multifunctional Materials
https://doi.org/10.1016/B978-0-12-813934-9.00003-7

semiconductor nanocrystal [9, 10]. In nanocrystals, the *sp-d* interactions between the host electrons and the impurity magnetic atom become stronger due to confinement and lead to new effects that have never been observed in bulks.

We will study the magneto-PL of Mn^{2+}-doped semiconductor core-shell colloidal quantum dots and investigate the effects of confinement structures on the PL of the nanocrystals. The general plan of the present chapter is to consider two effects that could alter the luminescence processes. First, the coated quantum dots with the absence of inversion symmetry in the structure of wurtzite nanocrystals possess a spontaneous polarization, a large internal dipole moment, and an electric field due to the piezoelectric effect [7, 13, 14]. This internal field mixes energy levels and changes the transition selection rules, leading to different results in PL pictures. Second, the Coulomb and exchange interactions of the *d*-electrons of the impurity centers with the confined electrons (or holes) existing inside the nanocrystals could change the PL properties in general, and could be an explanation for the observed behaviors [15]. In the strong confinement approximation, the boundary conditions enhance the coupling, and the effect on PL will be large.

The details will be considered in the following sections.

2 THEORY OF POLARIZATION IN POLAR NANOCRYSTALS

The spherical quantum confinement model with the effective mass approximation and an approximate treatment of the Coulomb interaction [1] is the most convenient model to describe quantum dots and the size dependence of their optical properties. The absorption features and their size dependence are well described by this model [2]. Also, the observable fine structure of the band-edge exciton was predicted as a result of the additional effects such as the nonspherical shape of the dot and the exchange interaction between the electron and the hole [7, 13, 16].

However, some observations give some results indicating that there should be some additional modification in the spherical model of nanocrystals. The Stark measurement [17] suggested the presence of a large change in the dipole moment during optical excitation. The detailed comparison of one-photon and two-photon band edge spectra of CdSe nanocrystals [18, 19] suggested the presence of the internal polarization that lifted the inversion symmetry [18–23]. Studies of colloidal CdSe nanocrystals by coating the CdSe surface with a layer of ZnS before capping with an organic base show a significant increase of PL quantum efficiency. The increase of PL partly is due to the difference of the band gaps, but might imply the existence of some "internal electric field," that mixes the optical energy levels as well.

2.1 Existence of a Permanent Dipole Moment in CdSe Quantum Dots

Guyotsionnest et al. [22] researched the one-photon and two-photon excitation transitions and observed the similarity of one-photon and two-photon spectra. However, in the two-photon spectrum, the transition $1S_{3/2}-1S_e$, which is parity-forbidden in bulks, has been found. They suggested that the observation of the parity-forbidden transition strongly indicates that a spontaneous polarization exists in the CdSe nanocrystal, or some internal electric field exists that mixes the odd and even band-edge hole states. The existence of the polarization also implies the lack of inversion symmetry of these nanocrystals.

In their later papers [18, 19], the authors have measured the magnitude of this dipole moment by the method of dielectric constants. CdSe nanocrystals were synthesized by injection of dimethyl cadmium and trioctylphosphine selenide in a hot bath of trioctylphosphine and trioctylphosphine oxide with size dispersion between 5% and 10%. Two different kinds of obtained nanocrystals were the diameters of 34 Å and 46 Å. The obtained dipole moments were of 25 debye (D) for the 17 Å radius and 47 D for the 23 Å radius nanocrystals. In the later paper [24], even larger values of the dipole moments have been reported: 66 D for the 34 Å radius dot and the dipole moment varies between 41 and 98 D for different dot sizes.

From the measurements of the conductivity, this dipole moment cannot account for the nanocrystal charge. The measured dipole must be an intrinsic property of an uncharge nanocrystal and exist permanently by itself in the nanocrystals. No thermal effect on the dipole moment is found.

By using the formula for the internal polarization [25]

$$P_0 = \mu(2\epsilon_1 + \epsilon_2)/4\pi R_0^3 \epsilon_1 \tag{1}$$

we found corresponding values of the internal polarization of 0.7 and 0.9 $\mu C/cm^2$ for the 23 and 17 Å radius nanocrystals, respectively. It has also been shown that the dipole moment is increasing linearly with the dot radius.

Different explanations have been offered for the origin of the permanent dipole moment in the nanocrystal: lifting of the inversion symmetry of the hexagonal lattice, laser photoionization [26], and thermal induce charging [18, 19].

2.2 Structure of CdSe Nanocrystals

The structure and optical properties of nanocrystals have been described by the spherical confinement approximation. The real structure and symmetry of the nanocrystals and their deviation from spherical symmetry would be some small perturbation to the model but might play an important role and lead to important changes in energy spectra or the selection rules of optical transitions.

The micrograph of the CdSe nanocrystal from high-resolution electron microscopy (HREM) and resonance spectroscopy has been obtained [20] and the HREM data were collected on a sample of average diameter 65 A. The nanocrystals from different directions have been viewed, and as a result, the hexagonal shape of the nanocrystal is claimed. The top and bottom surfaces terminate in polar (001) and $(00\overline{1})$. The (001) surface is purely cadmium terminated whereas the $(00\overline{1})$ is purely selenium. Resonant Raman depolarization data showed that the point group symmetry of the electronic states of the CdSe nanocrystal is not T, but rather both A and E. The fact that the triply degenerate state splits into A and E states supports the point group assignment of C_{3v} for the wurtzite CdSe nanocrystal. The CdSe nanocrystal has a hexagonal structure with a top purely cadmium terminated surface and a bottom purely selenium surface. It means that the CdSe nanocrystallite itself has no inversion symmetry. And of course, the coated CdSe nanocrystallite has no inversion symmetry as well.

2.3 Model for the Polarization: The Piezoelectric Effect

The fact that the CdSe nanocrystal has the top surface of purely Cd while the bottom surface of purely Se is very important. Namely, when the CdSe nanocrystal is capped by another semiconductor shell, the Cd and Se in the top and bottom surface of the nanocrystal will couple with the ions of the shell to form thin layers of different materials in the top and bottom surface of the nanocrystal.

For a clear picture, we consider a CdSe nanocrystal capped by a ZnS shell and model the absence of inversion symmetry by the layer model in Fig. 1. The ZnS semiconductor also can be described by the layer structure: a layer of Zn is continued by a layer of S, and a gain Zn, and so on. For CdSe coated by ZnS, we draw the layers of Cd, Se, Zn, and S in detail. In the top and bottom surfaces of the CdSe nanocrystal, the layers of the elements are in inverse order. The Cd layer (top layer of the nanocrystal surface) and S layer (the bottom layer of the ZnS shell) in the interface of CdSe and

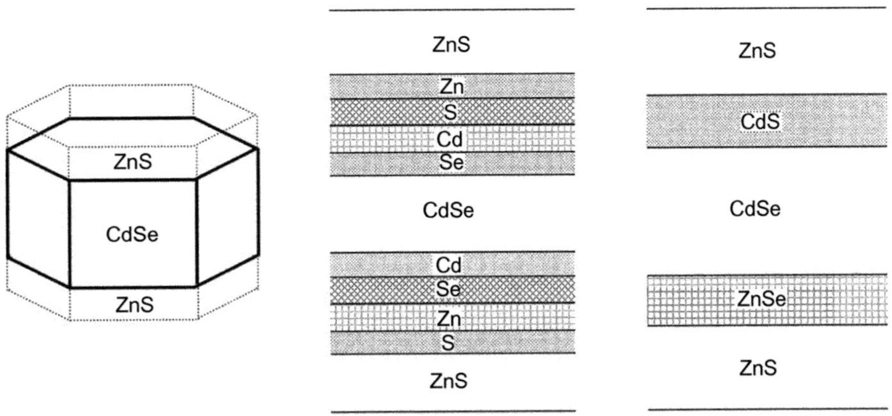

Fig. 1 Layer model for CdSe dot-coated ZnSe to show makeup layer (CdS) on *top* and (ZnS) on *bottom* in the interface CdSe-ZnS surfaces.

ZnS at the top surface are next to each other and can be considered as being coupled together, and so we have a thin layer of CdS in the top surface of the dot. With the same assumption, the Se layer (bottom layer of the nanocrystal surface) and the Zn layer (the top layer of the shell) in the CdSe-ZnS interface at the bottom surface are next to each other and can be considered as coupled together to form a thin layer of ZnSe in the bottom surface of the nanocrystal. The materials of those thin layers on top and bottom are different from each other as an indication for the absence of the inversion symmetry. Materials of those thin layers are difference from the material of the semiconductor shell, and the difference between the lattice constants of neighbor materials leads to strain on the surface of the nanocrystal.

Based on this model, we propose [14] a simple theory to explain the existence of the spontaneous polarization in the interface surface of CdSe and ZnS. In the interlayer bounding surface of a nanocrystallite, where different materials interface, there should exist a strain caused by the mismatch of lattice constants of different materials. Because of this strain, the piezoelectric effect produces an electric polarization in the surface layer of the crystallite, which is proportional to the difference of the lattice constants. The lack of inversion symmetry is a condition for the existence of the piezoelectric effect, and all these wurtzite structure materials are piezoelectric. So, originating from the piezoelectric effect, the spontaneous polarization in the surface has the form:

$$\mathbf{P} \sim e_{14} \frac{\Delta u}{u} \tag{2}$$

where e_{14} is the piezoelectric constant, u is the lattice constants, and Δu the difference of the lattice constants of materials in surface layers. For simplicity, we take Δu as a constant; in reality the strain will be distributed, that is, Δu will be a function of z. For CdSe coated by ZnS, the value $\frac{\Delta u}{u}$ is about 10%, and the polarization is rather large.

2.4 Calculation of the Internal Electric Field

To demonstrate how this surface polarization can cause the internal electric field inside the coated dots, we use a very simple geometric model, which is very close to the hexagonal structure of CdSe dots but with higher symmetry (Fig. 2).

Let us imagine that we have a spherical dot of CdSe coated by ZnS. In the surface of this dot, where CdSe and ZnS interface, there exists a thin layer composed of CdS in the upper half of the spherical surface and ZnSe in the

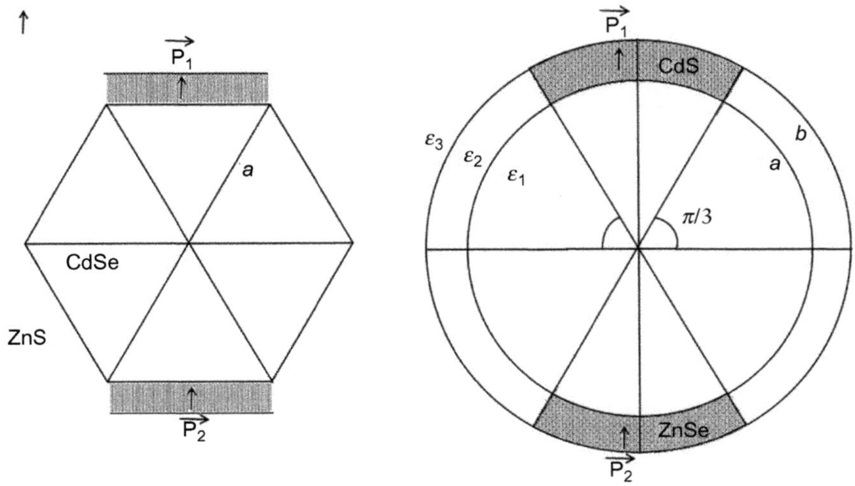

Fig. 2 Spherical model illustrating the hexagonal structure of a ZnS-coated CdSe nanocrystal. Surface polarization is different from zero only when it exists in the hexagonal structure.

lower half. To be similar to the hexagonal structure of the crystallite, where the polarization due to the piezoelectric effect is nonzero only in the top and bottom horizontal planes, in our spherical model we assume that the piezoelectricity causes a nonzero $\mathbf{P}_1$ only in the part of the CdS surface (top) layer with angle θ satisfying $\Pi/3 < \theta < 2\Pi/3$, and a nonzero $\mathbf{P}_2$ only in the part of the ZnSe surface (bottom) where $-2\Pi/3 < \theta < -\Pi/3$. In all other parts of the surface, the polarization equals zero.

In other words, we write the expression for the polarization in the dot surface:

$$\begin{aligned} \mathbf{P} &= \mathbf{P}_1 \Theta(\theta - \pi/3)\Theta(2\pi/3 - \theta) \\ \mathbf{P} &= \mathbf{P}_2 \Theta(\theta - 4\pi/3)\Theta(5\pi/3 - \theta) \end{aligned} \tag{3}$$

The dielectric constants of the dot, the thin layer, and the coat are ϵ_1, ϵ_2, and ϵ_3, respectively. Because of the lack of inversion symmetry, $\mathbf{P}_1$ and $\mathbf{P}_2$ will have the same direction. We use the usual electrostatic method to find the internal electric field inside the dot due to the surface spontaneous polarization. The surface charges due to the polarization in the inside (σ_a) and outside (σ_b) surfaces of the thin layer obey:

$$\begin{aligned} \sigma_a &= \mathbf{P} \cdot \mathbf{n} = P\cos\theta \\ \sigma_b &= \mathbf{P} \cdot \mathbf{n}' = -P\cos\theta \end{aligned} \tag{4}$$

with **P** given in Eq. (3). Due to the symmetry of our model, it is easy to choose the electrostatic potentials and the boundary conditions for this coated dot system. The electrostatic potential for regions inside the dot in the thin surface layer and in the coat is Φ_1, Φ_2, and Φ_3, respectively, which can be expanded in spherical polynomials:

$$\begin{aligned} \Phi_1 &= \sum_l \alpha_l r^l P_l(\cos\theta) \\ \Phi_2 &= \sum_l \left(\beta_l r^l + \frac{\gamma_l}{r^{l+1}}\right) P_l(\cos\theta) \\ \Phi_3 &= \sum_l \frac{\delta_l}{r^{l+1}} P_L(\cos\theta) \end{aligned} \tag{5}$$

where r is the distance to the center of the dot, P_l is a Legendre polynomial, α_l, β_l, γ_l, and δ_l are coefficients, and l are nonnegative integers. The boundary conditions at the surface a and b of the coated dot require the continuity of tangential components of the electric field

$$\begin{aligned} -\frac{1}{b}\partial\Phi_3/\partial\theta|_{r=b} &= -\frac{1}{b}\partial\Phi_2/\partial\theta|_{r=b} \\ -\frac{1}{a}\partial\Phi_2/\partial\theta|_{r=a} &= -\frac{1}{a}\partial\Phi_1/\partial\theta|_{r=a} \end{aligned} \tag{6}$$

and the discontinuity of the normal components of the displacement

$$\begin{aligned} \epsilon_2\partial\Phi_2/\partial\theta|_{r=a} - \epsilon_1\partial\Phi_1/\partial\theta|_{r=a} &= -P\cos\theta \\ \epsilon_3\partial\Phi_3/\partial\theta|_{r=b} - \epsilon_2\partial\Phi_2/\partial\theta|_{r=b} &= P\cos\theta \end{aligned} \tag{7}$$

Eqs. (6), (7) give the system of equations for the coefficients α_l, β_l, γ_l, and δ_l. To satisfy all equations simultaneously, all α_l, β_l, γ_l, and δ_l with $l \neq 1$ must be zero. Then the electrostatic potential inside and outside the quantum dot will be described by these coefficients as the following:

$$\begin{aligned} V_{\text{in}}(r) &= \alpha_1 r\cos\theta \\ V_{\text{out}}(r) &= \frac{\delta_1}{r^2}\cos\theta \end{aligned} \tag{8}$$

where $V_{\text{in}}(r)$ is the electrostatic potential at point r inside the quantum dot, and V_{out} is the electrostatic potential outside the dot. From Eq. (8) we have a solution for α_1, the expansion coefficient of the potential inside the nanocrystallite, as the following

$$\alpha_1 = -\alpha(P_1 + P_2) \tag{9}$$

where

$$\alpha = \left(1 - \frac{3\sqrt{3}}{8}\right) \frac{2(\epsilon_2 - \epsilon_3)a^3 b^3 - \epsilon_2 b^6 - 2\epsilon_3 a^6}{2(\epsilon_2 - \epsilon_1)(\epsilon_2 - \epsilon_3)a^3 b^3 - (\epsilon_2 b^3 + 2\epsilon_3 a^3)(\epsilon_1 A^3 + \epsilon_2 b^3)} \tag{10}$$

Then the electrostatic potential inside the nanocrystal is equal to:

$$V = -\alpha(P_1 + P_2)\cos\theta \tag{11}$$

and the internal electric field caused by the spontaneous polarization in the nanocrystallite dot is linearly proportional to the sum of $\mathbf{P}_1$ and $\mathbf{P}_2$:

$$\mathbf{E} = \alpha(\mathbf{P}_1 + \mathbf{P}_2) \tag{12}$$

From Eqs. (3), (12), we have the polarization in the surface and the electric field inside the nanocrystal due to the piezoelectric effect caused by the difference of the lattice constant of the material layers in the dot surface. We can estimate the polarization and the electric field numerically to compare with the experimental results [18, 19].

The calculation was made for the coated dot. We expect that for the pure semiconductor dot, there should exist the strain on the dot surface, also due to the mismatch of the lattice distances between the dot's material and the organic coat, but this requires further research.

2.5 Numerical Estimation and Comparison With the Experiments

In order to examine the amplitude of the polarization and the internal electric field, we can evaluate **P** and **E** with standard parameters of CdSe and ZnS. The piezoelectric constant $e_{14} = 4.2 \times 10^4(esu)$. For the lattice constants of CdSe, CdS, ZnS, and ZnSe, we take $u(CdSe) = 6.05\ A$, $u(CdS) = 5.83\ A$, $u(ZnS) = 5.40\ A$, $u(ZnSe) = 5.66\ A$ so that in the upper surface (interface of CdS-ZnS) $\frac{\Delta u_1}{u_1} = 0.08$, and the lower surface (interface of ZnSe-ZnS) $\frac{\Delta u_2}{u_2} = 0.05$. Hence, respectively, the polarization in the upper and lower surfaces equals $P_1 = 3.3 \times 10^3(esu)$ and $P_2 = 2.1 \times 10^3(esu)$ (or $P_1 = 1.1\ \mu C/cm^2$ and $P_2 = 0.7\ \mu C/cm^2$).

Dielectric constants $\epsilon_1 = 6.1$, $\epsilon_2 = 5.3$, and $\epsilon_3 = 5.2$ for a CdSe nanocrystal of radius $a = 20$ Å and the surface thin layer $b - a = 5$ Å, the coefficient $\alpha = 0.06$, and the magnitude of the internal electric field is about $3.3 \times 10^2(esu)$ (or 4.8×10^4 V/cm). The field is rather large, and the effect is considered important. Our model is supported later by experimental data [27] that confirm the presence of the strain in the core/shell surface.

For CdSe/CdS core/shell nanocrystals with the lattice mismatch between the core and shell of 3.9%, the presence of a strain up to 50 Å has been observed. The obvious effect of strain, such as the peak shift toward larger *d*-spacing, has been seen.

In summary, the piezoelectric effect has been considered as the origin of the spontaneous polarization and the large dipole moment in nanocrystallites, also obtaining a rather large internal electric field existing inside the nanocrystallite. This electric field exists not only in the coated dots but also in all nanocrystallites, which are put into other materials because of the strain in their surfaces. The existence of the internal electric field and the absence of inversion will change the electronic structure, mix the optical levels, and give new results in one- and two-photon transitions, in the Stark effect of excitons, in PL, and in other optical processes in quantum dots.

3 THEORY OF LUMINESCENCE IN MANGANESE-DOPED NANOCRYSTALS

In this section we study Mn-doped nanocrystals ZnS and consider the effect of an extra electron in the doped quantum dot ZnS:Mn^{2+} [15].

Impurity states in doped nanocrystals play an important role in the electronic structure, transition probabilities, and optical properties of the nanocrystal. In recent years, Mn-doped ZnS nanoparticles have been intensively studied. Among many wide band gap compounds, manganese is well known as an activator for PL and EL and the Mn^{2+} ion *d*-electron states act as efficient luminescent centers while doped by a semiconductor. In order to find a way to improve the efficiency, people are trying to reduce the size of the doped nanocrystals.

In 1994 Bhargava et al. [28, 29] obtained a ZnS semiconductor nanocrystal doped with Mn isoelectronic impurities and reported that Mn-doped ZnS nanocrystal can yield both the high luminescence efficiency and lifetime shortening. The doped nanocrystals ZnS:Mn were prepared by an organometallic reaction at room temperature and these nanocrystal were coated with polymer or monomer to maintain the separation of the particles. Their EXAFS experiments confirmed that the Mn^{2+} ion was occupying the site of the Zn^{2+} ion. In this ZnS:Mn nanocrystal system the significant increase of the quantum efficiency has been observed when the system was exposed to the 300 nm UV light; the room temperature quantum efficiency is increasing gradually but rather significantly with the decreasing size of the nanocrystals, from <1% for the nanocrystal of radius 70 Å to 18% for the nanocrystal of radius 35 Å. Size dependence of the quantum efficiency is obtained. The spectra of nanocrystals has been measured under identical conditions, so the ratio of the PL intensities reflects the ratio of their luminescence efficiencies.

The yellow emission characterized for Mn^{2+} in bulk ZnS [30–32], what is associated with the transition ${}^4T_1-{}^6A_1$, has been claimed to be observed in PL spectra for the Mn^{2+} in nanocrystal ZnS. In nanocrystals, however, the PL peak for the yellow emission is reported to be slightly shifted toward a lower energy (in bulk ZnS:Mn it peaks around 2.12 eV, and in nanocrystal ZnS:Mn it peaks at 2.10 eV). The line width of the yellow emission in the PL spectrum for the nanocrystal is larger than for the bulk. The luminescence lifetime of the Mn^{2+} ${}^4T_1-{}^6A_1$ transition was reported to decrease by five orders of magnitude from 1.8 ms in bulk to 3.7 and 20.5 ns in nanocrystals while maintaining the high (18%) quantum efficiency.

The authors suggested that the increase in quantum efficiency as well as the lifetime shortening is the result of strong hybridization of *s-p* electrons of the ZnS host and *d*-electrons of the Mn impurity due to confinement, and also of the modification of the crystal field near the surface of the nanocrystals.

Stimulated by these dramatic results, many other laboratories are trying to synthesized the Mn-doped ZnS nanocrystals and considerable attention has been paid to the optical properties of these kinds of materials [33–35]. Yu et al. [33] reports obtaining the yellow characteristic emission peak of ${}^4T_1-{}^6A_1$ where the peak is slightly shifted toward the higher energy. The increase in the luminescence efficiency of the nanocrystal ZnS:Mn in comparison with the Mn-doped ZnS bulk has been obtained. Decay time shortening to nanoseconds for nanocrystals has been reported for ZnS:Mn^{2+} by Sooklai et al. [36] while a lifetime shortening to microns for ZnTe:Mn^{2+} quantum is also Ito et al. [37].

Also, several different theories have been modeled to try to explain the processes happening with the impurity centers inside the confined nanocrystals [38, 39]. But up to now, the effects of the Mn-impurity on the electronic structure of the nanocrystal, or put in different way, the effects of the confinement on the energy structure and transitions of the Mn impurity in the nanocrystals, still are a controversy.

The aim of our work in this section is to study more about the doped ZnS:Mn nanocrystals. We consider a new mechanism to the Mn-doped ZnS nanocrystal, namely the effect of an extra electron in the doped nanocrystals, and will find out how the mechanism will affect the electronic structure and optical properties of the nanocrystals.

3.1 Model of Existence of an Extra Electron in a Nanocrystal

Our model assumes the existence of an extra electron in the quantum dot (or with the injection of an extra electron into the quantum dot). The extra electron couples with the electrons involved in the transition, splitting energy levels and mixing the wave functions, which so strongly breaks the former selection rules. Because the dot is small, the boundary conditions make the coupling very strong, and by controlling the presence or absence of the extra electron in the dot, one could control the optical transitions of the dot. This could provide the explanation to the observed shortening lifetime and enhancement of quantum efficiency.

The transition of our interest is ${}^4T_1{-}{}^6A_1$ of the Mn^{2+} ion in the crystal field of the ZnS nanocrystal. The luminescence transition from the lowest excited level ${}^4T_1$ (spin 3/2) to the ground state ${}^6A_1$ (spin 5/2) is forbidden. Only the weak spin-orbital interaction makes the transition slightly allowed.

If an extra electron is in the dot, the electron will couple with electrons of Mn^{2+} in each state, namely, it will couple with ${}^4T_1$ and ${}^6A_1$. For the coupling Hamiltonian the electron-electron exchange interaction will be used. The wave function of the d^5 electrons in ${}^4T_1$ and ${}^6A_1$ will be derived. For the extra electron that is confined in the dot, we use the wave function for a confined electron in the sphere

$$\phi_{1S}^{\mathrm{Dot}} = R_{1S}^{\mathrm{Dot}} Y_0^0 \left|\frac{1}{2}, \sigma\right\rangle \tag{13}$$

where R_{1S}^{Dot}, Y_0^0, $|\frac{1}{2},\sigma\rangle$ are the radial, angular, and spin functions for the confined electron, respectively. R_{1S}^{Dot} is the envelope function for the electron confined in a sphere of radius R.

In the next sections we will use the strong field limit to obtain the wave functions for d^5 electrons in ${}^4T_1$ and ${}^6A_1$ states.

3.2 A Single 3*d* Impurity Electron in Crystal Field-Strong Coupling Limit

Impurity centers are formed by foreign ions substituting for host ions or placed interstitially in the lattice. When the activator ions such as Mn are placed in the crystal field of the semiconductor lattice, the crystal field would affect the wave functions and energy structure of the impurity to form the so-called crystal field states of the impurity ions. For the transition metal ions with the outer $3d^n$ electron configuration, the crystal field energy and the interelectron Coulomb interaction are comparable [40], so these d^n electrons can be treated as an intermediate crystal field or a strong crystal field limit, and they are more often treated in the strong field approximation. Before presenting the application of strong crystal field approximation for the case of the $3d^5$ electron of Mn^{2+} ions, we recall some well-known results for a single $3d$ electron. Let us consider an ion having a single $3d$ electron outside the closed shell placed in an octahedral crystal field. In terms of spherical harmonics, the octahedral crystal field Hamiltonian can be expressed as the following [40]:

$$\mathcal{H}_c^{O_h}(r) = \frac{Ze^2}{4\pi\epsilon_0} C_0^{(4)}(\theta,\phi) + \left(\frac{5}{14}\right)^{1/2} C_4^{(4)}(\theta,\phi) + C_{-4}^{(4)}(\theta,\phi) + r^6 \text{ terms} + \cdots \tag{14}$$

where $\vec{r}\ (r,\theta,\phi)$ is the position of the d-electron,

$$C_t^{(k)}(\theta,\phi) = \left(\frac{4\pi}{2k+1}\right)^{1/2} Y_k^t(\theta,\phi) \tag{15}$$

To evaluate the effect of $H_c^{O_h}$ on the $3d$ electron states, one needs to evaluate the matrix elements of the type:

$$\langle n'l'm_l'|r^k C_t^{(k)}|nlm_l\rangle = \langle R'_{n'l'}|r^k|R'_{nl}\rangle\langle Y_{l'}{}^{m_l'}|C_t^{(k)}|Y_l{}^{m_l}\rangle \tag{16}$$

The radial integral can cause the configuration mixing, but because of the large separation between the energies of d, the mixing can be ignored and it is assumed that $nl = n'l'$. Then the matrix elements of $H_c^{O_h}$ will remain inside the states of the $3d$ configuration. The nonzero matrix elements will be the following [40]:

$$\begin{aligned} &\langle 3d0|H_c^{O_h}|3d0\rangle = 6Dq \\ &\langle 3d1|H_C^{O_h}|3d1\rangle = \langle 3d-1|H_c^{O_h}|3d-1\rangle = -4Dq \\ &\langle 3d2|H_c^{O_h}|3d2\rangle = \langle 3d-2|H_c^{O_h}|3d-2\rangle = Dq \\ &\langle 3d2|H_c^{O_h}|3d-2\rangle = \langle 3d-2|H_c^{O_h}|3d2\rangle = 5Dq \end{aligned} \tag{17}$$

where

$$D = \frac{1}{4\pi\epsilon_0}\frac{35Ze^2}{4a^5}, \quad q = \frac{2}{105}\langle r^4\rangle_{3d} \tag{18}$$

Due to Hamiltonian matrix diagonalization, the fivefold degenerate $3d$ level will split into two levels, the twofold degenerate level with the additional energy $+6Dq$, and the threefold degenerate with the additional energy of $-4Dq$. In Fig. 3 the splitting diagram for a $3d$ electron in the octahedral crystal field is shown [40].

The eigenstates of the twofold degenerate level (the e-orbitals) are written in the following form:

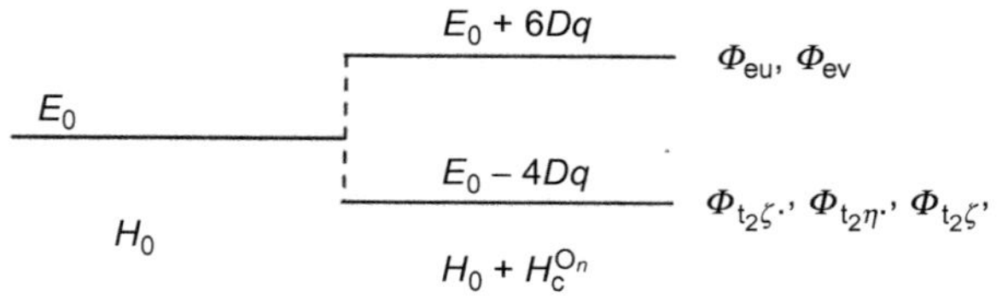

Fig. 3 Splitting diagram of a single 3*d* electron in an octahedral crystal field. The fivefold degenerated electron with energy E^0 splits in an octahedral crystal field into two levels.

$$
\begin{aligned}
\phi_{eu} &= |3d0\rangle = R'_{3d}(r)\left(\frac{5}{4\pi}\right)^{1/2}\left(\frac{3z^2 - r^2}{2r^2}\right) \\
\phi_{ev} &= \left(\frac{1}{2}\right)^{1/2}(|3d2\rangle + |3d-2\rangle) = R'_{3d}(r)\left(\frac{5}{4\pi}\right)^{1/2} 3^{1/2}\left(\frac{x^2 - y^2}{2r^2}\right)
\end{aligned}
\tag{19}
$$

And the eigenstates for the threefold degenerate level (the t_2-orbital) are written as:

$$
\begin{aligned}
\phi_{t_2\xi} &= \left(\frac{i}{2}\right)^{1/2}(|3d1\rangle + |3d-1\rangle) = R'_{3d}(r)\left(\frac{5}{4\pi}\right)^{1/2} 3^{1/2}\left(\frac{yz}{r^2}\right) \\
\phi_{t_2\eta} &= -\left(\frac{1}{2}\right)^{1/2}(|3d1\rangle - |3d-1\rangle) = R'_{3d}(r)\left(\frac{5}{4\pi}\right)^{1/2} 3^{1/2}\left(\frac{xz}{r^2}\right) \\
\phi_{t_2\zeta} &= -\left(\frac{i}{2}\right)^{1/2}(|3d2\rangle - |3d-2\rangle) = R'_{3d}(r)\left(\frac{5}{4\pi}\right)^{1/2} 3^{1/2}\left(\frac{xy}{2r^2}\right)
\end{aligned}
\tag{20}
$$

For simplicity, from now on we will denote theses eigenstates as u, v, ξ, η, and ζ, respectively.

3.3 3d^5 Electron in Crystal Field: Mn^{2+} Energy Levels and Wave Functions

For a multi-3*d*-electron system, the orbital Hamiltonian in the crystal field will be the following

$$
H = \sum_i H_0^0(r_i) + \sum_{i>j} H'(r_i, r_j) \tag{21}
$$

where $H_0^0(r_i)$ is the Hamiltonian of a single *d*-electron interacting with the crystal field we considered in the last section, and $H'(r_i, r_j)$ is electron-electron interaction between *a*-electrons. For the strong crystal field limit, the Hamiltonian $H_0^0(r_i)$ will be solved by first obtaining the wave functions of a single-3*d* electron, as in the last section, then these states will interact by the Hamiltonian $H'(r_i, r_j)$. The wave functions of 3d-electron orbitals are u and v, which belong to the E irreducible (in O_h) representation, and η, ξ, and θ (which belong to the T_2 reducible representation. So, in the strong crystal field case, the wave functions of the multi-3d-electron Hamiltonian will be the products of these one-electron orbitals. In an octahedral crystal field, the Hamiltonian Eq. (21) is invariant under all the rotation operators of the group O, and the wave functions will transform according to irreducible representation of the O group. The products of irreducible representations of the O group follow the rules:

$$
\begin{array}{ll}
A_1 \times A_1 = A_1 & E \times E = A_1 + A_2 + E \\
A_1 \times A_2 = A_2 & E \times T_1 = T_1 + T_2 \\
A_1 \times E = E & E \times T_2 = T_1 + T_2 \\
A_1 \times T_1 = T_1 & T_1 \times T_2 = A_2 + E + T_1 + T_2 \\
A_1 \times T_2 = T_2 & T_1 \times T_1 = A_1 + E + T_1 + T_2 \\
A_2 \times A_2 = A_1 & T_2 \times T_2 = A_1 + E + T_1 + T_2 \\
A_2 \times T_1 = T_2 & \\
A_2 \times T_2 = T_1 &
\end{array}
$$

The d^n electrons are treated in the strong field approximation where the wave functions of multi-3d-electron Hamiltonian are written as the products of these one-electron orbitals. The system Hamiltonian is invariant under all the rotation operators of the group O_h and the wave function transforms according to the irreducible representation of the O_h group. The reduction process for the two 3d-electron configurations gives the products $E \times E$, $E \times T_2$, and $T_2 \times T_2$, where A_1 and A_2 are 1D representations (A_1 is symmetric and A_2 is antisymmetric), E is the 2D representation, and T_1 and T_2 are 3D representations of the group O. Beginning the reduction process or the two 3d-electron configuration, we will have the products $E \times E$, $T_2 \times T_2$, and $E \times T_2$. The $E \times E$ product functions will belong to A_1, A_2 and E irreducible representation. The product functions of the $E \times E$ orbitals will be the following:

$$
\begin{aligned}
|e^2, A_1\rangle &= \frac{1}{\sqrt{2}}[u(r_1)u(r_2) + v(r_1)v(r_2)] \\
|e^2, A_2\rangle &= \frac{1}{\sqrt{2}}[u(r_1)v(r_2) - v(r_1)u(r_2)] \\
|e^2, Eu\rangle &= \frac{1}{\sqrt{2}}[u(r_1)u(r_2) - v(r_1)v(r_2)] \\
|e^2, Ev\rangle &= \frac{1}{\sqrt{2}}[u(r_1)v(r_2) + v(r_1)u(r_2)]
\end{aligned} \tag{22}
$$

The $|e^2, A_1\rangle$ function is symmetric under the interchange of T_1 and T_2. Due to the Pauli principle, to obtain the antisymmetric function, its spin function must be the antisymmetric spin function with the total spin $S = O$. The total wave function is the product of the orbital and the spin functions and if we use the notation for the Slater determinant

$$\frac{1}{\sqrt{2}}\begin{vmatrix} u(r_1)\alpha(\sigma_1) & u(r_2)\alpha(\sigma_2) \\ u(r_1)\alpha(\sigma_1) & u(r_2)\alpha(\sigma_2) \end{vmatrix} = |u^+u^-| \tag{23}$$

the final full wave function $|e^2, A_1\rangle$ is written in the short form

$$|e^2, A_1, M_S = 0\rangle = \frac{1}{\sqrt{2}}([u^+u^-] + [v^+v^-]) \tag{24}$$

Similarly we will have all the functions of $|e^2, A_2\rangle$, $|e^2, A_2\rangle$, $|e^2, E\rangle$, $|t_2^2, A_1\rangle$, $|t_2^2, A_2\rangle$, $|t_2^2, E\rangle|t_2^2, T_2\rangle$, $|t_2^2, T_1\rangle|t_2e, T_1\rangle$, and $|t_2e, T_2\rangle$.

The reduction gives all the possible wave functions and energies t^n, e^n, and $t^m e^k$ for the d^5 configuration of Mn^{2+} in the octahedral crystal field [15].

The ground state of the $3d^5$ configuration is the sextet ${}^6A_1$ with the total spin 5/2. The first excited level is the triplet states ${}^4T_1$. We are interested in the optical transition between the first excited state and the ground state; this is the transition ${}^4T_1 - {}^6A_1$. The wave function of the ground state ${}^6A_1$ is derived from $t_2^3 \times e_2$ [15]:

$$|{}^6A_1\rangle = |t_2^3({}^4A_2)e^2({}^3A_2), {}^6A_1 a_1\rangle \tag{25}$$

which gives us

$$|{}^6A_1\rangle = -[-\xi^+\eta^-\zeta^+u^+v^+] \tag{26}$$

and the triplet wave functions of the lowest excited state are derived from $t_2^4({}^3T_1) \times e^1({}^2E)$ [15]

$$\begin{aligned}
|{}^4T_1x\rangle &= -\frac{1}{2}[\xi^+\xi^-\eta^+\zeta^+u^+] - \frac{\sqrt{3}}{2}[\xi^+\xi^-\eta^+\zeta^+v^+] \\
&\quad + \frac{\beta}{\sqrt{2}}[\xi^+\eta^{+2}\zeta^+p_z^+] - \frac{\beta}{\sqrt{2}}[\xi^+\eta^+\zeta^{+2}p_y^+] \\
|{}^4T_1y\rangle &= \frac{1}{2}[-\xi^+\eta^+\eta^-\zeta^+u^+] + \frac{\sqrt{3}}{2}[\xi^+\eta^+\eta^-\zeta^+v^+] \\
&\quad + \frac{\beta}{\sqrt{2}}[\xi^{+2}\eta^+\zeta^+p_z^+] - \frac{\beta}{\sqrt{2}}[\xi^+\eta^+\zeta^{+2}p_x^+] \\
|{}^4T_1z\rangle &= [-\xi^+\eta^-\zeta^+\zeta^+u^+] - \frac{\beta}{\sqrt{2}}[\xi^{+2}\eta^+\zeta^+p_y^+] \\
&\quad - \frac{\beta}{\sqrt{2}}[\xi^+\eta^{+2}\zeta^{+2}p_x^+]
\end{aligned} \tag{27}$$

where the brackets denote Slater determinants of the 3d wave functions.

3.4 Coulomb and Exchange Interaction

The exchange interaction between the extra electron and the impurity configuration has the form

$$H_{\text{ex}} = -JS_{\text{Mn}}s_e \tag{28}$$

where S_{Mn} and s_e are the total spin of the d^5 configuration in the states of interest and the spin of the extra electron, respectively. j is the exchange matrix element, which has the form:

$$j_{ii} = \sum_{k,\lambda} \langle \overline{\phi_k \phi_\lambda} | V | \phi_\lambda \phi_k \rangle \tag{29}$$

ϕ_k is the electron wave function, V is a two-electron interaction operator

$$V(r_1, r_2) = \frac{q_1 q_2}{r^2} \tag{30}$$

that can be expanded in terms of spherical harmonics

$$V(r_1, r_2) = \frac{q_1 q_2}{r_>} \sum_{k=0}^{\infty} \frac{4\pi}{2k+1} \frac{r_<^k}{r_>^k} \sum_m Y_{km}(1) \overline{Y}_{km}(2) \tag{31}$$

and with the wave functions of the electrons, the matrix elements consist of two parts—the radial and the angular parts:

$$\langle \phi_i \phi_j | V | \phi_m \phi_t \rangle = \sum_{k=0}^{\infty} \rho^k(n^i l^i, n^j l^j, n^m l^m, n^t m^t) \times A_k \sigma(s^i s^t) \sigma(s^j s^m) \tag{32}$$

where

$$\begin{aligned} \rho^k(n^i l^i, n^j l^j, n^m l^m, n^t m^t) &= \left\langle R_{n^i l^i} R_{n^j l^j} \left| \frac{e^2 r_<^k}{r_>^k} \right| R_{n^m l^m} R_{n^t l^t} \right\rangle \\ A_k \sigma(s^i s^t) \sigma(s^j s^m) &= \frac{4\pi}{2k+1} \sum_{p=-k}^{k} \langle \overline{Y}_{l^i p^i} Y_{kp} Y_{l^t p^t} \rangle \langle \overline{Y}_{l^j p^j} \overline{Y}_{kp} Y_{l^m p^m} \rangle \end{aligned} \tag{33}$$

where $\langle \overline{Y}_{l^i p^i} Y_{kp} Y_{l^t p^t} \rangle$ is a Clebsch-Gordan coefficients that is different from zero only when the following conditions are satisfied

$$\begin{aligned} p^i &= p + p^t \\ k + l^i + l^t &= \text{even} \\ l^i - l^t &\le l^i + l^t \end{aligned} \tag{34}$$

These conditions will reduce nonzero terms in the sum.

3.5 Interaction of an Mn^{2+} Ion With the Extra Electron in the Nanocrystal and ${}^4T_1-{}^6A_1$ Transition

Now we suppose that, due to the confinement, the extra electron inside the nanocrystal is close enough to the Mn^{2+} ion at the center so that the electron can couple via the Coulomb and exchange interactions with the d^5 electrons in both the ground state ${}^6A_1$ and the excited state ${}^4T_1$. The Coulomb and exchange matrix elements of the interaction $V\,(r_1,\ r_2)$ between the wave functions of $3d^5$ electrons of the Mn^{2+} center and the extra electron have been evaluated as:

$$
\begin{aligned}
K &= \left\langle \Psi^e_{1s}(r_1)\begin{Bmatrix} {}^6A_1(r_2) \\ {}^4T_1(r_2) \end{Bmatrix} | V(r_1,r_2) | \Psi^e_{1s}(r_1)\begin{Bmatrix} {}^6A_1(r_2) \\ {}^4T_1(r_2) \end{Bmatrix} \right\rangle \\
J &= \left\langle \begin{Bmatrix} {}^6A_1(r_1) \\ {}^4T_1(r_1) \end{Bmatrix} \Psi^e_{1s}(r_2) | V(r_1,r_2) | \Psi^e_{1s}(r_1)\begin{Bmatrix} {}^6A_1(r_2) \\ {}^4T_1(r_2) \end{Bmatrix} \right\rangle
\end{aligned}
\tag{35}
$$

The matrix element of the exchange interaction of the injected electron and the d^5 electrons in the state ${}^6A_1$ (or ${}^4T_1$) will have the form:

$$
\begin{aligned}
&\langle {}^6A_1(\vec{r})\phi^{\text{Dot}}_{1s}(\vec{r'}) | V(r-r') | \phi^{\text{Dot}6}_{1s}A_1(\vec{r})\rangle \\
&= \langle [\xi^+\eta^+\zeta^+u^+v^+]\ \ (r) R^{\text{Dot}}_{1s}(r') Y^0_0 | V(r-r') | R^{\text{Dot}}_{1s}(r) \\
&\quad Y^0_0[\xi^+\eta^+\zeta^+u^+v^+]\ \ (r')\rangle
\end{aligned}
\tag{36}
$$

The wave function of ${}^6A_1$ $[\xi^+\eta^+\zeta^+u^+v^+]$ is a Slater determinant with five single d-electron wave functions with five different coordinates. In fact, the extra electron has an exchange interaction with each of the five d-electrons. Because these five coordinates are independent, the integrals will be taken separately. In the calculation of the determinantal matrix elements of Eq. (36), we have to deal with the single-electron matrix element of the type, for example:

$$
\begin{aligned}
&\langle \zeta(r)\phi(r') | V(r-r') | \phi(r)\zeta(r')\rangle \\
&= \frac{4\pi q_1 q_2}{j_1(\pi)^2}\frac{1}{R^3}\sum_{k=0}^{\infty}\frac{1}{2k+1}\left\langle \overline{R}_{3d}(r)(\overline{Y}^1_2+\overline{Y}^{-1}_2)\overline{j}_0\left(\pi\frac{r'}{R}\overline{Y}^0_0\right)\right| \\
&\quad \times \frac{r^k}{r'^{k+1}}\sum_{m=-k}^{k} Y^m_k(1)Y^m_k(2)|\overline{j}_0\left(\pi\frac{r'}{R}\overline{Y}^0_0\overline{Y}^{-1}_2\right)
\end{aligned}
\tag{37}
$$

The conditions for nonzero Clebsch-Gordan coefficients give the selection rules for k and m. And the radial part of the matrix element becomes:

$$\rho^{\text{exchange}}(3d,s,k=2)=\int r^2dr\int r'^2dr' R_{3d}(r)R_{10}^{\text{Dot}}(r')\frac{r^2}{r'^3}R_{3d}(r')R_{10}^{\text{Dot}}(r) \tag{38}$$

Here R_{10}^{Dot} is the envelope function for the electron confined in a dot. R_{3d} is the orbital function of the $3d$-electron that can be approximated as Slater function $R-3d=r^2e^{-1.87r/a}$, where $a=\frac{\hbar^2}{me^2}$. For the Coulomb interaction:

$$\rho^{\text{Coulomb}}(3d,s,k=2)=\int r^2dr\int r'^2dr' R_{3d}(r)R_{10}^{\text{Dot}}(r')\frac{1}{r'}R_{3d}(r)R_{10}^{\text{Dot}}(r') \tag{39}$$

Then the matrix element (37) has the form:

$$\langle\zeta(r)\phi(r')|V(r-r')|\phi(r)\zeta(r')\rangle=\frac{2q_1q_2}{j_1(\pi)^2}R^3\rho(3d,s,k) \tag{40}$$

The integrals are taken over the dot radius.

As the result of the calculations, the matrix elements of the Coulomb and exchange interactions between the extra electron and the $^6A_1$ state are obtained in the form [15]:

$$\begin{aligned} K(^6\text{A}_1,{}^1S)&=5\text{e}^2\rho_{\kappa=0}^{\text{Coulomb}}=K \\ K(^6\text{A}_1,{}^1\text{S})&=\frac{\text{e}^2}{25}\rho_{\kappa=2}^{\text{Coulomb}}=J \end{aligned} \tag{41}$$

And for the Coulomb and exchange interactions between the extra electron and the $^4T_1$ state:

$$\begin{aligned} K(^4T_1,{}^1S)&=5\text{e}^2\rho_{\kappa=0}^{\text{Coulomb}}=K \\ J(^4T_1xy,{}^1S)&=\frac{\text{e}^2}{25}\rho_{\kappa=2}^{\text{Coulomb}}=J \\ J(^4T_1z,{}^1S)&=\frac{\text{e}^2}{25}\rho_{\kappa=2}^{\text{Coulomb}}=-J \end{aligned} \tag{42}$$

As the effect of the Coulomb and exchange interactions with the injected electron, the $^6A_1$ level splits into two sublevels, $^6A_1\phi(1sC)$ and $^6A_1\phi(1s\downarrow)$ with total spin $S=3$ and $S=2$ with corresponding exchange energies (Fig. 4)

$$\begin{aligned} E_{\text{ex-C}}(^6A_1\phi(1s\uparrow,S=3)&=K-7/4J \\ E_{\text{ex-C}}(^6A_1\phi(1s\downarrow,S=2)&=K+5/4J \end{aligned} \tag{43}$$

and corresponding wave functions are linear combinations of the products of the wave function of the $3d^5$ electrons with spin 3/2 and 5/2 (Eqs. 1, 2) and the extra electron.

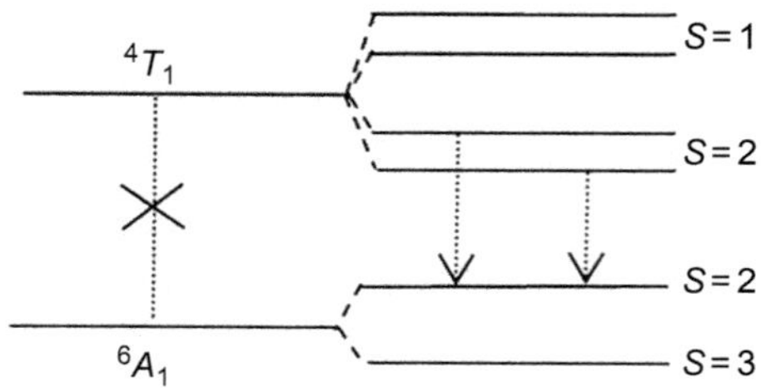

Fig. 4 Energy splitting of $^4T_1$ and $^6A_1$ states of Mn^{2+} and allowable transitions due to Coulomb and exchange interaction with the extra electron.

$$\begin{aligned} |^6A_1\phi(1s\uparrow, S=3\rangle &= [-\xi^+\eta^+\zeta^+u^+v^+]\ |s+\rangle \\ |^6A_1\phi(1s\downarrow, S=3\rangle &= [-\xi^+\eta^+\zeta^+u^+v^+]\ |s-\rangle \end{aligned} \tag{44}$$

The $^4T_1$ level splits into four sublevels: two states $^4T_{1z}\phi_{1s}(\downarrow)$, $^4T_{1xy}\phi_{1s}(\downarrow)$ of the total spin $S = 1$ with the energies corrections

$$\begin{aligned} E_{ex-C}(^4T_1z\phi(1s\downarrow, S=1) &= K - 5/4J \\ E_{ex-C}(^4T_1xy\phi(1s\downarrow, S=1) &= K + 5/4J \end{aligned} \tag{45}$$

and corresponding wave functions:

$$\begin{aligned} |^4T_1z\phi(1s\downarrow, S=1)\rangle &= [\zeta^+\eta^+\xi^+\xi^-u^+]\ |s-\rangle + [\zeta^2\eta^+\xi^+p_y^+]\ |s-\rangle \\ &\quad - [\zeta^+\eta^2\xi^+p_x^+]\ |s-\rangle \end{aligned} \tag{46}$$

$$\begin{aligned} &|^4T_1xy\phi(1s\downarrow, S=1)\rangle = \\ &\quad \frac{\sqrt{2}}{4}[\zeta^+\zeta^-\eta^+\xi^+u^+]\ |s-\rangle + \frac{\sqrt{6}}{4}[\zeta^+\zeta^-\eta^+\xi^+v^+]\ |s-\rangle \\ &\quad -\frac{\sqrt{1}}{2}[\zeta^+\eta^+\eta^-\xi^+u^+]\ |s-\rangle - \frac{\sqrt{1}}{2}[\zeta^+\eta^+\eta^-\xi^+v+^+]\ |s-\rangle \\ &\quad +\frac{\beta}{2}[\zeta^+\eta^2\xi^+p_z^+]\ |s-\rangle - \frac{\beta}{2}[\zeta^+\eta^2\xi^2p_y^+]\ |s-\rangle \\ &\quad +\frac{\beta}{2}[\zeta^2\eta^2\xi^+p_z^+]\ |s-\rangle - \frac{\beta}{2}[\zeta^+\eta^2\xi^2p_x^+]\ |s-\rangle \end{aligned} \tag{47}$$

and two states $^4T_{1z}\phi_{1s}(\uparrow)$, $^4T_{1xy}\phi_{1s}(\uparrow)$ with the value of the total spin $S = 2$ with the exchange energies

$$\begin{aligned} E_{ex-C}(^4T_1z\phi(1s\uparrow, S=2) &= K + 3/4J \\ E_{ex-C}(^4T_1xy\phi(1s\downarrow, S=2) &= K - 3/4J \end{aligned} \tag{48}$$

and corresponding wave functions:

$$
\begin{aligned}
|^4T_1z\phi(1s\uparrow, S=2)\rangle = [\zeta^+\eta^+\xi^+\xi^-u^+]\ |s+\rangle + \frac{\beta}{\sqrt{2}}[\zeta^2\eta^+\xi^+p_y^+]\ |s+\rangle \\
-[\zeta^+\eta^2\xi^+p_x^+]\ |s-\rangle
\end{aligned}
\tag{49}
$$

$$
\begin{aligned}
&|^4T_1xy\phi(1s\uparrow, S=2)\rangle = \\
&\frac{\sqrt{2}}{4}[\zeta^+\zeta^-\eta^+\xi^+u^+]\ |s+\rangle + \frac{\sqrt{6}}{4}[\zeta^+\zeta^-\eta^+\xi^+v^+]\ |s+\rangle \\
&-\frac{\sqrt{1}}{2}[\zeta^+\eta^+\eta^-\xi^+u^+]\ |s+\rangle - \frac{\sqrt{1}}{2}[\zeta^+\eta^+\eta^-\xi^+v+^+]\ |s+\rangle \\
&+\frac{\beta}{2}[\zeta^+\eta^2\xi^+p_z^+]\ |s+\rangle - \frac{\beta}{2}[\zeta^+\eta^2\xi^2p_y^+]\ |s+\rangle \\
&+\frac{\beta}{2}[\zeta^2\eta^2\xi^+p_z^+]\ |s+\rangle - \frac{\beta}{2}[\zeta^+\eta^2\xi^2p_x^+]\ |s+\rangle
\end{aligned}
\tag{50}
$$

The matrix elements K and J of the Coulomb and exchange interactions between the $3d^5$ electrons and the extra electron are the radial integrals that are inversely proportional to the dot radius.

The two sublevels $^4T_{1z}\phi_{1s}(\uparrow)$, $^4T_{1xy}\phi_{1s}(\uparrow)$ of the 4_1^T state of the Mn^{2+} ion and the sublevel $^6A_1\phi(1s\downarrow)$ of the ground state $^6A_1$ now have the same total spin $S = 2$ and the transition between them is allowable. Fig. 4 shows the exchange energy splitting of the 4_1^T and $^6A_1$ levels and the transitions allowable between them.

3.6 Luminescence and Lifetime of Transition

As the result of the Coulomb and exchange interactions with an extra electron, both the $^6A_1$ and $^4T_1$ have the states of the same spin $S = 2$ and the interaction between the electron existing in the dot and the d-electrons of the Mn-impurity center make the previously forbidden transition allowable.

The matrix element of the electric dipole transition [41] between the sublevels of the state $^6A_1$ and $^4T_1$ again have the form of the determinant matrix element [15]

$$
\begin{aligned}
&\langle^4T_1z, \phi_{1s+}^{\text{Dot}}, S=2|\ \vec{r}\ \cdot \hat{e}|^6A_1\phi_{1s-}^{\text{Dot}}\rangle \\
&= \langle[\zeta^+\eta^+\xi^+\xi^-u^+]\ \langle s+|r|[\zeta^+\eta^+\xi^+\xi^-u^+]\ |s-\rangle \\
&= \langle R_{3d}R_{1s}^{\text{Dot}}\rangle\langle(i)Y_1^q(j)\rangle
\end{aligned}
\tag{51}
$$

where the angular and radial parts are separated. Similar to the calculation of the exchange matrix element in the last section, the angular part is calculated using the Clebsch-Gordan coefficients. The radial part is calculated using the envelope function as the Bessel function and an approximation function for the radial function of the d-electron of the Mn-ion. The radial part shows a complicated dependence on the inverse of the dot radius and its powers. The numerical calculation for the case of a quantum dot of Mn-doping ZnS with standard parameters of Mn and ZnS shows the splitting of 1.0−1.5 eV for the doped ZnS:Mn quantum dot of radius 50 Å. The probability of the transition approximately equals $5.4 \times 10^4\ s^{-1}$. It results in a lifetime, what is inversely proportional to the transition probability, of 2×10^{-5} s. The integral strongly depends on the radius of the nanocrystal. So the presence of an extra electron in the quantum dot really makes the spin-forbidden transition allowable and shortens the lifetime. The theory supports the results obtained in the experimental works [28, 29, 33–37, 42, 43].

4 MN^{2+} EMISSION IN MN-DOPED QUANTUM DOTS

A magnetophotoluminescence study of a strongly confined diluted magnetic semiconductor (DMS) in Mn^{2+}-doped ZnSe/CdSe core-shell colloidal nanocrystals leads has reported interesting results. Unlike the cases in bulks or in other conventional DMS materials, the yellow emission characterized for Mn^{2+}, which is associated with the $d-d$ internal transition ${}^4T_1-{}^6A_1$, was reported as not suppressed in applied $B\|\vec{z}$ magnetic fields and circularly polarized as usual, and instead, developed a circular polarization. More interestingly, the Mn^{2+} PL has been found to have a large splitting between the σ^+ and σ^- components that depend on the applied field. This again has not been found in the characteristics of the Mn^{2+} PL in bulks and other conventional DMS materials and is the result of the strong confinement of the nanocrystals.

In the PL spectra for Mn^{2+} in the wide band-gap semiconductor bulk, the characterized yellow emission is associated with the internal transition ${}^4T_1-{}^6A_1$ within the excited $3d^5$ shell of the Mn^{2+} ion. Because the two impurity levels are inside the band gap, a band-edge exciton can either recombine radiatively (band-edge exciton PL) or transfer its energy to an Mn^{2+} ion through a recombination Auger process, which results in the excitation of the Mn^{2+} ion and is accombined by emitting a photon and a transition from ${}^4T_1$ to ${}^6A_1$ (Mn^{2+} PL). At low temperature and under a magnetic field in bulks, this Auger recombination process is dramatically suppressed due to the spin projection conservation selection rule of the exciton-Mn system [11, 12]. The ground state of Mn^{2+} of the $3d^5$ configuration is the sextet $^6A_1$ with the total spin $S^{Mn} = 5/2$ while its first excited state is the triplet $^4T_1$ that has the total spin $S^{Mn} = 3/2$. Under an applied magnetic field, the ground state $^6A_1$ splits into six Zeeman levels with spin projection $S_z^{Mn} = \pm 5/2, \pm 3/2, \pm 1/2$ and the $^4T_1$ splits into four levels with spin projection $S_z^{Mn} = \pm 3/2, \pm 1/2$. Because the paramagnetic Mn^{2+}polarizes in the B field, the $-5/2$ state is predominantly occupied. There are two kinds of excitons existing in the system: the bright exciton with the total (spin and orbital) angular momentum projection $J = \pm 1$, which has spin projection $s_z^{ex} = 0$, and the dark exciton, with the total angular momentum projection $J \pm 2$ and the spin projection $s_z^{ex} = \pm 1$. With the two main orientations of the magnetic field, there are two geometries with different emission intensities. When $B \| \vec{z}$, in Faraday geometry, the bright exciton is predominantly occupied. The Auger recombination, which excites the Mn^{2+} from its $^6A_1$ to its $^4T_1$, requires total spin ($S^{Mn} + s^{ex}$) and its projection conservation [44, 45], therefore the bright exciton ($s^{ex} = 0$) cannot excite the Mn^{2+} out of its spin $\pm 5/2$ $^6A_1$ ground state into the spin 3/2 $^4T_1$ lowest excited state. As a result, the Mn PL is suppressed and the exciton PL is increased, and only very weak PL is allowed due to spin-orbit coupling for the bulk materials.

But the results are quite different for the case of Mn^{2+}-doped ZnSe/CdSe core-shell colloidal nanocrystals. The two mechanisms that we study in this chapter could change the Auger process in Faraday geometry, when the magnetic field $B \| \vec{z}$, and explain the results obtained in [46]. If there exists an extra electron in the quantum dot, the electron would interact with the Mn^{2+} center by an exchange Coulomb interaction. The exchange interaction mixes the states and changes the total spin of the system. On the other hand, the existence of the permanent polarization in the dot due to stress and mismatch of the lattice constances on the dot surface [14] causes an internal electric field inside the dot that mixes the exciton states, allowing the excitation of the Mn^{2+} from the spin 5/2 to the spin 3/2 state. Both the two mechanisms are rather strong due to the small size of the dot, leading to the Auger recombination and explaining why in nanocrystals the yellow emission band $^6A_1$–$^4T_1$ is not suppressed in the magnetic field. Depending on the magnitude of the applied magnetic field and the size of the dot, there is a competition between the exchange interaction, the polarization, and the magnetic field splitting to decide the relieving of the suppression of the Mn^{2+} PL under the magnetic field.

5 CONCLUSION

In conclusion, we have studied the two effects that could affect optical processes in quantum dots. The existence of an electron (or a hole) inside a doped dot leads to the Coulomb and exchange interactions between this confined electron (hole) with the ground state $^6A_1$ and the first excited $^4T_1$, producing new Mn mixed states that satisfy the spin and spin projection conservation law for the transition involving the bright exciton. The existence of an intrinsic electric field, on the other hand, modifies the exciton wave functions and produces a mixture of the bright exciton with the higher excitons. The first effect is larger for the smaller dots while the second effect is larger for the larger dots and there is a competition between these two effects depending on the sizes of the nanocrystals.

REFERENCES

[1] L. Brus, L.E. Brus, J. Chem. Phys. 80 (1982) 4403.
[2] J.J. Shiang, A.V. Kadavanich, R.K. Grubbs, A.P. Alivisatos, J. Phys. Chem. 99 (1995) 17417.
[3] J.B. Xia, Phys. Rev. B 40 (1989) 8500.
[4] L. Banyai, S.W. Koch, Semiconductor Quantum Dot, Series on Atomic, Molecular and Optical Physics, vol. 2, World Scientific, Singapore, 1993.
[5] A.L. Efros, A.V. Rodina, Phys. Rev. B 47 (1993) 10005(R).
[6] A.L. Efros, Phys. Rev. B 46 (1992) 7448.
[7] M. Nirmal, D.J. Norris, M. Kuno, M.G. Bawendi, A.L. Efros, M. Rosen, Phys. Rev. Lett. 75 (1995) 3728.
[8] U. Wogon, Optical Properties of Semiconductor Quantum Dots, Springer Verlag, Berlin Heidelberg, 1996.
[9] R.N. Bhargava, et al., Phys. Rev. Lett. 72 (1994) 416.
[10] R.N. Bhargava, J. Lumin. 70 (1996) 85.
[11] L. Besombes, et al., Phys. Rev. Lett. 93 (2004) 207403.
[12] A.V. Chernenko, et al., J. Phys. Cond. Mat. 22 (2010) 355306.
[13] D.J. Norris, A. Sacra, C.B. Murray, M.G. Bawendi, Phys. Rev. Lett. 72 (1994) 2612.
[14] N.Q. Huong, J.L. Birman, J. Chem. Phys. 108 (1998) 1769.
[15] N.Q. Huong, J.L. Birman, Phys. Rev. B 69 (2005) 085321.
[16] A.L. Efros, M. Rosen, M. Kuno, M. Nirmal, D.J. Norris, M. Bawendi, Phys. Rev. B 54 (1996) 4883.
[17] V.L. Colvin, A.P. Alivisatos, J. Phys. Chem. 100 (1996) 13226.
[18] M.E. Schmidt, S.A. Blanton, M.A. Hines, P. Guyot-Sionnest, J. Chem. Phys. 106 (1997) 12.
[19] S.A. Blanton, R.L. Lebeny, M.A. Hines, P. Guyot-Sionnest, J. Chem. Phys. 99 (1995) 17417.
[20] J.J. Shiang, A.V. Kadavanich, R.K. Grubbs, A.P. Alivisatos, J. Phys. Chem. 99 (1995) 17417.
[21] C.B. Murray, D.J. Norris, M.G. Bawendi, J. Am. Chem. Soc. 115 (1993) 8706.
[22] M. Hines, P. Guyot-Sionnest, J. Phys. Chem. 100 (1996) 468.
[23] S.A. Empedocles, D.J. Norris, M.G. Bawendi, Phys. Rev. Lett. 77 (1996) 3873.

[24] M. Shim, P. Guyot-Sionnest, J. Chem. Phys. 11 (1999) 6955.
[25] W. Cady, Piezoelectricity, McGraw-Hill Book Co., New York, NY, 1946
[26] L. Brus, J. Phys. Chem. 105 (2001) 1725.
[27] X. Peng, M.C. Schlamp, A.V. Kadavanich, A.P. Alivisatos, J. Am. Chem. Soc. 119 (1997) 7019.
[28] D. Gallagher, W.E. Heady, J.M. Racz, R.N. Bharagava, J. Mat. Res. 10 (1995) 870.
[29] R.N. Bhargava, J. Lumin. 70 (1996) 85.
[30] R. Parrot, C. Naud, F. Gendron, Phys. Rev. B 13 (1976) 3748.
[31] H.E. Gumlich, J. Lumin. 23 (1981) 73.
[32] L.A. Hemstreet, Phys. Rev. B 22 (1980) 4590.
[33] T. Il Yu, M.S. Isobe, Takahashi, Mater. Sci. Eng. B 38 (1996) 177.
[34] M. Tanaka, J. Qi, Y. Masumoto, unpublished.
[35] A.A. Bol, A. Meijerink, Phys. Rev. B 548 (1998) R15997.
[36] K. Sooklai, B.S. Collum, S.M. Angel, M.J. Murphy, Phys. Chem. 100 (1996) 4551.
[37] H. Ito, T. Tanako, T. Kuroda, F. Minami, H. Akinaga, J. Lumin. 72–74 (1997) 342.
[38] V. Albe, C. Jouanin, D. Bertho, Phys. Rev. Let. B 57 (1996) 373.
[39] T.A. Kennedy, E.R. Glaser, P.B. Klein, R.N. Bhargava, Phys. Rev. 52 (1995) R14356.
[40] B. Henderson, G.S. Imbursch, Optical Spectroscopy of Inorganic Solids, Oxford University Press, Oxford, 1989.
[41] T.P. Sokrates, Deep Center in Semiconductor Physics: A State-of-the-Art Approach, Gordon and Breach Science Publisher, New York, NY, 1992.
[42] R.N. Bhargava, D. Gallagher, T. Welker, J. Lumin. 60–61 (1994) 275.
[43] R.N. Bhargava, D. Gallagher, X. Hong, A. Nurmiko, Phys. Rev. Lett. 72 (1994) 416.
[44] M. Nawrocky, et al., Phys. Rev. B 52 (1995) R2241.
[45] R. Viswanatha, J.M. Pietryga, V.I. Klomov, S.A. Crooker, Phys. Rev. Lett. 107 (2011) 067402.
[46] R. Beaulac, et al., Nano Let. 8 (2008) 1197.

CHAPTER 4

Silica-Based Nanostructures in Biomedicine

Timur Sh. Atabaev*,†, Nguyen Hoa Hong†
*Department of Chemistry, School of Science and Technology, Nazarbayev University, Astana, Kazakhstan
†Department of Physics and Astronomy, Seoul National University, Seoul, Republic of Korea

1 INTRODUCTION

Over the last several decades, various nanomaterials with unique mechanical, chemical, physical, optical, magnetic, and electrical properties have attracted the considerable attention of many scientists [1]. Among them, silicon dioxide or silica (SiO_2) is an optically transparent, nontoxic material and has a great chemical and thermal resistance, which may make it useful for various applications. Additionally, the fabrication process of SiO_2 nanostructures is simple, controllable, and cost-effective. For example, silica has been widely employed as an air moisture absorbent due to its mesoporous characteristics. The high specific surface area of silica nanoparticles (NPs) up to $1200\,m^2/g$ also makes them suitable for controllable drug delivery applications. The surface of silica nanoparticles can be easily functionalized with specific functional groups for targeting purposes [2]. Furthermore, it is possible to create multifunctional silica nanoparticles with the help of various fluorophores, dopants, and assembly techniques. Thus, the use of silica-based nanostructures in clinical settings has been raised exponentially (Fig. 1) due to their wide range of biomedical applications (drug delivery, cell tracking and imaging, etc.) [3–6]. In this chapter, we review and discuss recent developments related to the usage of silica-based nanostructures for nanomedical applications. A special focus will be given to the preparation, surface functionalization, cell-tracking ability, sensing, and drug delivery properties of multifunctional silica-based nanostructures. We also provide our opinion on some challenges that need to be addressed to start the usage of silica-based nanostructures in our daily life.

Nano-sized Multifunctional Materials
https://doi.org/10.1016/B978-0-12-813934-9.00004-9

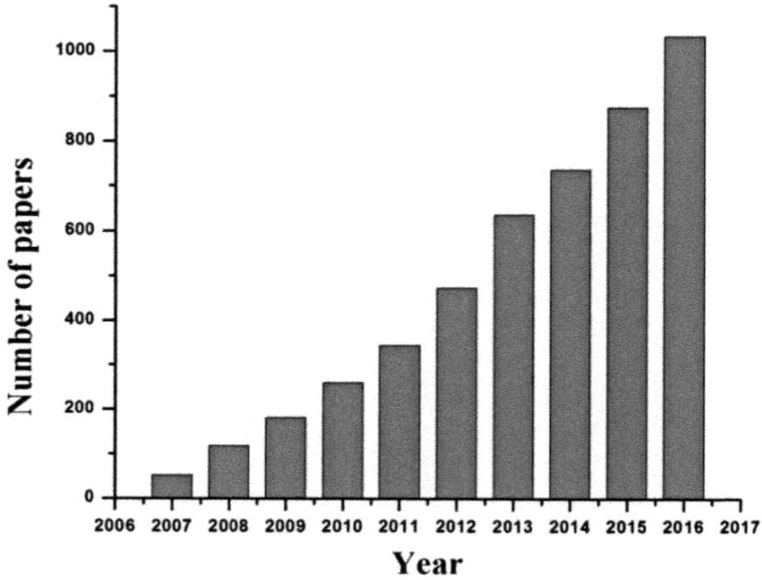

Fig. 1 Number of published papers according to the Web of Science core collection (topics "mesoporous silica" and "drug delivery").

2 SYNTHESIS

The nucleation, growth, and aggregation processes must be strongly considered during the synthesis process of silica-based nanostructures. Basically, synthesis strategies of silica nanoparticles need to satisfy several conditions: (a) low-cost and user-friendly production, (b) uniform and narrow size distribution of nanoparticles, (c) controllable surface characteristics, and (d) long-term colloidal stability. The Stöber method and the microemulsion method are among the several well-known approaches for the synthesis of silica nanoparticles [7,8]. The Stöber method was first reported in 1968 by a research team of Werner Stöber [7]. This method can be used for the production of monodisperse silica particles with sizes ranging from 0.05 to 2 μm, depending on the synthesis conditions. In principle, the Stöber method represents the hydrolysis and condensation reactions of the organosilane precursors in alcohol media with the help of basic media. Basic conditions give silica a highly negative surface charge, allowing the production of stable colloidal suspension. Thus, basic and highly diluted conditions are usually employed during the synthesis to maintain the colloidal stability of silica nanoparticles [9].

Microemulsion is another well-known method for the production of silica nanoparticles [8,10]. Usually, two or more immiscible phases with/without surfactants are used to form a thermodynamically stable dispersion. In the case of silica nanoparticles, a surfactant stabilized water-in-oil microemulsion (reverse microemulsion) is commonly used. To decrease the interfacial energy, the dispersed phase (water) may form spherical droplets. The nucleation and growth of silica occur within an aqueous droplet. The size and shape of the aqueous droplet can be easily adjusted by the fabrication conditions. Thus, the final size of silica nanoparticles can also be easily adjusted. This method is more suitable for the synthesis of silica nanoparticles in the size range of tens to hundreds of nanometers. Silica nanoparticles produced by the microemulsion method have better monodispersity compared to that produced by the Stöber method [11]. However, this method is more costly and not environmentally friendly because it requires large amounts of various hazardous substances to be used in the reaction. Additionally, silica nanoparticles prepared by the microemulsion method need to be cleaned from the remaining surfactants on their surfaces.

3 MESOPOROUS SILICA

Direct synthesis of various mesoporous silica sieves using cationic surfactants was first reported by scientists of Mobil Oil Research and Development in 1992 [12,13]. Among all mesoporous silica types, MCM-41 (obtained in basic media) and SBA-15 (obtained in acid media) became very important materials in many areas of science and engineering. The easiness of the synthesis method, the ability to control the size and shape of pores, large surface areas, thermal stability, and nontoxicity make them potentially applicable in adsorption [14], catalysis [15,16], chemical sensing [17], drug delivery [18,19], etc. Recently, Polshettivar and coworkers developed novel fibrous silica nanospheres (KCC-1) [20]. Compared to MCM-41 and SBA-15, KCC-1 nanospheres have large pore sizes, three-dimensional (3D) dendric pore networks, and high pore volume (Fig. 2). Furthermore, the fiber density and sizes of KCC-1 can be easily controlled [21], which makes it ideal for use in catalysis [22], solar cells [23], drug delivery [24], sensing and removal of metal ions from water [25], etc.

4 BIOCOMPATIBILITY

The US Food and Drug Administration (FDA) categorized the silica as a "generally recognized as safe" material [26]. Hudson and coworkers

Fig. 2 Morphology images of (A) MCM-41, (B) SBA-15, and (C) KCC-1 silica. *Reprinted with permission from J. Lai, X. Mu, Y. Xu, X. Wu, C. Wu, C. Li, J. Chen, Y. Zhao, Light-responsive nanogated ensemble based on polymer grafted mesoporous silica hybrid nanoparticles. Chem. Commun. 46 (2010) 7370–7372; M.A. Ballem, F. Soderlind, P. Nordblad, P.-O. Kall, M. Oden, Growth of Gd_2O_3 nanoparticles inside mesoporous silica frameworks. Micropor. Mesopor. Mater. 168 (2013) 221–224; V. Polshettivar, D. Cha, X. Zhang, J.M. Basset, High-surface-area silica nanospheres (KCC-1) with a fibrous morphology, Angew. Chem. Int. Ed. 49 (2010) 9652–9656, respectively.*

performed in vitro toxicology tests of various mesoporous silica types [27]. According to their study, mesoporous silica nanostructures at concentrations less than 100 μg/mL could be well tolerated by different cell lines. In another study, Lu and coworkers showed that mesoporous silica nanostructures can be safely used in vivo at dosages less than 200 mg/kg [28]. Moreover, a number of studies suggested that mesoporous silica nanoparticles have good hemocompatibility, which is a requirement for their potential intravenous administration [29,30]. However, Yu and coworkers have showed that the hemolytic activity of silica nanoparticles was dependent on geometry, porosity, and surface charge [31]. In addition, a recent study highlights that silica nanoparticle toxicity depends on its size, dose, and cell type [32]. Therefore, all these factors should be considered to avoid undesirable toxicity and other side effects caused by the usage of silica nanostructures.

5 DYE-DOPED SILICA NANOPARTICLES

Fluorescence microscopy has been recognized as one of the most versatile optical imaging methods. In fluorescence imaging, the fluorophores located near the site of interest are excited by an external source of light and then reemited at another wavelength that can be detected. Thus, the optical signal in fluorescence microscopy strongly depends on the fluorophore properties. Organic dyes, quantum dots (QDs), and lanthanide compounds are typical fluorophores that can be employed for the optical imaging of living cells [33]. To date, organic dyes such as fluorescein isothiocyanate (FITC), Alexa Fluor, and DyLight Fluor are commonly used in biotechnology. However, most of these organic dyes have poor photostability and environment-dependent optical features, which limits their practical bioapplications [34]. To address this issue, these organic dyes can be encapsulated or doped into a silica matrix. These dye-doped silica nanoparticles produce an enhanced optical signal compared to a single dye molecule [4,35]. An organic dye with desired excitation and emission wavelengths can be used, depending on particular applications. Moreover, thanks to well-established synthetic protocols, fluorescent silica nanoparticles with particular sizes can be easily prepared. For example, Santra and coworkers have used inorganic tris (2,2′-bipyridyl) dichlororuthenium(II) hexahydrate luminescent dye for doping inside the silica [36]. In this work, authors employed a water-in-oil microemulsion procedure to prepare various sized (~5, 63, and 400 nm) luminescent silica nanoparticles. In all cases, the authors observed that prepared samples emitted a strong visible light with a maximum at 608 nm. In addition, the authors used human leukemia cells to demonstrate that their prepared samples could be very effective for membrane staining. Conventional green emitting Fluorescein isothiocyanate (FITC) dye has also been doped into a silica matrix [4,37]. In both cases, prepared nanoparticles showed no obvious emission peak shift compared to that of the pure FITC emission in an aqueous solution. Fig. 3 shows that these FITC-doped silica nanoparticles can be effectively used as optical nanoprobes for high-resolution cellular imaging. Another study [38] suggested that it is also possible to use the near-infrared (NIR) dye-doped silica for sensitive clinical imaging. At the same time, dye-doped silica NPs can be also effectively used in other areas. Kim and coworkers [39] have demonstrated the usage of fluorescent silica nanoparticles for latent fingerprint detection on hydrophobic and hydrophilic substrates. The authors of another study [40] employed dye-doped silica NPs to increase the detection sensitivity in flow cytometry analysis. Furthermore, dye-doped silica NPs have been widely employed for various sensing and detection applications [41,42].

6 INORGANIC MAGNETIC CORE COATED WITH SILICA SHELL

In recent years, studies on the use of nanoparticles such as Fe_3O_4, MnO, Gd_2O_3, Dy_2O_3, and Ho_2O_3 for MRI purposes have increased significantly thanks to the unique magnetic properties of those nanoparticles [43–47]. However, the toxicity of the above-mentioned magnetic nanoparticles

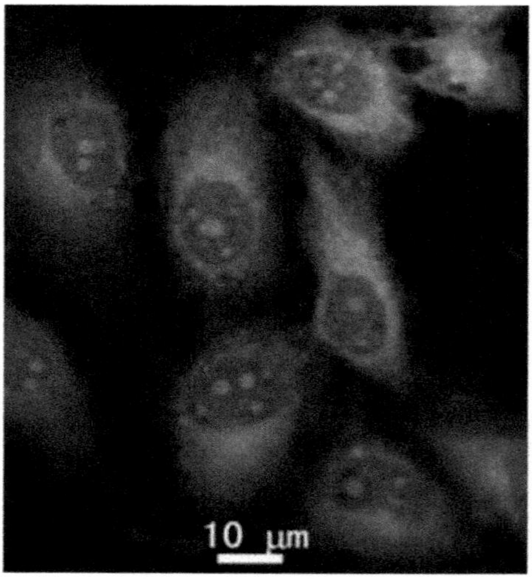

Fig. 3 Confocal laser scanning microscope image of nonsmall cell lung cancer line labeled with FITC-doped silica NPs. Reprinted with permission from [37].

has greatly reduced their bioapplications. The observed toxicity is associated with the so-called "Trojan horse" mechanism in which metal ions can be leached from the nanoparticles and produce reactive oxygen species [47]. Therefore, in order to reduce the toxic effects that might exist in magnetic NPs, some researchers cover them with a thin shell of inert silica layer (Fig. 4). Additionally, these core-shell nanostructures can improve the

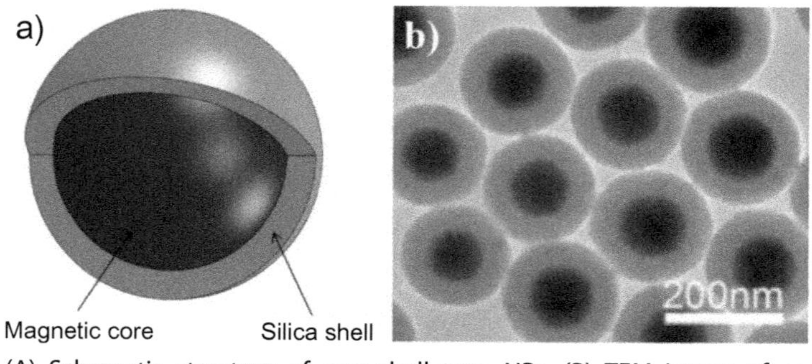

Fig. 4 (A) Schematic structure of core-shell type NPs, (B) TEM image of core-shell Fe_3O_4@SiO_2 NPs. *Reprinted with permission from L. Cheng, Y. Liu, B. Zou, et al., Template-etching route to construct uniform rattle-type Fe_3O_4@SiO_2 hollow microspheres as drug carrier. Mater. Sci. Eng. C 75 (2017) 829–835.*

colloidal stability of magnetic nanoparticles. For example, Zhang and coworkers reported the preparation of multifunctional $Fe_3O_4@SiO_2$ NPs functionalized with FITC fluorescent dye [48]. In addition to the obvious magnetic and luminescent properties, these nanostructures can be used for optical pH sensing in cells. Another study reports on the preparation of uniform rattle-type $Fe_3O_4@SiO_2$ hollow microspheres for controlled drug delivery applications [49]. In this study, doxorubicin (DOX) was used as a model drug to demonstrate the controlled drug delivery properties of prepared rattle-type $Fe_3O_4@SiO_2$. It was also shown that rattle-type $Fe_3O_4@SiO_2$ can slowly release the DOX (83% in 72h) at pH 5.7 in a PBS solution. Thus, prolonged and controlled drug delivery can be carried out to reach the place of interest. A recent study [50] showed that intravenously injected $Fe_3O_4@SiO_2$ NPs mainly accumulated in the liver and could still be found there after 84 days. This study highlights that silica-based core-shell NPs can be used for controlled drug delivery applications, especially for liver tissues.

$MnO@SiO_2$ is another type of magnetic core-silica shell nanostructure that is mainly used for magnetic resonance (MRI) visualization. Lee and coworkers have developed silica-coated MnO NPs (~20 nm) to control the MRI relaxivity rate in response to some specific physiological changes [51]. The main goal of this research is to develop a new MRI contrast agent that shows the low relaxivity at normal physiological conditions (pH = 7.4). The relaxivity value of these $MnO@SiO_2$ NPs increases after cell uptake by endocytosis and exposure to low pH = 5. Thus, a contrast agent that can "switch on" after uptaking of the cells may eventually enhance the specificity of cell labeling. In another study, it was shown that silica-coated MnO NPs were less prone to Mn leaching compared to MnO NPs coated with conventional organic coatings [52]. The authors have shown that the dye encapsulation into the silica shell layer can make these nanostructures fluorescent for potential optical detection and visualization. In addition, prepared $MnO@SiO_2$ NPs were highly soluble and stable in some aqueous solutions, including the physiological saline due to the attachment of hydrophilic polyethylene glycol (PEG) chains on their surface. On the other hand, mesoporous silica coating can also be quite useful in some other cases, as was shown in a recent study [53]. The nanosized channels of a mesoporous silica layer can increase the water permeability for improved T_1-weighted MRI imaging while enhancing the diffusion of the oxygen singlet from the iridium complex for photodynamic therapy.

7 INORGANIC OPTICAL CORE COATED WITH A SILICA SHELL

Semiconductor quantum dots (QDs) have a broad absorption band and a narrow tunable emission band, depending on their sizes. As was shown earlier, semiconductor QDs have a great potential for live cell visualization and early cancer detection [54,55]. On the other hand, semiconductor QDs are considered to be toxic, mainly due to the atoms of heavy elements in their composition [56]. Therefore, a similar core-shell coating strategy can be adopted to reduce the QD toxicity. In particular, the toxicity of CdSe/CdS/ZnS QDs was significantly reduced after a silica coating [57]. Furthermore, the resultant silica-coated QDs showed excellent long-term photo and colloidal stability, even in some harsh environments. Nearly similar results were also reported by another research group [58]. According to their results, the silica-coated CdSe QDs were nontoxic even at relatively high loadings (up to 1000 nM). At the same time, silica coating can be used to enhance the optical properties of some optical materials. For example, Zhang and coworkers showed that the quantum yield of some QDs with appropriate structure and composition could be enhanced after a silica coating [58]. Therefore, silica coating can highly reduce the QD toxicity and enhance their fluorescent properties. The luminescence enhancement can also be observed in the case of phosphor materials. In this case, the inert and optically transparent silica coating can protect the phosphors from luminescence killers such as water and air. For instance, Gd_2O_3-based phosphors used for biomedical applications should always be coated with silica, for two reasons. First, leaching of free Gd^{3+} ions is quite dangerous for living cells [59]. Second, the luminescent signal of Gd_2O_3-based phosphors can be reduced in the presence of water molecules [60]. Lü and coworkers showed that the two-photon upconversion emission spectra of Y_2O_3:Eu^{3+} can be improved after a silica coating [61]. Atabaev and coworkers also observed the luminescence enhancement of Y_2O_3:Eu^{3+}, Co^{2+} NPs after a silica coating process [62]. Thus, the phosphors with improved luminescent properties can be used for optical devices and biotechnology. In some cases, researchers combined both magnetic and optical NPs in one silica nanoparticle [63]. The fluorescent silica shell layer on the surface of other optical materials can be used in sensing applications. For example, the FITC-doped silica shell was coated on the surface of Y_2O_3:Eu^{3+} [42] and rhodamine B dye-doped Fe_3O_4@SiO_2 [64] nanostructures. These nanostructures can be used for ratiometric pH detection, where the inner optical core serves as a reference and the outer shell layer serves as a pH indicator. Thus, thanks to the simple silica coating design, it has become possible to obtain biocompatible NPs with a large range of functionalities.

8 RARE-EARTH DOPED SILICA

Rare-earth metals can be directly doped into a silica matrix, which represents another promising method for magneto-optical nanoprobe manufacturing. For example, silica microspheres doped with lanthanides such as praseodymium (Pr) and erbium (Er) were prepared using the acid-catalyzed hydrolysis of tetraethyl orthosilicate (TEOS) [65,66]. Prepared lanthanide-doped SiO_2 microspheres have shown good luminescent properties in both the visible and the IR regions, which can be used for bioimaging. The Er-doped silica nanoparticles with a size range of 20–25 nm were also prepared using a reverse micelle and sol-gel methods [67]. In another report, authors used Eu-doped SiO_2 nanospheres as efficient labels in DNA microarrays [68]. Fedorenko and coworkers showed that amino-modified green emitting terbium (Tb)-doped silica NPs could be adopted for cellular imaging [69]. Besides that, authors have shown that the toxicity of the prepared samples depends on their sizes. In particular, the toxicity was lower for larger SiO_2 NPs, which could be considered as a common trend for SiO_2 [70]. On the other hand, magnetic ions can also be incorporated or doped into a silica to produce magneto-responsive nanoprobes. For example, manganese chelates can be immobilized on the silica nanospheres for potential MRI applications [71]. Cytotoxicity and MRI relaxivity studies showed that Mn-chelated SiO_2 nanospheres exhibited a good longitudinal relaxivity value of $r_1 = 7.18\,mM^{-1}\,s^{-1}$ and can be safely used at concentrations up to 500 μg/mL. In another report, a gadolinium (Gd) was used as a dopant for silica to prepare a potential T_2-weighted MRI nanoagent [72]. A near-infrared indocyanine green dye was also incorporated into silica NPs to attain bifunctionality. Fig. 5 shows that these NPs can be detected optically, in addition to excellent MRI visualization capability (relaxivity value of $r_2 = 67\,mM^{-1}\,s^{-1}$ at 4.7 T).

9 SILICA-BASED SENSORS

Silica-based nanomaterials can be widely used for various sensing and biosensing applications. For example, surface-modified fluorescent mesoporous silica can be exploited for selective colorimetric detection of

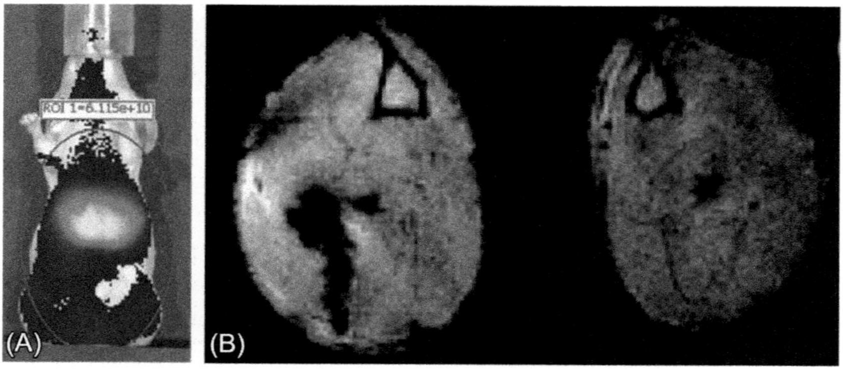

Fig. 5 (A) NIR fluorescence signal in mouse from injected Gd-doped ICG-SiO_2 NPs, and (B) in vivo T_2*-weighted gradient-echo MRI image of mouse calf muscle injected with Gd-doped ICG-SiO_2 NPs. *Reprinted with permission from P. Sharma, N.E. Bengtsson, G.A. Walter, et al., Gadolinium-doped silica nanoparticles encapsulating indocyanine green for near infrared and magnetic resonance imaging, Small 8 (2012) 2856–2868.*

mercury(II) cations [73]. Authors of another report have shown that surface-modified silica NPs doped with Eu(III) complex were suitable for trace determination of Cu(II) in water or biological environments [74]. It was shown that the prepared samples have high selectivity and sensitivity in the range of 10–100 μM. As was shown recently [75], amine groups anchored on the silica NPs can capture probes for 2,4,6-trinitrotoluene (TNT). Thus, these functionalized silica NPs can be used for colorimetric-based detection of TNT. Silica NPs in the form of aerogel films on Au-patterned alumina substrates could be used as a humidity sensor [76].

In the case of biomedical research, silica NPs can be utilized to detect the viability of amoebas [77]. In particular, authors have investigated the amoeba phagocytosis using the fluorescent detection methods. Kell and coworkers reported a development of label-free DNA biosensors based on SiO_2 NPs [78]. It was shown that these biosensors have a limit of detection of about 2 nM ssDNA using a standard fluorometer. Accordingly, developed biosensors may be used for highly sensitive label-free DNA detection without the need for polymerase chain reaction. In another report, a quantitative fluorescent bioassay was developed based on Eu(III) chelate-doped SiO_2 NPs [79]. Authors showed that fluorescent SiO_2 NPs labeled with streptavidin can be applied for the time-resolved detection of the human hepatitis B surface antigen.

10 SUMMARY AND OUTLOOK

This chapter has mainly focused on prospective biomedical applications of silica-based nanostructures. We have briefly discussed the synthesis methods of silica, silica types, and toxicity issues related to the medical use of silica. Some examples of the use of silica-based nanostructures for biomedical applications have been provided. Plenty of examples have shown that the various SiO_2-based NPs could be effectively employed as multifunctional therapy-diagnostic nanoprobes and biosensors. However, it should be mentioned that developed SiO_2-based nanoprobes are still in an early stage of research. There are some remaining tasks to be solved before SiO_2-based nanoprobes can be widely used for biomedical applications. For real clinical applications, it is required that these nanoprobes have excellent biocompatibility, good colloidal stability, long circulation time, and controllable size range. In fact, some of these problems can be easily solved by surface functionalization while the remaining ones require a long-term approach. In particular, the issues with silica NP uptake and the toxicity dependence on particle size, morphology, composition, and cell type have not been fully solved yet [80]. Furthermore, the possible health consequences for the uptake of multifunctional NPs should also be considered. Additionally, functional ligands can be used for theranostic NPs to specifically reach the diseased site. All these standing issues can be solved if one complete nanoparticle-cell interaction problem could be addressed by the cowork of scientists from different disciplines.

Despite the lack of research data, silica-based nanomaterials provide excellent opportunities for future biomedicine. As more clinical trials become available, the silica-based nanomaterials can be significantly improved to meet the requirements for safe clinical applications. In this context, it should be noted that the FDA recently approved silica-based NPs (Cornell Dots) for clinical applications in cancer diagnostics and therapy [84, 85]. Therefore, further research activity in this direction could help us to better understand the nanoparticle-cell interaction and to design a versatile all-in-one nanoprobe.

REFERENCES

[1] T.S. Atabaev, Multimodal inorganic nanoparticles for biomedical applications, in: A. M. Grumezescu (Ed.), Nanobiomaterials in Medical Imaging: Applications of Nanobiomaterials, Elsevier Inc., Oxford, UK, 2016, pp. 253–278

[2] T.S. Atabaev, J.H. Lee, J.J. Lee, D.W. Han, Y.H. Hwang, H.K. Kim, N.H. Hong, Mesoporous silica with fibrous morphology: a multifunctional core-shell platform for biomedical applications, Nanotechnology 24 (2013) 345603.
[3] W. Stöber, A. Fink, E. Bohn, Controlled growth of monodisperse silica spheres in the micron size range, J. Colloid Interface Sci. 26 (1968) 62–69.
[4] K.S. Finnie, J.R. Bartlett, C.J.A. Barbe, L. Kong, Formation of silica nanoparticles in microemulsions, Langmuir 23 (2007) 3017–3024.
[5] S.H. Wu, Y. Hung, C.-Y. Mou, Mesoporous silica nanoparticles as nanocarriers, Chem. Commun. 47 (2011) 9972–9985.
[6] R.P. Bagwe, C. Yang, L.R. Hillard, W. Tan, Optimization of dye-doped silica nanoparticles prepared using a reverse microemulsion method, Langmuir 20 (2004) 8336–8342.
[7] F.J. Arriagada, K.J. Osseo-Asare, Synthesis of nanosize silica in a nonionic water-in-oil microemulsion: effects of the water/surfactant molar ratio and ammonia concentration, J. Colloid Interface Sci. 211 (1999) 210–220.
[8] C.T. Kresge, M.E. Leonowicz, W.J. Roth, J.C. Vartuli, J.S. Beck, Ordered mesoporous molecular sieves synthesized by a liquid-crystal template mechanism, Nature 359 (1992) 710–712.
[9] P. Van Der Voort, M. Mathieu, F. Mees, E.F. Vansant, Synthesis of high-quality MCM-48 and MCM-41 by means of the GEMINI surfactant method, J. Phys. Chem. B 102 (1998) 8847–8851.
[10] J. Fan, C. Yu, F. Gao, J. Lei, B. Tian, L. Wang, Q. Luo, B. Tu, W. Zhou, D. Zhao, Cubic mesoporous silica with large controllable entrance sizes and advanced adsorption properties, Angew. Chem. Int. Ed. 42 (2003) 3146–3150.
[11] X. Wang, K.S.K. Lin, J.C.C. Chan, S. Cheng, Direct synthesis and catalytic applications of ordered large pore aminopropyl-functionalized SBA-15 mesoporous materials, J. Phys. Chem. B 109 (2005) 1763–1769.
[12] S.P. Naik, V. Bui, T. Ryu, J.D. Miller, W. Zmierczal, Al-MCM-41 as methanol dehydration catalyst, Appl. Catal. A Gen. 381 (2010) 183–190.
[13] V.S.Y. Lin, C.Y. Lai, J. Huang, S.A. Song, S. Xu, Molecular recognition inside of multifunctionalized mesoporous silicas: toward selective fluorescence detection of dopamine and glucosamine, J. Am. Chem. Soc. 123 (2001) 11510–11511.
[14] W. Zeng, X.-F. Qian, J. Yin, Z.-K. Zhu, The drug delivery system of MCM-41 materials via co-condensation synthesis, Mater. Chem. Phys. 97 (2006) 437–441.
[15] S.W. Song, K. Hidajat, S. Kawi, Functionalized SBD-15 materials as carriers for controlled drug delivery: influence of surface properties on matrix-drug interactions, Langmuir 21 (2005) 9568–9575.
[16] V. Polshettivar, D. Cha, X. Zhang, J.M. Basset, High-surface-area silica nanospheres (KCC-1) with a fibrous morphology, Angew. Chem. Int. Ed. 49 (2010) 9652–9656.
[17] N. Bayal, B. Singh, R. Singh, V. Polshettivar, Size and fiber density controlled synthesis of fibrous nanosilica spheres (KCC-1), Sci. Rep. 6 (2016) 24888.
[18] N. Bayal, R. Singh, V. Polshettivar, Nanostructured silica-titania hybrid using dendritic fibrous nanosilica as a photocatalyst, ChemSusChem 10 (2017) 2182–2191.
[19] J.S. Kang, J. Lim, W.Y. Rho, J. Kim, D.S. Moon, J. Jeong, D. Jung, J.W. Choi, J. K. Lee, Y.E. Sung, Wrinkled silica/titania nanoparticles with tunable interwrinkle distances for efficient utilization of photons in dye-sensitized solar cells, Sci. Rep. 6 (2016) 30829.
[20] T.S. Atabaev, G. Urmanova, N.H. Hong, Highly mesoporous silica nanoparticles for potential drug delivery applications, Nano Life 04 (2014) 1441003.
[21] Z. Sun, H. Li, D. Guo, J. Sun, G. Cui, Y. Liu, Y. Tian, S. Yan, A multifunctional magnetic core-shell fibrous silica sensing for highly sensitive detection and removal of Zn^{2+} from aqueous solution, J. Mater. Chem. C 3 (2015) 4713–4722.

[22] J.E. Lee, N. Lee, T. Kim, J. Kim, T. Hyeon, Multifunctional mesoporous silica nanocomposite nanoparticles for theranostic applications, Acc. Chem. Res. 44 (2011) 893–902.
[23] Y. Zhao, X. Sun, G. Zhang, B.G. Trewyn, I.I. Slowing, V.S.Y. Lin, Interaction of mesoporous silica nanoparticles with human red blood cell membranes: size and surface effects, ACS Nano 5 (2011) 1366–1375.
[24] I.Y. Kim, E. Joachim, H. Choi, K. Kim, Toxicity of silica nanoparticles depends on size, dose, and cell type, Nanomedicine 11 (2015) 1407–1416.
[25] E.B. Cho, D.O. Volkov, I. Sokolov, Ultrabright fluorescent mesoporous silica nanoparticles, Small 6 (2010) 2314–2319.
[26] B. Quan, K. Choi, Y.-H. Kim, K.W. Kang, D.S. Chung, Near infrared dye indocyanine green doped silica nanoparticles for biological imaging, Talanta 99 (2012) 387–393.
[27] Y.J. Kim, H.S. Jung, J. Lim, S.J. Ryu, J.K. Lee, Rapid imaging of latent fingerprints using biocompatible fluorescent silica nanoparticles, Langmuir 32 (2016) 8077–8083.
[28] M.C. Estévez, M.B. O'Donoghue, X. Chen, W. Tan, Highly fluorescent dye-doped silica nanoparticles increase flow cytometry sensitivity for cancer cell monitoring, Nano Res. 2 (2009) 448–461.
[29] L. Xue, B. Li, Q. Fei, G. Feng, Y. Huan, Z. Shi, Carboxylate-modified squaraine dye doped silica fluorescent pH sensors, Nanotechnology 21 (2010) 215502.
[30] T.S. Atabaev, H.H.T. Vu, Y.H. Hwang, H.K. Kim, Ratiometric pH sensor based on fluorescent core-shell nanoparticles, J. Nanosci. Nanotechnol. 17 (2017) 8313–8316.
[31] R.Y. Hong, B. Feng, L.L. Chen, G.H. Liu, H.Z. Li, Y. Zheng, D.G. Wei, Synthesis, characterization and MRI application of dextran-coated Fe_3O_4 magnetic nanoparticles, Biochem. Eng. J. 42 (2008) 290–300.
[32] H.B. Na, J.H. Lee, K. An, et al., Development of a T_1 contrast agent for magnetic resonance imaging using MnO nanoparticles, Angew. Chem. Int. Ed. 46 (2007) 5397–5401.
[33] T.S. Atabaev, J.H. Lee, D.W. Han, et al., Ultrafine PEG-capped gadolinia nanoparticles: cytotoxicity and potential biomedical applications for MRI and luminescent imaging, RSC Adv. 4 (2014) 34343–34349.
[34] K. Kattel, J.Y. Park, W. Xu, et al., Paramagnetic dysprosium oxide nanoparticles and dysprosium hydroxide nanorods as T_2 MRI contrast agents, Biomaterials 33 (2012) 3254–3261.
[35] T.S. Atabaev, Y.C. Shin, S.J. Song, et al., Toxicity and T_2-weighted magnetic resonance imaging potentials of holmium oxide nanoparticles, Nanomaterials 7 (2017) 216.
[36] Y. Zhang, S.W.Y. Gong, L. Jin, et al., Magnetic nanocomposites of Fe_3O_4/SiO_2-FITC with pH-dependent fluorescence emission, Chin. Chem. Lett. 20 (2009) 969–972.
[37] L. Cheng, Y. Liu, B. Zou, et al., Template-etching route to construct uniform rattle-type Fe_3O_4@SiO_2 hollow microspheres as drug carrier, Mater. Sci. Eng. C 75 (2017) 829–835.
[38] S. Smulders, A. Ketkar-Atre, K. Luyts, et al., Body-distribution of SiO_2-Fe_3O_4 core-shell nanoparticles after intravenous injection and intratracheal instilation, Nanotoxicology 10 (2016) 567–574.
[39] Y.C. Lee, D.Y. Chen, S.J. Dodd, et al., The use of silica coated MnO nanoparticles to control MRI relaxivity in response to specific physiological changes, Biomaterials 33 (2012) 3560–3567.
[40] T.D. Schladt, K. Koll, S. Prüfer, et al., Multifunctional superparamagnetic MnO@SiO_2 core/shell nanoparticles and their applications for optical and magnetic resonance imaging, J. Mater. Chem. 22 (2012) 9253–9262.

[41] Y.K. Peng, C.W. Lai, C.L. Liu, et al., A new and facile method to prepare uniform hollow MnO/functionalized $mSiO_2$ core/shell nanocomposites, ACS Nano 5 (2011) 4177–4187.
[42] M.A. Walling, J.A. Novak, J.R.E. Shepard, Quantum dots for live cells and in vivo imaging, Int. J. Mol. Sci. 10 (2009) 441–491.
[43] P. Pericleous, M. Gazouli, A. Lyberopoulou, et al., Quantum dots hold promise for early cancer imaging and detection, Int. J. Cancer 131 (2012) 519–528.
[44] K.T. Yong, W.C. Law, R. Hu, et al., Nanotoxicity assessment of quantum dots: from cellular to primate studies, Chem. Soc. Rev. 42 (2013) 1236–1250.
[45] Y. Ma, Y. Li, S. Ma, X. Zhong, Highly bright water-soluble silica coated quantum dots with excellent stability, J. Mater. Chem. B 2 (2014) 5043–5051.
[46] M. Vibin, R. Vinayakan, A. John, et al., Cytotoxicity and fluorescence studies of silica-coated CdSe quantum dots for bioimaging applications, J. Nanopart. Res. 13 (2011) 2587–2596.
[47] B. Zhang, X. Gong, L. Hao, et al., A novel method to enhance quantum yield of silica-coated quantum dots for biodetection, Nanotechnology 19 (2008) 465604.
[48] T.S. Atabaev, J.H. Lee, D.W. Han, et al., Multicolor nanoprobes based on silica-coated gadolinium oxide nanoparticles with highly reduced toxicity, RSC Adv. 6 (2016) 19758–19762.
[49] T.S. Atabaev, Z. Piao, Y.H. Hwang, et al., Bifunctional $Gd_2O_3:Er^{3+}$ particles with enhanced visible upconversion luminescence, J. Alloys Compd. 572 (2013) 113–117.
[50] Q. Lü, A.H. Li, F.Y. Guo, et al., The two-photon excitation of SiO_2-coated $Y_2O_3:Eu^{3+}$ nanoparticles by a near-infrared femtosecond laser, Nanotechnology 19 (2008) 205704.
[51] T.S. Atabaev, O.S. Jin, J.H. Lee, et al., Facile synthesis of bifunctional silica-coated core-shell $Y_2O_3:Eu^{3+}$, Co^{2+} composite particles for biomedical applications, RCS Adv. 2 (2012) 9495–9501.
[52] S.T. Selvan, Silica-coated quantum dots and magnetic nanoparticles for bioimaging applications, Biointerphases 5 (2010) F110–F115.
[53] A. Lapresta-Fernandez, T. Doussineau, S. Dutz, et al., Magnetic and fluorescent core-shell nanoparticles for ratiometric pH detection, Nanotechnology 22 (2011) 415501.
[54] C.E. Moran, G.D. Hale, N.J. Halas, Synthesis and characterization of lanthanide-doped silica microspheres, Langmuir 17 (2001) 8376–8379.
[55] F. Enrichi, Luminescent amino-functionalized or erbium-doped silica spheres for biological applications, Ann. N. Y. Acad. Sci. 1130 (2008) 262–266.
[56] H. Park, D.S. Bae, Synthesis and characterization of erbium-doped SiO_2 nanoparticles fabricated by using reverse micelle and sol-gel processing, J. Korean Phys. Soc. 61 (2012) 1481–1484.
[57] F. Enrichi, R. Ricco, A. Meneghello, et al., Luminescent dye-doped or rare-earth-doped monodisperse silica nanospheres as efficient labels in DNA microarrays, Proc. SPIE 7393 (2009) 73930M.
[58] S.V. Fedorenko, A.R. Mustafina, A.R. Mukhametshina, et al., Cellular imaging by green luminescence of Tb(III)-doped aminomodified silica nanoparticles, Mater. Sci. Eng. C 76 (2017) 551–558.
[59] E. Barron, I. Passagne, A. Auger, et al., Monitoring biological effects of 20 nm versus 100 nm silica nanoparticles induced on a human renal cell line using Fourier transform infrared spectroscopy, Anal. Methods 8 (2016) 2233–2242.
[60] M. Palmai, A. Petho, L.N. Nagy, et al., Direct immobilization of manganese chelates on silica nanospheres for MRI applications, J. Colloid Interface Sci. 498 (2017) 298–305.
[61] P. Sharma, N.E. Bengtsson, G.A. Walter, et al., Gadolinium-doped silica nanoparticles encapsulating indocyanine green for near infrared and magnetic resonance imaging, Small 8 (2012) 2856–2868.

[62] G. Sanchez, D. Curiel, I. Ratera, et al., Modified mesoporous silica nanoparticles as a reusable selective chromogenic sensor for mercury(II) recognition, Dalton Trans. 42 (2013) 6318–6326.
[63] X. Zheng, J. Pan, L. Gao, et al., Silica nanoparticles doped with a europium(III) complex and coated with an ion imprinted polymer for rapid determination of copper(II), Microchim. Acta 182 (2015) 753–761.
[64] N. Idros, M.Y. Ho, M. Pivnenko, et al., Colometric-based detection of TNT explosives using functionalized silica nanoparticles, Sensors 15 (2015) 12891–12905.
[65] C.T. Wang, C.L. Wu, I.C. Chen, Y.H. Huang, Humidity sensors based on silica nanoparticle aerogel thin films, Sens. Actuators B Chem. 107 (2005) 402–410.
[66] G. Patonay, M. Henary, G. Chapman, et al., Fluorescent silica nanoparticles containing covalently bound dyes for reporter, marker and sensor applications, Proc. SPIE 9723 (2016) 97230Q.
[67] A.J. Kell, L. Page, S. Tan, et al., The development of a silica nanoparticle-based label-free DNA biosensor, Nanoscale 3 (2011) 3747–3754.
[68] Z. Ye, M. Tan, G. Wang, J. Yuan, Novel fluorescent europium chelate-doped silica nanoparticles: preparation, characterization and time-resolved fluorometric application, J. Mater. Chem. 14 (2004) 851–856.
[69] B. Korzeniowska, R. Nooney, D. Wencel, C. McDonagh, Silica nanoparticles for cell imaging and intracellular sensing, Nanotechnology 24 (2013) 442002.
[70] M. Benezra, O. Penate-Medina, P.B. Zanzonico, et al., Multimodal silica nanoparticles are effective cancer-targeted probes in a model of human melanoma, J. Clin. Invest. 121 (2011) 2768–2780.
[71] Z. Xu, X. Ma, Y.E. Gao, et al., Multifunctional silica nanoparticles as a promising theranostic platform for biomedical applications, Mater. Chem. Front. 1 (2017) 1257–1272.
[72] A. Liberman, N. Mendez, W.C. Trogler, A.C. Kummel, Synthesis and surface functionalization of silica nanoparticles for nanomedicine, Surf. Sci. Rep. 69 (2014) 132–158.
[73] S.P. Hudson, R.F. Padera, R. Langer, D.S. Kohane, The biocompatibility of mesoporous silicates, Biomaterials 29 (2008) 4045–4055.
[74] J. Lu, M. Liong, Z. Li, J.I. Zink, F. Tamanoi, Biocompatibility, biodistribution, and drug-delivery efficiency of mesoporous silica nanoparticles for cancer therapy in animals, Small 6 (2010) 1794–1805.
[75] L. Yildirimer, N.T.K. Thanh, M. Loizidou, A.M. Seifalian, Toxicology and clinical potential of nanoparticles, Nano Today 6 (2011) 585–607.
[76] M. Joglekar, R.A. Roggers, Y. Zhao, B.G. Trewyn, Interaction effects of mesoporous silica nanoparticles with different morphologies on human red blood cells, RSC Adv. 3 (2013) 2454–2461.
[77] T. Yu, A. Malugin, H. Ghandehari, Impact of silica nanoparticle design on cellular toxicity and hemolytic activity, ACS Nano 5 (2011) 5717–5728.
[78] T.S. Atabaev, G. Urmanova, M. Ajmal, N.H. Hong, Fabrication of nontoxic dye-embedded silica particles for live cells imaging purposes, BioNanoSci 3 (2013) 132–136.
[79] P. Sharma, S. Brown, G. Walter, S. Santra, B. Moudgil, Nanoparticles for bioimaging, Adv. Colloid Interf. Sci. 123-126 (2006) 471–485.
[80] R. Xu, L. Huang, W. Wei, X. Chen, X. Zhang, X. Zhang, Real-time imaging and tracking of ultrastable organic dye nanoparticles in living cells, Biomaterials 93 (2016) 38–47.

FURTHER READING

[81] F. Torney, B.G. Trewyn, V.S.-Y. Lin, K. Wang, Mesoporous silica nanoparticles deliver DNA and chemicals into plants, Nat. Nanotechnol. 2 (2007) 295–300.

[82] S. Santra, K. Wang, R.T.W. Tan, Development of novel dye-doped silica nanoparticles for biomarker application, J. Biomed. Opt. 6 (2001) 160–166.
[83] N. Zhang, E. Ding, X. Feng, Y. Xu, H. Cai, Synthesis, characterizations of dye-doped silica nanoparticles and their application in labeling cells, Colloids Surf. B 89 (2012) 133–138.
[84] D. Luo, W.M. Saltzman, Thinking of silica, Gene Ther. 13 (2006) 585–586.
[85] J. Lai, X. Mu, Y. Xu, X. Wu, C. Wu, C. Li, J. Chen, Y. Zhao, Light-responsive nanogated ensemble based on polymer grafted mesoporous silica hybrid nanoparticles, Chem. Commun. 46 (2010) 7370–7372.
[86] M.A. Ballem, F. Soderlind, P. Nordblad, P.-O. Kall, M. Oden, Growth of Gd_2O_3 nanoparticles inside mesoporous silica frameworks, Micropor. Mesopor. Mater. 168 (2013) 221–224.

CHAPTER 5

Conceptualization of Self-Assembled Nanobiomimicry at the Intersection of Biology and Material Science: New Routes to Smarter Quantum-Scale Materials

Sanjiv Sonkaria, Varsha Khare
Institute of Advanced Machinery and Design, Seoul National University, Seoul, Republic of Korea

1 INTRODUCTION

Nanobiomimetic processes are gaining recognition as important routes to acquiring novel materials with new functional properties. Metal and semiconductor intercalation at the soft-hard nanomaterial interface has the potential to drive energy applications for soft robotics and drug delivery applications (Fig. 1). Despite progress over the decades, there is an increasing need to adopt more strategic processes to fabricate low-dimensional energy materials that have the ability to operate as low-power devices. Titanium dioxide as an exceptionally important semiconductor material is being underutilized because of current synthetic barriers to accessing its quantum state. The considerable difficulty in obtaining quantum-sized materials will undoubtedly reduce the technological impact in addressing energy, environmental, and potential biomedical applications. In the case of TiO_2, much remains to be explored as a quantum-sized material. Engineering new material properties through the application of synthetic protocols for acquiring TiO_2 to the quantum scale has been dismally poor with the exception of one or two successful attempts. It is important to stress early that there have been numerous claims to the successful synthesis of quantum-sized TIO_2. However, these particles greatly exceed the true quantum confinement range of $\leq 3\,nm$.

Nano-sized Multifunctional Materials
https://doi.org/10.1016/B978-0-12-813934-9.00005-0

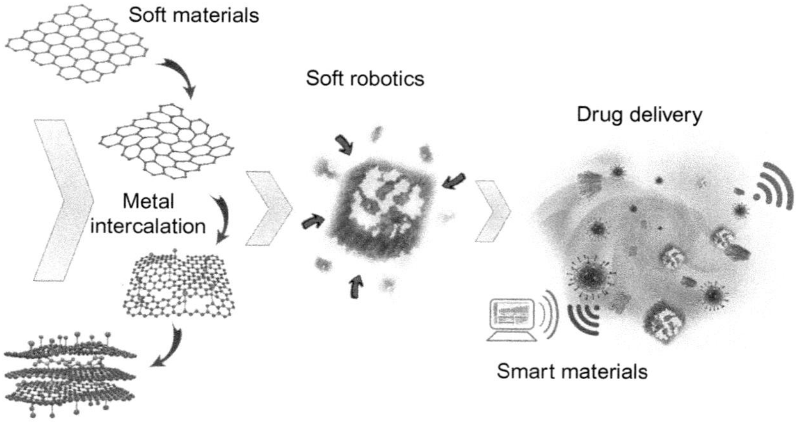

Fig. 1 Metal intercalation of nanomaterials. Metal incorporation of nano and quantum materials as potential multifunctional hybrid structures as energy materials for soft robotic and drug delivery applications.

Understanding the coordination chemistry [1] of growth parameters that guides synthetic events in the self-assembly of complex molecular nanoarchitectures from structurally simpler bulk precursor materials is of enormous multidisciplinary interest. For example, reduced precursor activity has been shown to improve the size distribution of particles in the quantum-scale range [2]. The exploitation of diverse nucleating growth patterns that reveal strikingly different nanostructural shape dominating features in self-stabilized structures is being pursued by integrating metals in polymer systems to form supramolecular nanoassemblies composed of metal organic frameworks (MOFs) [3] and supramolecular coordination complexes (SCCs) [4]. The effect of organic-inorganic conjugates on energy-driven charge transfer processes at the polymer-metal oxide interface is limited by nanoparticle morphology, degree of porosity, and accessibility to metal oxide binding sites using both a polymer-filled metal oxide composite or a polymer-metal oxide comixture approach [5]. This suggests that charge transfer processes might be better aided by applying more conceptually effective biomimetic synthetic processes [6] to control physical and chemical parameters.

From a biological perspective, the shape and size selection of biomineralized particles at the nano and quantum scales [7] (Table 1) may preferentially occur via an amorphous trapped state to liberate geometrically controlled crystals that may have both a chemical and functional significance to changing environments. This point is best emphasized by cooperativity

Table 1 Examples of representative biomediated synthesis of important biomineralized metallic salts and semiconductor metals

Organism/biological entity	Biomediated synthetic material	Size range	Ref.
Tenotrophomonas maltophilia Cystathionine γ-lyase (smCSE)	Cadmium sulfide (CdS)	Quantum conferment	[7a]
Stenotrophomonas maltophilia	PbS, PbS-CdS	Quantum conferment	[7b]
Plant cystoliths	Amorphous calcium carbonate (ACC)	Micrometer	[7c]
Modified collagen template	ZnO nanowires	Nanometer	[7d]
Silicateins	SiO_2 (silica) or TiO_2 (titania)	Micrometer	[7e]
Yeast cells	Cadmium telluride (CdSe)	Quantum conferment	[7f]
Candida glabrata and *Schizosaccharomyces pombe*	Cadmium sulfide (CdS)	Quantum conferment	[7g]
Fungus, Verticillium sp.	Gold $[AuCl_4]^-$	Nanometer	[7h]
Rhodobacter sphaeroides	ZnS	Nanometer	[7i]
F. oxysporum fungus	Bimetallic Au—Ag Alloy	Nanometer	[7j]
Shewanella algae	Platinum, $PtCl_6^{2-}$	Nanometer	[7k]
Desulfovibrio desulfuricans NCIMB 8307	Pd(II) to Pd(0) reduction	Nanometer	[7l]
RNA Pdase	Pd from $Pd2(DBA)_3$ (DBA = dibenzylideneacetone)	Nanometer	[7m]

effects between biological entities in the assembly of nanoparticles. For examples, two RNA phases have been reported to assemble palladium nanoparticles from the organometallic precursor complex Pd_2(dibenzylideneacetone)$_3$ into crystals that are otherwise confined to the amorphous phase in the absence of one or the other RNA pdase [7m]. The potential to cease opportunities to trap a tunable amorphous metastable phase at the synthetic level is of considerable developing interest.

The metastable arrest phase is often marked by changes in the structural configuration on energy scales that allow landscape dynamics to form a materially disordered, quenched-like state analogous to nonnative folds in proteins [8] aligned with dysfunctional conformations [9] and nanoscale confinement in biomineralization [10]. However, such nucleating shape-size relationships in synthetic systems are potentially tunable to deliver new chemical functionalities. Synthetic accessibility to such transient metastable configurations found in nature has been challenging to assemble and has seldomly been addressed. Further, the dynamic nature of an amorphous state is an important element of a mechanism that allows the shape- and size-dependent changes to occur in response to environmental and chemical signals. The control of spacing and size parametrics in amorphous precursors by synthetic intervention [11] via the introduction of biodriven-like cavities should predictably affect thermodynamic considerations and the kinetic behavior of crystallization [12].

Dimensionality, which signifies the property of surrounding space as it applies to quantum materials, is also proving to be an important criterion to templating approaches in controlling the parametrics of size in nucleating materials. Ionic liquids (ILs) [13] composed of loosely coordinated ions exhibit low symmetry due to weak ionic interactions. The spatial arrangement between and around the ions is sufficiently feeble to generate degrees of structural regularity, interrupting the asymmetric behavior at the hard-soft interface of the interacting charged surfaces. Further, tailoring the templating properties of organic ionic liquids to control the shape and size of the metal oxide nanoparticles via the process of molecular self-assembly driven by thermodynamic and kinetic processes is an exploitable route to generate polymer-coated organometallic structures that exert surface passivation properties to effectively modulate the band gap characteristics of semiconductors as energy materials. Self-assembly events at the polymer-metal oxide interface can take advantage of the flexibility around a templated anionic-cationic reaction medium to drive the association and convergence of complementary chemical structures, enabling the supply of guest structures to facilitate host-guest supramolecular interactions binding a network of hydrocarbon chains. The availability of homogenously structured and well-distributed patterned surfaces formed at the solid-liquid interface from core materials provides a rich source of nucleating sites that strongly defines the physical and chemical nature of stabilized nanoparticles during and after the growth process. This characteristic has demonstrated considerable utility in directing structure-induced shape modifications during nucleation events. Surface heterogeneity at the solid-liquid interface during carbonization of ionic liquids is a potential route to acquire materials of low dimensionality.

This objective is of considerable importance to the assembly of semiflexible polymers and their correlation with biopolymers on a synthetic level as well as processes that impart new properties at the soft-hard interface of contrasting materials. As biomimetics begins to take shape, polymer structural engagements around nucleating sites that favor unusual bending and strain energetics of atypical structures can be driven to self-associate into metastable arrangements compensated through entropic loss. This could have far-reaching implications in mimicking the flexibility of biopolymer frameworks to include stretching-bending characteristics, strained geometries, and a template scaffold precursor that supports multidimensional cross-linked networks conjugated into a supramolecular lattice. Such chemical contacts are an important criteria for flow characteristics for charge transfer events, for example.

In this chapter, we explore the emerging role of "synthetic nanobiomimicry" in controlling both the shape and size characteristics of hard-to-configure quantum-scale materials, exemplified by TiO_2. The focal point of this investigation converges on enabling bulk precursor chemistry to drive a biomimetic synthetic mechanism of confinement directing the quantum "trapped" growth of the rutile phase of TiO_2. The quantum arrest of TiO_2 is synthetically difficult and biogenically unreported, and is reputed to be exceptionally hard to achieve on the low-dimensional scale. Initially, the correlation between bulk precursor surface chemistry and the control of interfacial surface properties at heterogeneous interfaces is examined by reference to its catalytic chemistry and the self-assembly of functional groups modulated by key elemental doping events triggering an interdependence between polymer and surface ligand chemistry occurring on different time scales in a single reaction. Parallels are drawn between the folding trajectory of biologically driven assemblies on a transient scale and geometrically trapped nonequilibrium synthetic structures intermediate between stable and unstable energy states. This is conceptually illustrated by the energy landscape folding trajectory of proteins [14] en route to forming stable folded structures from their unfolded states (Fig. 2). Experimental investigations of such nonequilibrium systems will inevitably contribute to a better appreciation of theoretical problems in understanding the transition between equilibrium and nonequilibrium driven self-assemblies [15]. Further, we discuss the relevance of such materials to the technological development of energy materials and their utilization as alternative sources for harnessing electrical power from solar energy.

2 TUNING SURFACE CHEMISTRY TOWARD CHEMICAL SELECTIVITY: ACCESSIBILITY TO HARD-TO-ACHIEVE ATOMIC CONFIGURATIONS

The concept of "atomically directed" nanofabrication allowing a precise balance to be established between controllability and applicability in the

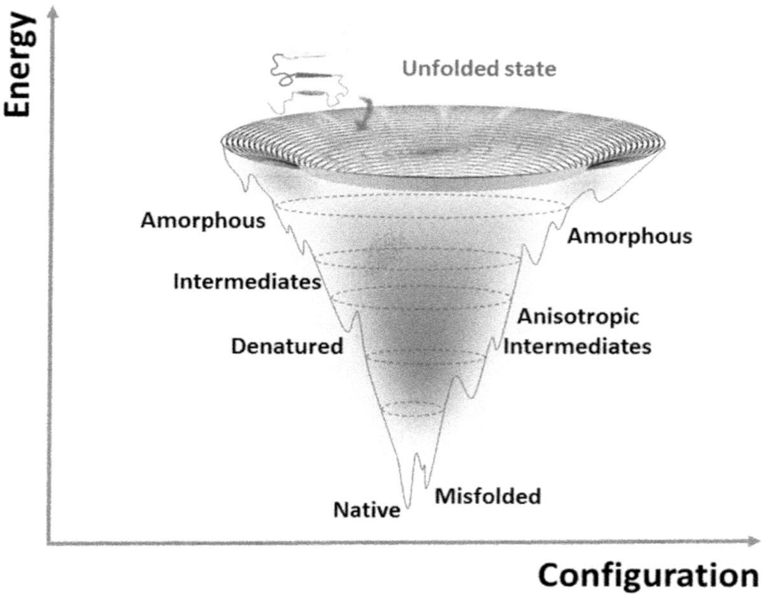

Fig. 2 Conceptual folding funnel energy landscape of protein folding. Energy landscape depicting possible folding trajectories of a protein from the unfolded state to its native state. Kinetic pathways can lead to the thermodynamic entrapment of nonequilibrium structures as amorphous and intermediate states with varying transitory time scales.

framework of nanomaterial design is far from being achieved. However, analogous to early reports highlighting the limitations of conventional chemical synthesis as a strategic route to access complex synthetic structures in the nanometer range [16], this possibility is being examined with semiconductor colloidal particles forming larger structures from localized short-range interactions between competing excited and ground-state structures [17] through the recognition and convenience of self-assembly processes [18]. Central to this realization that distinguishes the hallmark characteristics of self-assembly processes over chemical approaches in the utilization of complementary geometrical interactions between reactant molecules of a covalent nature, the hierarchical ordering of nanostructures and the effective use of building blocks in the construction of supramolecular hybrid complexes is enabling these key steps to proceed spontaneously through natural chemical selection and differential binding of chemical ligands favoring the incorporation of otherwise inaccessible functional properties. This approach might help understand and predict the untapped potential of thermodynamically driven molecular forces to form stable shape-dominant structures, broadening the relationship between molecular patterns and function.

A considerable geometrical challenge in semiconductor materials chemistry has been ways to understand and support the assembly processes leading to the quantum confinement of TiO_2. As an exceptionally important semiconductor material, TiO_2 is being underutilized because of current synthetic barriers to accessing its quantum state. Considerable difficulty in obtaining quantum-sized materials will undoubtedly reduce the technological impact in addressing energy, environmental, and potential biomedical applications. The change in optoelectronic properties under quantum confinement is intrinsically related to size that affects their optical absorption band energy range, exciton energies, recombination properties, and density of states (DOS) at the band edge that imparts discrete particle wave function behavior. These parameters are surface-state dependent and the mobility of defect states that show temperature-dependent behavior is largely observed within the quantum region [19].

The unmatched potential to develop TiO_2 as a multifunctional biocompatible material has been limited by considerable challenges to enhance the untapped properties of TiO_2 at the quantum scale to exploit shape- and size-dependent photoluminescence characteristics accessible through discrete forbidden quantum band regimes. This nontriviality in the quantum scaling of TiO_2 suggests that conventional growth patterns that are sensitive to interfacial energies at surface edges or defect sites and the composition of starting materials that influence temperature-dependent enthalpies of interfacial surfaces require alternative strategies. A key criteria to implement novel approaches to organize matter is by targeting structure-induced shape modifications during nucleation. Better manipulation of protein caged-nanoparticle assemblies with novel architectures has been achieved using moderate to low reaction rates by lowering temperature, precursor concentrations, and the stoichiometry of reactants. Under these conditions, the possibility of directing the controlled diffusion of inorganic precursors through protein-shell barriers into hollow cavities imitating dispersed nanoreactor-like centers offers much potential to broaden the diversity of nucleation events under nanoscale confinement assisting the self-assembly of more novel architectures. Such self-assembly processes, however, proceed in driving equilibrium structures that are associated with more predictable nucleation and growth patterns forming complex supramolecular hybrid structures that may favor geometries capable of hosting guest structures in bonded cavities.

Indeed, hard-to-achieve atomic configurations epitomized by the quantum confinement of TiO_2 are being fueled by the desire to address approaches to induce more shape adaptive mechanisms to inhibit nucleation growth as the primary phase in governing size and shape. This is decisively much harder to achieve at the quantum scale. While chemical paths to equilibrium self-assembly phases share more stable thermodynamic boundaries occupying deep energy minima's restricting morphological transformations resulting from heightened energy barriers, the relative ease in adaptability of nonequilibrium phases renders host structures more shape compliant to congruent guests, increasing the possibility of confinement. Because shape and size flexibility is a primary function of intermolecular interactions, a direct route to self-program desired dimensions at the quantum nanoscale can be achieved through the use of structural templates.

3 CONCEPTUALIZATION OF SYNTHETIC NANOBIOMIMICRY: LESSONS FROM BIOLOGY

3.1 Energy Landscapes and Biomimicry

Multicomponent colloidal systems originating from self-assembly processes exhibit complex phase transitions at the liquid-solid interface. Efforts to decipher and understand the delicate balance between the multiplex interplay of enthalpically driven pathways and their underlying relationship with surface chemistries have been slow. The nucleation and growth dynamics of naturally occurring biomolecules and those of synthetic origin directed by self-assembly processes proceed via their respective transient conformational and anisotropic states governed by an energy landscape defined by thermodynamic and kinetic energy barrier constraints. The perception that relates the organization of folded structures into task-specific proteins or synthetic nanomaterials with synthetically adapted morphologies as an elementary function of atomic-scale interactions at the level of nucleation and prenucleation must also influence their ability to evolve into amorphous-phase or crystalline states of matter. The natural selection of folding transitions in proteins conceptually adheres to a folding funnel-like trajectory of trapped events en route to their native structural conformations. These folding transitions conceptually proceed through multiple pathways in which folding events are dictated by the energy landscape of interfacial surface interactions between the bulk environment and the folding protein exterior. Conformational precision of structure-function processes makes use of folding patterns that occur through a number of enthalpically trapped configurations distributed along a path of thermochemical barriers distinguishable by a succession of local energy minima of misfolds, aggregates, nonnative structures, or amorphous and interchangeable conformational states.

Here, we extend the analogy that funnel-shaped energy landscapes are equally applicable to nonbiological crystallization processes. The existence of multiple pathways along the funnel route is modulated by kinetic and thermodynamic pathways that can often lead to protein misfolding and subsequent entrapment of metastable species. These exhibit a scale of shape dependency upon confinement to local energy minima when descending the funnel trajectory. An interesting development in elucidating characteristic folding patterns of proteins along their corresponding enthalphic routes is the work on heteropolymers. Such polymer compositions derived from multiple monomers can significantly interrupt and delay native three-dimensional configurations [20] influenced by key differences in the nature of kinetic traps and their pathways. In extending this notion, if we envision the amorphous or crystallization phases of nonbiological processes to proceed through a funnel-type ensemble of events that is essentially shape-dependent, it may be feasible to influence and subsequently arrest novel hard to achieve atomic configurations to tunable geometries such as spaced confined crystalline cages of metal-organic complexes. While protein cages such as ferritin have been used as templates to immobilize and enhance the electronic localization of semiconductor photocatalyts [21], self-assembled groups might also be directed to form nanochemical geometries capable of confining and inhibiting metal nucleation growth at the organic-metal interface. We draw an analogy with efforts to control protein nucleation under conditions of confinement that act to alter nucleation energy barriers to growth or diminishment, suppressing or allowing alternative kinetic pathways to crystallization and metastable phases [22].

The shape dynamics of metal-conjugated polymeric materials of unusual amorphous types are interesting in their biomimicking characteristics of protein folding patterns that arise during mismatch in topological alignments of protein chains and their comparison to the self-organization of chemically driven morphologies. Such crystallization processes are gaining increasing importance, particularly in the formation of nanocavities or nanocages that leads to the trapping of metastabilized protein cores or metal oxides. In certain instances, for example, the growth dynamics within nanoconfined spaces might drive the assembly of unexpected or unconventional shape configurations.

Compelling evidence supporting multisubunit crystallization initiated from a monomeric fold via an amorphous precursor phase pathway [23] is equally accessible to folding funnel-like energy landscapes assumed to be exclusive to singular folding events. The cornerstones to these breakthrough discoveries have relied on in situ AFM imaging techniques that have been used to describe monomer growth by adsorption measurements. In view of the soft matter perspective on protein crystal assembly [24] and our current understanding of protein architecture, the dependency of three-dimensional biomolecular structure on physical and chemical parameters has provided much insight on synthetic material behavior correlating to mineralization processes and phase change characteristics. Valuable clues to novel synthetic material development at the quantum-nano scale are derived from a mixed phase concurrence of amorphous and crystalline states of biogenic origin [25]. This work has important implications in distinguishing the persistence of a water-rich bionanomineral phase that precedes a dehydration and crystallization phase change.

3.2 Bioadaptive Mimicry of Synthetic Self-Assemblies

Adaptation of synthetic assemblies at the level of nucleation in biologically controlled environments occurs at the atomic level, enabling exquisite control over structural geometries through selective kinetic pathways. Ordered scaling of complex hierarchal morphologies of varying shape and size has both a chemical and functional significance that is often adaptive to environmental signals and operational needs. Hence there exists a fine relationship between the flexibility and diversity of materials that transcends physical, chemical, and functional properties. The intrinsic nature of biosynthetic materials to escalate from disordered to ordered structural states, for example as stimuli-responsive multiinterface bioassemblies with the ability to function intelligently on different scales (Fig. 3), is a fundamental aim of biomimicry. One example of such materials is those that can be made responsive to magnetic fields (Fig. 4) through shape-responsive properties by exploiting attractive and repulsive forces in magnetized hybrid materials. As we approach the ability to match the atomic-scale manipulation of matter that is inherent in biological systems, challenges remain in understanding precise mechanisms that allow the fabrication of such materials solely by synthetic means.

Adaptive behavior is executed through host-guest interactions that proceed via recognition-based interactions. The versatility of chemical bonds

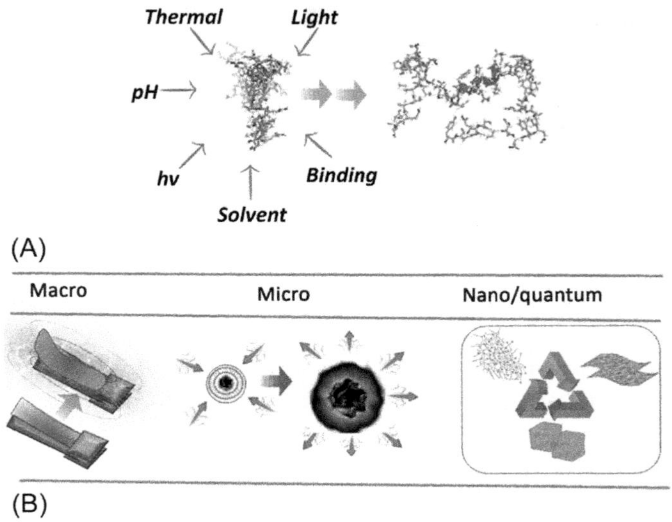

Fig. 3 Biomimicking intelligent materials. (A) Biomolecules with built-in intrinsic stimuli responsive properties that may be multifunctional in nature triggered by pH, photons, light, solvent, heat, or binding interactions. (B) Fabrication of stimuli-responsive materials on the macro, micro, and nano/quantum scale.

and functional groups that determines shape, size, geometry, and selectivity on different scales offers the potential to be manipulated at the molecular and atomic level to controllably engineer task-specific structures that biology has perfected. Examples of biologically triggered events include conformational binding, signal transduction, catalysis, cooperative binding, and triggered release mechanisms that are essentially stimuli responsive [26] but also include nonproteins such as polymers [27]. Probably, the most significant examples of adaptive biology are those based on evolving recognition mechanisms against antigenic structures through immune responses generating a large number of antibody structural variants.

Biology has access to vast control mechanisms that permit the synthesis of structured materials by self-assembly routed through bottom-up processes. It is fascinating and thought provoking that such mechanisms are applicable to the assembly of nonbiological materials for both natural [28] and engineered organisms [29], including titanium dioxide (titania) [7e]. From an application perspective, the logical aim of using natural biological machineries to synthesize metal oxide semiconductors is geared toward improving biointerfacing between contrasting material properties, leading to better hierarchical control of structures [30]. To achieve this purely by synthetic approaches

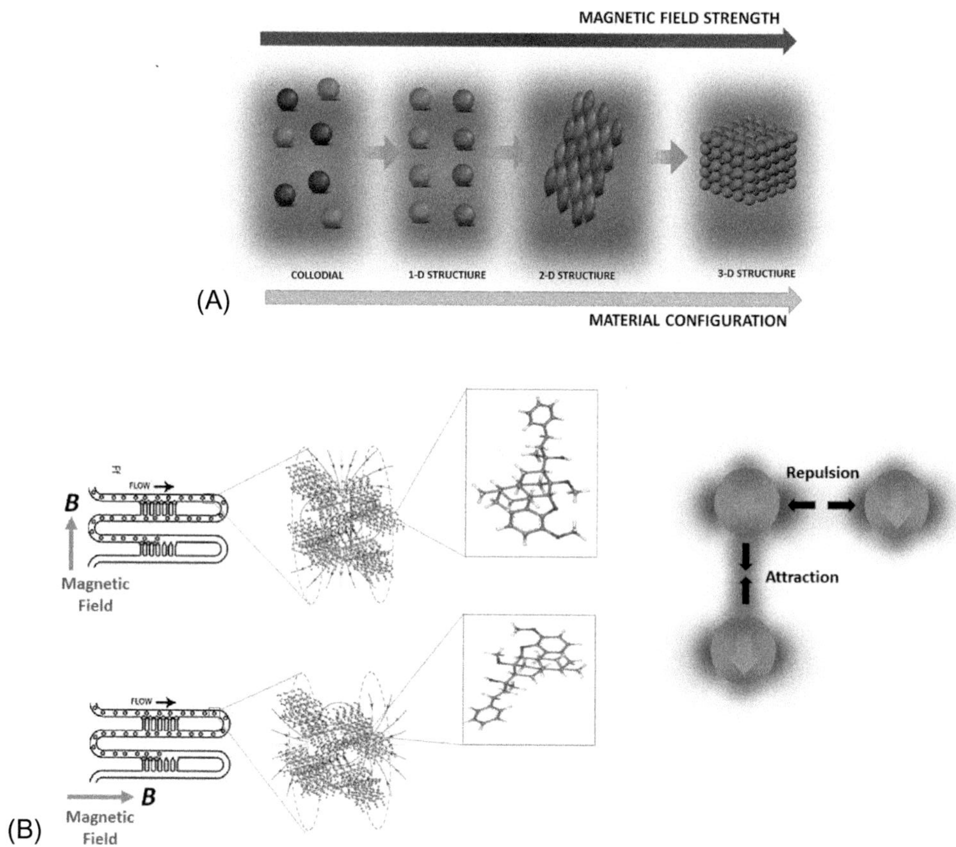

Fig. 4 Magnetic hybrid nanomaterials as stimuli-responsive materials. (A) Shape-responsive magnetized colloidal materials undergoing structural change induced by magnetic fields. (B) Exploiting attractive and repulsive forces of microfluidic encapsulated magnetic hybrid materials.

requires the use of templated structures that sufficiently mimic biodriven templates [31]. Current challenges depend on knowledge-based strategies to decipher chemical codes to design templates that combine functional group selectivity and energy landscapes to impart biological properties in resulting material behavioral characteristics. This might necessarily make use of contrasting multicomponent interfaces such as metal and polymer growth nucleation from bulk templates affecting morphology and dimensionality to induce new properties. Here, flexibility associated with self-assembled processes has shown much promise in tuning material geometrical designs [32], The programmable scale of events can be structured akin to enthalpically driven biological folding funnel-type trajectories [14, 33] steered through selective kinetic pathways and thermodynamic traps similar to peptide blocks [34]. Hence, the principles of synthetic biology [35] correctly applied to a synthetic template blueprint might make possible the accessibility to new structures via specific-template design [36]. The current focus in synthetic materials fabrication, however, relates to the interrelationship between structural diversity and control over shape and size, particularly in view of rising interest in low-dimensional materials at the subnanometer or quantum scale as low-consumption energy materials.

Organisms have evolved sophisticated routes to self-assemble semiconductor materials mechanistically with crystal nucleation patterns embodied at the nanoquantum range [7i, 37]. The fabrication of organometallic structures that precedes the nanometer scale afar from the quantum confinement range is accessible by chemical synthetic approaches, but growth restriction within the quantum dimensional space is still challenging for promising semiconductor materials such as TiO_2 that exhibit quantum properties below 2 nm. As the evolving parallel between guest molecules (e.g., metals or metal oxides) exposed to biological environments or guest-host interactions with chemical templates in synthetic systems unfolds, a concurrent theme is emerging of adaptability in which chemical systems obey and conform to the behavioral characteristics of biomolecules. This is possible if molecular chemistry makes use of flexible bonding associations that are synonymous with nature's spontaneity of noncovalent interactions. Some of the features important to self-assembly [38] require cooperativity and synergistic binding but not necessarily associations that are complementary in nature. Hence, transitions between disordered and ordered networks tend to utilize (1) weak binding affinities, (2) multiple binding events that occur under different time scales, (3) directional steering at heterogeneous surfaces, (4) intrinsically driven binding orientations, (5) geometric constraints in space (e.g., nanospaces influencing dimensional scale, stress-strain and rigidity-fragility relationships, confinement, water dynamics and glassy states), (6) reversibility, (7) shape and size convergence, and (8) kinetic pathway selectivity and thermodynamic trapping of nonequilibrium metastable states.

Decoding the intricate steps of enthalpically driven surface associations that define spontaneous self-assembly at the biointerface is a cornerstone to successfully mimicking such complimentary surface interactions chemically to access size-restricted growth parameters. Because biology is perplexed by the complexities of multiphase and mixed interactions of contrasting surfaces in compartmentalized areas of the cell, our ability to biomimic and access novel architectures of subnanometer and quantum states depends on the nature and availability of chemical bulk precursors or their synthetic tunability (bulk precursor with catalytic surface properties) as the building blocks for self-assembly. However, the shape and size persistence of unfolded protein-peptide chains during the folding process, for example, might be advantaged by the structural sequence of amino acids as elemental building blocks guiding the pattern of folding. This suggests an inherent spatial order locked within the inner structure in biological systems that provides directional significance from the onset of nucleation. This points toward a functional role for chemically ordered charged templated environments bearing spatial and chemical orientation of freely mobile building units generated from bulk precursors under thermally driven reaction conditions. The shape and size selection of host structures such as semiconductor metals by the self-assembly of polymers and their escalation in scale from molecular to supramolecular [39] level artificially can be generated by enabling the biomimicry characteristics of biopolymer catalysis. Such properties are inspired by biocatalysts of nonequilibrium polymerization [40] that use dynamic noncovalent interactions with molecular forces that allow the build up of scalable compartmentalized polymer scaffolds adaptable to the growth features of a nucleating host.

This opens the possibility of utilizing polymerization cross-linking strategies for precise monomer assembly that provide cross-linked bonding networks for building coordination networks composed of coordination cages for low-dimensional colloidal synthesis. Cytoskeleton proteins that shuffle from monomer to polymerized states responsive to signaling and enzyme regulation are just one example of nature's molecular forces at work, intelligently shaping the needs of the environment. The biological ordering of hieratical polymers in living systems that use space and geometry to adapt and control polymer structures to their environment is an emerging area of focus. True synthetic mimics in complex self-assembly systems are not restricted to single polymer events during nucleation, but materialize as metastable multicomponent copolymer blends. For example, they have been shown to predictably adopt thermodynamic characteristics at copolymer buffering ratios of 1:1 and are strongly temperature dependent [41]. We propose that the imitation of synthetically driven biomimetic routes to metastable structures can open routes to tunable material properties of shape, size, and function through chemically designed building blocks that participate in cooperative nucleation processes at complex interfaces. This is possible for metastable phases due to the small enthalphic and free energy differences among surfaces states [42] with interchangeable morphological structures.

An area that is currently occupying much importance in the field of biomimetic self-assembly that connects geometrical size and shape and control over the complexity of fabricated functionally relevant structures converges upon supramolecular chemistry. The principles that drive the self-assembly of supramolecular structures that are perhaps intrinsically unfavorable in design from a biological perspective are those that can be stabilized and altered at will and that lie outside the thermodynamic equilibrium. The transition from molecular to supramolecular assembly occurs readily with organometallics and becomes feasible when shape and structural complimentary occur with ligand-receptor type positioning. This is possible for larger building blocks permitting a supramolecular architecture. This scenario of recognition-based binding in which metallic elements or complexes are accommodated as stabilizing centers to larger scaffolds are hallmark characteristics of polymers that claim their own functional significance to the trapped metallic cores. Cooperativity involving copolymer growth between monomers around guest molecules under favorable conditions that promote reactive cross-linking is becoming increasingly programmable, given the ease by which polymeric molecules [38,39] adapt to both chain and branch chain-like growth mechanisms. It is conceivable that the multitude of shapes and polymer pores across different scale lengths is possible through the availability of bonding sites along the polymer. This may cater to induced flexibility under geometric constraints, influencing both the shape and size of polymer scaffolds and their effect of metallic growth nucleation. A more interesting approach to gain deeper access to unreported supramolecular assembly types could be better achieved by allowing the natural selection of polymer growth by self-assembly via nucleation mechanisms at metal oxide catalytic sites through monomer assembly that may otherwise be impeded by the reaction of preassembled polymers. The conceptual idea of biomimetic self-assembly is not new but the disconnect between how nature assembles biomolecules and how we as chemists and material scientists conceive material fabrication by synthetic means has been considerably hard to decipher. Surveying and establishing common elements to both processes has been blurred because the interest has largely been on generating novel synthetic materials as end products, which has diverted much of the attention away from the relevance of fundamental concerns on the dynamics of bulk to nano- or quantum-scale chemistry. The change, however, has been marked by efforts to identify elements of biomimetics of self-assembly and supramolecular chemistry with natural biological processes governing self-assembly of quantum state biomineralization [6]. A quantum leap in this direction by the group of De Cola [43] that provided unprecedented insight into the dynamic equilibrium of multiassemblies, supporting indirect evidence of overlapping or converging kinetic pathways. The interlocking of dynamics of kinetic pathways and thermodynamic traps is reminiscent of energy landscapes that conceptually define protein folding trajectories from unstable (high energy) to stable (low energy) conformations. The fact that such assemblies undergo further structural refinement where the interconversion is controllable by a predefined solvent to water ratio demonstrates the intermediatory nature of the kinetic products and the shift toward nonequilibrium states. The stimuli-responsive behavior shows, however, that this trend is reversible and tunable in shape and size. Manipulation of structural features and physical properties by understanding the nature of bonds in self-assembled complexes has the potential for behavior as smart materials selectively in response to specific stimuli exposure of a particular energy and type (e.g., photon, pH, thermal, solvent, etc.). This is particularly prominent at constrained interfaces between reaction surfaces such as polymers and semiconductors in which different material constituents exhibit stimuli responsive properties unhindered by overlapping energy bands.

One of the most complex and unquestionably cryptic routes to crystallization organisms use to adapt to environments requires some species to adopt an amorphous pathway, essentially arresting biomineralized growth to a precrystallization phase. The time scale of biomimetic processes is infinitesimally small in comparison to biologically driven mineralization and its evolutionary origin over millions of years. Thus, through strategies that have allowed hierarchical structures to take a variety of shapes and sizes from macro to quantum scales mediated via an amorphous phase by delaying crystallization, organisms can control not only dimension but also mechanical properties with ease. Calcium carbonate biomineralization is the most well-studied area of scalar growth [44]. An important observation in nanomaterials chemistry is the diversity of hierarchical structures accessible by synthetic means. Although low-dimensional morphologies at the subnanometer scale are significantly outnumbered by successes on the nanometer and micrometer scale, lessons from biology provide knowledge-based evidence that hard to achieve atomic configurations of lower dimensions may inevitably be coupled to unexplored kinetic routes intermediate between stable and unstable states of matter. Access to the miniaturization of bulk state materials through pathways that are novel to synthetic systems but have long evolved as established routes in biological systems requires chemically driven self-assembly to be tunable for alternative ways to achieve minimization of interfacial energies. This may require new approaches to allow the adaptability of biomimetic processes to be more favorable toward (1) template driven effects, (2) amorphicity, (3) polymer stress-induced confinement at catalytic sites, (4) water and glass transition state dynamics, and (5) tunable polymorphism of nonequilibrium states. Fig. 5 shows that

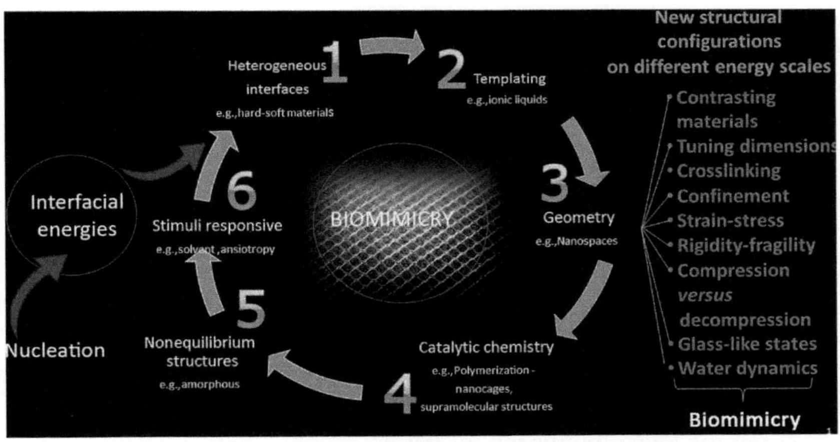

Fig. 5 Dissecting the important elements to synthetic nanobiomimicry. Directing interfacial energies at the level of nucleation at complex heterogeneous interfaces in nanobiomimicry defines the basic morphological blueprint akin to biodriven assemblies by kinetic selection steered by structural templates en route to crystallization. New geometries that pass through a stable transitory amorphous phase provide a further level of morphological selection of closely related nonequilibrium structures. Synthetic templating using the principles of biology by mimicking biointerfaces such as those observed for biomineralization provide much scope for controlling events for both pre- and postnucleation events on different energy scales.

establishing a well-understood correlation between surface chemistry and its impact on steering interfacial energies at heterogonous interfaces could provide hidden clues to imitate biological pathways, improving our approach to biomimetic self-assembly.

This path to discovery entails a more systematic understanding of template induced programmability to access heterointerfaces with energy delivered on a scale to drive precise shape, size, and geometrical outcomes of materials with tunable functionalities. The evolution of biomimicry lies within exploratory nanomaterials chemistry. Its new direction and progress toward the next generation of materials requires stronger convergence with biological strategies in their aim of generating and utilizing energy materials.

We initially aim to draw a parallel between biological systems and synthetically driven processes that drive the assembly of metal complexation at the organic and inorganic interface. This will provide a basis to identify commonalities and key differences in the assembly of biological and synthetic scaffolds, naturally leading to a discussion of biomimetic systems and promising routes to adopt nucleation events that define these processes at the structural and molecular level.

3.3 The Amorphous-Glass Phase—Adopting a Biogenic Route to Host-Guest Confinement

A key feature associated with biomineralization that permits organisms to synchronize crystal characteristics to adapt to their environmental needs such as shape and size equates short-range order that operates as a dominating characteristic of the amorphous phase. As in plant cell wall mechanics, the porosity of networks that contributes to the attainment of a hydroskeleton-like composition of networks, mechanical stiffness, rigidity, and deformation characteristics of the polymer plays a key role in the anisotropic nature of the polymer shell. In the context of current findings, these behavioral properties might resemble plant cell wall pectins forming a gel-like composition of an amorphous nature in which the matrix provides an assembly of interlinking networks for embedding polysaccharides. The mechanical rigidity of the polymer surroundings can impart important consequences in the growth parametrics of cells and tissues [45] on a tunable scale of shape and size. Recent investigations have explored the correlation between polymeric additives and their effect on crystal seed growth of amorphous synthetic formulations. Polymers have been shown to inhibit crystallization, driving new growth patterns by altering (a) crystal growth rate,

(b) crystal morphology, and (c) the size and shape in the presence of water [46]. This significantly affects the evolution of crystal maturation via an amorphous phase.

The dynamic nature of an amorphous state further is an important element of a mechanism that allows shape- and size-dependent changes to occur in response to environmental and chemical signals. The control of spacing and size parametrics in amorphous precursors via the introduction of cavities profoundly effects the thermodynamic consideration and kinetic behavior of crystallization [12]. This relates to the molecular imprint of the amorphous phase affecting the bond order and dictates the overall crystalline qualities in transitioning from the amorphous to crystalline phase. Some key elements of the amorphous phase that have important implications in material assembly are shown in Fig. 6.

With respect to nonbiological materials, the ability to control such dynamics synthetically from an amorphous phase is a rare occurrence but is indeed possible, as demonstrated by amorphous alumina in which changes induced in the short-range order of the material in the precursor phase are not only possible but have the potential to control crystallization akin to biomineralization [47]. Complex interfaces show characteristics of heterogeneous nucleation marked by the structural diversity of mixed nucleating compositions with discrete or overlapping assembly rates. Nucleation within

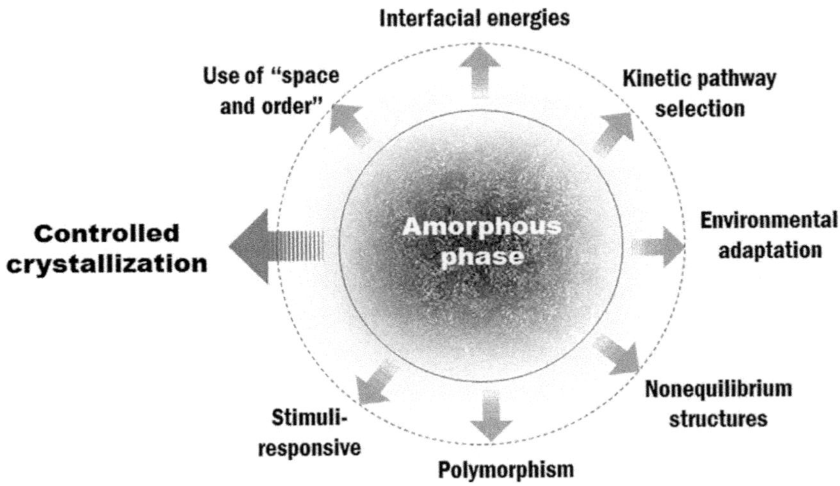

Fig. 6 Adaptability of structures to environments via the "amorphous" phase. Steps to controlled crystallization used by organisms via the "precrystallization arrest phase through an amorphous route.

the confines of a polymer matrix caters to geometrical features that contribute to the enhancement of monomer-monomer interactions in sync with preferential crystallographic orientations steered by the polymer shell by solute-adsorptive compartmentalization [48] using polymer microgels [49]. Further, a shape-dependent kinetic mechanism driving nucleation triggered by the resulting polymer nanopores can orientate molecules to favorable configurations for nucleation to occur at angular positions on the pores [50].

The notion that the conformational dynamics of intrinsically disordered proteins (IDPs) follow a pattern of folding into their respective native structures from the unfolded state where chemical and surface interactions dominate could be predictably comparable to postnucleation events in synthetic systems. In particular, intrinsically disordered proteins may be a special case in this respect due to thermodynamic and kinetic considerations [51] being equally applicable to nucleating species after the critical nucleation phase has occurred. Hence, this may bring important relevance to the growth nucleation of enthalpically separated soft and hard materials that might be considered to compete against nucleation sites between self-assembled polymer and TiO_2 surface interfaces. Soft interactions that define chemical interactions between IDPs in a polymer crowded environment can occur via electrostatic associations, hydrogen bonding through donor-acceptor sites, and weaker van der Waals forces. Geometrically designed supramolecular protein assemblies using proteins as individually aligned building blocks [52] have been artificially infused at the protein-protein interface to engineer nanocaged structures to reliably imitate porous networks formed from internal cavities.

The size control of metallic crystals under protein-templated growth within a protein cage is offset by the intrinsic nature of the material where the confinement of crystalline growth to smaller domains occurs independently of the caging effect resulting from defect behavior [53]. The intervention of nucleation and growth by polymer addition has the potential to shift the phase of material assembly that alters the dynamics between the different stages of crystal growth. Polymer blends through the association of self-assembled structures that kinetically originate from nonconventional routes have been predicted to hinder crystallization. This approach experimentally supports an analogous relationship that might exist between nonequilibrium driven assemblies and polymer stress relaxation processes that determine the extent of polymer distortion as a result of chain configuration connectivity and the degree of amorphicity.

The enrichment of polymer-melt interfaces from growing organic nanochains around nucleating metal centers caters not only to metal growth compartmentalization but also to shape-tunable organometallic photoresponsive amorphous to crystal transitory materials. Recent attention on cross-linked polymer colloids has revealed an emerging relationship in shaping the size and morphology of synthetic networks and cross-linker density of monomers chemically responsive to a structure-regulating environment. Block-copolymer templates as size-specific geometric templates composed of PS:PMMA have recently been combined with capillary force assembly to direct a PbSe QD formation of 5 nm [54], demonstrating the importance of directional control of monomer-monomer bonding to desired geometries at the larger scale. Further, the possibility of tailoring polymer nano or subnanocavities in porous coordination polymers by selective choice of ligands and adapting monomer reactivity intrinsically controlled by metal surface characteristics mediating stereo-, regio-, and chemical selectivity proposes to alter the geometrical scale of colloidal MOFs [55] by size and space effects. While polymers can alter the interfacial energies at adsorbed sites modulating growth of nuclei, the growth mechanisms that dominate during protein 2D nucleation adopt a "pinning growth" type morphology [56]. It is observed that primary crystal formation dictates morphological patterns and anisotropic features during nuclei evolution. In this context, the prediction of sequential deposition growth of polymer layers at the TiO_2 surface enforces the notion that the nature of complex coordination leading to ligand stabilization at the polymer metal interface is geometrically important to aspects of nuclei growth and inhibition. This is indicative of cosynergistic growth patterns dominating at the polymer-polymer and polymer-metal growth nucleation sites. This results in atomic configurations different to assembly events achieved via polymer addition. Hence, the competition between polymer and nuclei growth is chemically tunable. Due to the difference in temperature dependency expected for nucleation and polymerization at surface sites, an important parametric consideration is applying a thermal separation between both events where rapid polymerization is most effective in inhibiting nuclei growth at the nano to subnano scale. A consistency in the current result supports the generation of a class of materials that neither exhibits indistinguishable growth and nucleation processes or ones that show distinctly separated nucleation and growth characteristics. However, opposing crystal growth and nucleation patterns at low and high rates, respectively, in the limits of a temperature domain [57] are of growing interest here.

An interesting advancement to heterogeneous nucleation to polymer-metal interfacial chemistry concerns the stabilization of nonequilibrium states arising from contrasting bulk materials forming assemblies at the nano and quantum scales. Biogenically engineered bacterial strains have generated enzymatically driven metal sulfide nanocrystals under quantum confinement [7a,7b]. Recently, the addition of polystyrene sulfonate (PSS) was shown to arrest the amorphous precursor phase of calcium carbonate (ACC) by effectively interrupting crystallization by compartmentalization of Ca^{2+} ions forming instead the stabilized metastable phase [58] inhibiting ACC nucleation via calcium-PSS charge-to-charge complexation. Further, polyvinyl (alcohol) PVA and poly(ethylene oxide) (PEO) forming molecular nanocages of 1.5 and 0.6 nm diameter coordinated to Cu^{2+} show the metal to be weakly bound to PVA surfaces in smaller confinements and correlates to greater restrictive movement of the metal in contrast to larger cages [59].

With a core focus on developing methodologies that permit single phase metal-polymer blends, thermal decomposition processes have been applied at temperatures in which metal complex precursors and polymer additives were reacted in the molten state above the transition glass states (T_gs) of the polymer to improve polymer flexibility at the metal surface. Chain flexibility correlates to the length of the polymer repeat and characteristically alters the T_g [60]. The precipitation of $Cd(SC_{12}H_{25})_2$ with 1-dodecanethiol blended with amorphous polystyrene in chloroform showed that the long-chain polymer precursors were effective in confining the growth of CdS particles in the range of 2.2–2.8 nm [61], well below the quantum confinement limit of 4 nm. Although the T_g of complexed polystyrene has gone unreported in this work, the effect on the band edge properties of the semiconductor were significantly altered to a green-yellow intensity consistent with a trapped state emission phenomena normally associated with quantum-scale growth properties.

The noncrystallizable polymer chain that is characteristic of a disordered network of cross-links misaligned in an amorphous state form polymer composites with porous nanocavities for accommodating guest water molecules. The nature of the hydrated state is evidenced by tight or lose water binding associations by FTIR analysis, signifying a supramolecular state composed of weak interactions. This finding follows a similar sequence of events reported by Distefano et al. [62] using the host-guest polymerization of a vinyl polymer precursor (2,5-divinyl-terephalate or DVTP) ligand with Cu^{2+} cations driving cross-linking of olefin monomers confined in a porous coordination polymer (PCP) matrix.

From a dynamic and structural perspective, the role of water in confined states during protein folding is significantly different to its bulk behavior and protein behavioral characteristics distinctly relate to hydrogen bonding abilities of water that differ in the unconfined hydrated state to the internalized nonhydrated state [63]. Differences in the binding modes of water linked to functional group stability in the polymer-metal matrix suggest that water might play an analogous function in the geometrical stability of synthetic materials, as shown in the configurational ordering of proteins and protein recognition [64]. The internalization of water molecules that acts to confine the dynamic behavior of water in polymers has very interesting consequences from bulk water, not only in their binding arrangements but also in their solvent dependent motilities. The associated translational and rotational movement of water under confinement in a solvent-exposed environment, particularly for multilayered supramolecular structures, is likely to be influenced by the complexity of interfacial layers and may deviate from typical characteristic hydrogen-bonding patterns normally associated with water. The ice-like mobility of water resulting from the specific ordering of molecules into a stacking-type architecture at the nano and quantum scale suggests geometrical-based mechanisms to alter conducting properties at multiscale lengths. The coordination of molecular water in the metal ligand complex might account for the behavioral differences at the metal center from bulk to nano to quantum scale dimensions. A significant implication of polymer-induced stabilization of proteins is an appreciation of the disruption of folding pathways intercepting conventional nucleation routes by increasing the free energy barrier.

4 CONCLUDING COMMENTS

In biogenic materials, structural diversification of materials is exquisitely steered by pathway selection governed by free-energy landscapes of reacting surfaces. These events are intimately related with particle dynamics that are often contrasting in their properties. Conditioning synthetically driven self-assembly to unlock interfacial forces to adopt growth mechanisms akin to natural systems could have strong implications in defining shape, size, and structural complexity materials. In Chapter 11, some of these aspects are considered in detail in the polymorphic assembly of quantum-confined titanium dioxide as a tunable amorphous material. We aim to demonstrate how self-assembly can induce biomimetic characteristics, allowing accessibility to otherwise inaccessible material compositions.

REFERENCES

[1] T.R. Cook, Y.-R. Zheng, P.J. Stang, Metal–organic frameworks and self-assembled supramolecular coordination complexes: comparing and contrasting the design, synthesis, and functionality of metal–organic materials, Chem. Rev. 113 (1) (2013) 734–777.

[2] D.K. Harris, M.G. Bawendi, Improved precursor chemistry for the synthesis of III–V quantum dots, J. Am. Chem. Soc. 134 (50) (2012) 20211–20213.

[3] R.J. Kuppler, D.J. Timmons, Q.-R. Fang, J.-R. Li, T.A. Makal, M.D. Young, D. Yuan, D. Zhao, W. Zhuang, H.-C. Zhou, Potential applications of metal-organic frameworks, Coord. Chem. Rev. 253 (23–24) (2009) 3042–3066.

[4] X. Wu, D.J. Young, T.S.A. Hor, A molecular artisans guide to supramolecular coordination complexes and metal organic frameworks, J. Mol. Eng. Mater. 03 (01n02) (2015) 1540004.

[5] J. Boucle, P. Ravirajan, J. Nelson, Hybrid polymer-metal oxide thin films for photovoltaic applications, J. Mater. Chem. 17 (30) (2007) 3141–3153.

[6] S. Sonkaria, S.-H. Ahn, C.S. Lee, V. Khare, "On the dot"—the timing of self-assembled growth to the quantum scale, Chem. Eur. J. 23 (34) (2017) 8104–8117.

[7] (a)R. Dunleavy, L. Lu, C.J. Kiely, S. McIntosh, B.W. Berger, Single-enzyme biomineralization of cadmium sulfide nanocrystals with controlled optical properties, Proc. Natl. Acad. Sci. 113 (19) (2016) 5275–5280; (b)L.C. Spangler, L. Lu, C.J. Kiely, B.W. Berger, S. McIntosh, Biomineralization of PbS and PbS-CdS core-shell nanocrystals and their application in quantum dot sensitized solar cells, J. Mater. Chem. A 4 (16) (2016) 6107–6115; (c)A. Gal, A. Hirsch, S. Siegel, C. Li, B. Aichmayer, Y. Politi, P. Fratzl, S. Weiner, L. Addadi, Plant cystoliths: a complex functional biocomposite of four distinct silica and amorphous calcium carbonate phases, Chem. Eur. J. 18 (33) (2012) 10262–10270; (d)H. Bai, F. Xu, L. Anjia, H. Matsui, Low temperature synthesis of ZnO nanowires by using a genetically-modified collagen-like triple helix as a catalytic template, Soft Matter 5 (5) (2009) 966–969; (e)L.A. Bawazer, M. Izumi, D. Kolodin, J.R. Neilson, B. Schwenzer, D.E. Morse, Evolutionary selection of enzymatically synthesized semiconductors from biomimetic mineralization vesicles, Proc. Natl. Acad. Sci. 109 (26) (2012) E1705–E1714; (f)H. Bao, N. Hao, Y. Yang, D. Zhao, Biosynthesis of biocompatible cadmium telluride quantum dots using yeast cells, Nano Res. 3 (7) (2010) 481–489; (g)C.T. Dameron, R.N. Reese, R.K. Mehra, A.R. Kortan, P.J. Carroll, M. L. Steigerwald, L.E. Brus, D.R. Winge, Biosynthesis of cadmium sulphide quantum semiconductor crystallites, Nature 338 (1989) 596; (h)P. Mukherjee, A. Ahmad, D. Mandal, S. Senapati, S.R. Sainkar, M.I. Khan, R. Ramani, R. Parischa, P. V. Ajayakumar, M. Alam, M. Sastry, R. Kumar, Bioreduction of $AuCl_4^-$ ions by the Fungus, *Verticillium* sp. and surface trapping of the gold nanoparticles formed, Angew. Chem. Int. Ed. 40 (2001) 3585–3588; (i)H.-J. Bai, Z.-M. Zhang, J. Gong, Biological synthesis of semiconductor zinc sulfide nanoparticles by immobilized Rhodobacter sphaeroides, Biotechnol. Lett. 28 (14) (2006) 1135–1139; (j)S. Senapati, A. Ahmad, M.I. Khan, M. Sastry, R. Kumar, Extracellular biosynthesis of bimetallic Au–Ag alloy nanoparticles, Small 1 (5) (2005) 517–520; (k)Y. Konishi, K. Ohno, N. Saitoh, T. Nomura, S. Nagamine, H. Hishida, Y. Takahashi, T. Uruga, Bioreductive deposition of platinum nanoparticles on the bacterium Shewanella algae, J. Biotechnol. 128 (3) (2007) 648–653; (l)P. Yong, N.A. Rowson, J.P.G. Farr, I.R. Harris, L.E. Macaskie, Bioreduction and biocrystallization of palladium by *Desulfovibrio desulfuricans* NCIMB 8307, Biotechnol. Bioeng. 80 (4) (2002) 369–379; (m)J.L. Rouge, C.J. Ackerson, D.L. Feldheim, B. E. Eaton, Cooperativity between two selected RNA Pdases in the synthesis of Pd nanoparticles, J. Mater. Chem. 20 (38) (2010) 8394–8398.

[8] X.-L.J. Shu-Qun Liu, Y. Tao, D.-Y. Tan, K.-Q. Zhang, Y.-X. Fu, P. Kaumaya (Ed.), Protein Folding, Binding and Energy Landscape: A Synthesis, Protein Engineering,

InTech, 2012. ISBN 978-953-51-0037-9. Available from: http://www.intechopen.com/books/protein-engineering/protein-foldingbinding-and-energy-landscape-a-synthesis.

[9] T.R. Jahn, S.E. Radford, Folding versus aggregation: polypeptide conformations on competing pathways, Arch. Biochem. Biophys. 469 (1) (2008) 100–117.

[10] B. Cantaert, E. Beniash, F.C. Meldrum, Nanoscale confinement controls the crystallization of calcium phosphate: relevance to bone formation, Chem. Eur. J. 19 (44) (2013) 14918–14924.

[11] J.-P. Zhang, H.-L. Zhou, D.-D. Zhou, P.-Q. Liao, X.-M. Chen, Controlling flexibility of metal–organic frameworks, Natl. Sci. Rev. (2017) nwx127.

[12] P. Fratzl, F.D. Fischer, J. Svoboda, J. Aizenberg, A kinetic model of the transformation of a micropatterned amorphous precursor into a porous single crystal, Acta Biomater. 6 (3) (2010) 1001–1005.

[13] Z. Lei, B. Chen, Y.-M. Koo, D.R. MacFarlane, Introduction: ionic liquids, Chem. Rev. 117 (10) (2017) 6633–6635.

[14] J.N. Onuchic, N.D. Socci, Z. Luthey-Schulten, P.G. Wolynes, Protein folding funnels: the nature of the transition state ensemble, Fold. Des. 1 (6) (1996) 441–450.

[15] M. Nguyen, S. Vaikuntanathan, Design principles for nonequilibrium self-assembly, Proc. Natl. Acad. Sci. 113 (50) (2016) 14231–14236.

[16] G.M. Whitesides, J.P. Mathias, C.T. Seto, Molecular self-assembly and nanochemistry: a chemical strategy for the synthesis of nanostructures, Science 254 (5036) (1991) 1312–1319.

[17] Z. Zeravcic, V.N. Manoharan, M.P. Brenner, Size limits of self-assembled colloidal structures made using specific interactions, Proc. Natl. Acad. Sci. 111 (45) (2014) 15918–15923.

[18] K. Thorkelsson, P. Bai, T. Xu, Self-assembly and applications of anisotropic nanomaterials: a review, Nano Today 10 (1) (2015) 48–66.

[19] E. Kuçur, W. Bücking, R. Giernoth, T. Nann, Determination of defect states in semiconductor nanocrystals by cyclic voltammetry, J. Phys. Chem. B 109 (43) (2005) 20355–20360.

[20] H.S. Chan, K.A. Dill, Transition states and folding dynamics of proteins and heteropolymers, J. Chem. Phys. 100 (12) (1994) 9238–9257.

[21] X. Liu, W. Wei, Q. Yuan, X. Zhang, N. Li, Y. Du, G. Ma, C. Yan, D. Ma, Apoferritin-CeO_2 nano-truffle that has excellent artificial redox enzyme activity, Chem. Commun. 48 (26) (2012) 3155–3157.

[22] O. Galkin, P.G. Vekilov, Control of protein crystal nucleation around the metastable liquid–liquid phase boundary, Proc. Natl. Acad. Sci. 97 (12) (2000) 6277–6281.

[23] S. Chung, S.-H. Shin, C.R. Bertozzi, J.J. De Yoreo, Self-catalyzed growth of S layers via an amorphous-to-crystalline transition limited by folding kinetics, Proc. Natl. Acad. Sci. 107 (38) (2010) 16536–16541.

[24] D. Fusco, P. Charbonneau, Soft matter perspective on protein crystal assembly, Colloids Surf. B: Biointerfaces 137 (2016) 22–31.

[25] Y.U.T. Gong, C.E. Killian, I.C. Olson, N.P. Appathurai, A.L. Amasino, M.C. Martin, L.J. Holt, F.H. Wilt, P.U.P.A. Gilbert, Phase transitions in biogenic amorphous calcium carbonate, Proc. Natl. Acad. Sci. 109 (16) (2012) 6088–6093.

[26] H. Yang, B. Yuan, X. Zhang, O.A. Scherman, Supramolecular chemistry at interfaces: host–guest interactions for fabricating multifunctional biointerfaces, Acc. Chem. Res. 47 (7) (2014) 2106–2115.

[27] T.M. Reineke, Stimuli-responsive polymers for biological detection and delivery, ACS Macro Lett. 5 (1) (2016) 14–18.

[28] (a)M. Sarikaya, C. Tamerler, A.K.Y. Jen, K. Schulten, F. Baneyx, Molecular biomimetics: nanotechnology through biology, Nat. Mater. 2 (2003) 577; (b)A. Ahmad,

P. Mukherjee, D. Mandal, S. Senapati, M.I. Khan, R. Kumar, M. Sastry, Enzyme mediated extracellular synthesis of CdS nanoparticles by the fungus, *Fusarium oxysporum*, J. Am. Chem. Soc. 124 (41) (2002) 12108–12109; (c)P. Mukherjee, A. Ahmad, D. Mandal, S. Senapati, S.R. Sainkar, M.I. Khan, R. Parishcha, P.V. Ajaykumar, M. Alam, R. Kumar, M. Sastry, Fungus-mediated synthesis of silver nanoparticles and their immobilization in the mycelial matrix: a novel biological approach to nanoparticle synthesis, Nano Lett. 1 (10) (2001) 515–519.

[29] S.-K. Lee, D.S. Yun, A.M. Belcher, Cobalt ion mediated self-assembly of genetically engineered bacteriophage for biomimetic Co–Pt hybrid material, Biomacromolecules 7 (1) (2006) 14–17.

[30] M. Sytnyk, M. Jakešová, M. Litviňuková, O. Mashkov, D. Kriegner, J. Stangl, J. Nebesářová, F.W. Fecher, W. Schöfberger, N.S. Sariciftci, R. Schindl, W. Heiss, E.D. Głowacki, Cellular interfaces with hydrogen-bonded organic semiconductor hierarchical nanocrystals, Nat. Commun. 8 (1) (2017) 91.

[31] A. Kumar, V. Kumar, Biotemplated inorganic nanostructures: supramolecular directed nanosystems of semiconductor(s)/metal(s) mediated by nucleic acids and their properties, Chem. Rev. 114 (14) (2014) 7044–7078.

[32] M. Šuvakov, M. Andjelković, B. Tadić, Hidden geometries in networks arising from cooperative self-assembly, Sci. Rep. 8 (1) (2018) 1987.

[33] (a)M. Karplus, Behind the folding funnel diagram, Nat. Chem. Biol. 7 (2011) 401; (b)P. G. Wolynes, Folding funnels and energy landscapes of larger proteins within the capillarity approximation, Proc. Natl. Acad. Sci. 94 (12) (1997) 6170–6175; (c)J. N. Onuchic, Z. Luthey-Schulten, P.G. Wolynes, Theory of protein folding: the energy landscape perspective, Annu. Rev. Phys. Chem. 48 (1) (1997) 545–600.

[34] J. Wang, K. Liu, R. Xing, X. Yan, Peptide self-assembly: thermodynamics and kinetics, Chem. Soc. Rev. 45 (20) (2016) 5589–5604.

[35] K.R. MaryJoe, C.R. Warren, Creating biological nanomaterials using synthetic biology, Sci. Technol. Adv. Mater. 15 (1) (2014)014401.

[36] C. Toro, J.M. Buriak, Template synthesis approach to nanomaterials: Charles Martin, Chem. Mater. 26 (17) (2014) 4889–4890.

[37] (a)M. Shao, R. Zhang, C. Wang, B. Hu, D. Pang, Z. Xie, Living cell synthesis of CdSe quantum dots: manipulation based on the transformation mechanism of intracellular Se-precursors, Nano Res. (2017); (b)S.R. Stürzenbaum, M. Höckner, A. Panneerselvam, J. Levitt, J.S. Bouillard, S. Taniguchi, L.A. Dailey, R. A. Khanbeigi, E.V. Rosca, M. Thanou, K. Suhling, A.V. Zayats, M. Green, Biosynthesis of luminescent quantum dots in an earthworm, Nat. Nanotechnol. 8 (2012) 57; (c)H.-J. Bai, Z.-M. Zhang, Microbial synthesis of semiconductor lead sulfide nanoparticles using immobilized *Rhodobacter sphaeroides*, Mater. Lett. 63 (9) (2009) 764–766.

[38] O. Ikkala, G. ten Brinke, Functional materials based on self-assembly of polymeric supramolecules, Science 295 (5564) (2002) 2407–2409.

[39] T. Aida, E.W. Meijer, S.I. Stupp, Functional supramolecular polymers, Science 335 (6070) (2012) 813–817.

[40] A. Sorrenti, J. Leira-Iglesias, A. Sato, T.M. Hermans, Non-equilibrium steady states in supramolecular polymerization, Nat. Commun. 8 (2017) 15899.

[41] P.A. Korevaar, C. Grenier, A.J. Markvoort, A.P.H.J. Schenning, T.F.A. de Greef, E. W. Meijer, Model-driven optimization of multicomponent self-assembly processes, Proc. Natl. Acad. Sci. U. S. A. 110 (43) (2013) 17205–17210.

[42] A. Navrotsky, Energetic clues to pathways to biomineralization: precursors, clusters, and nanoparticles, Proc. Natl. Acad. Sci. U. S. A. 101 (33) (2004) 12096–12101.

[43] A. Aliprandi, M. Mauro, L. De Cola, Controlling and imaging biomimetic self-assembly, Nat. Chem. 8 (2015) 10.

[44] S. Weiner, L. Addadi, Design strategies in mineralized biological materials, J. Mater. Chem. 7 (5) (1997) 689–702.

[45] A.J. Bidhendi, A. Geitmann, Relating the mechanics of the primary plant cell wall to morphogenesis, J. Exp. Bot. 67 (2) (2016) 449–461.
[46] C.J. Schram, S.P. Beaudoin, L.S. Taylor, Polymer inhibition of crystal growth by surface poisoning, Cryst. Growth Des. 16 (4) (2016) 2094–2103.
[47] L. Bloch, Y. Kauffmann, B. Pokroy, Size effect on the short range order in amorphous materials, Cond. Mater. Mater. Sci. (2014) 1–12.
[48] Y. Diao, M.E. Helgeson, Z.A. Siam, P.S. Doyle, A.S. Myerson, T.A. Hatton, B. L. Trout, Nucleation under soft confinement: role of polymer–solute interactions, Cryst. Growth Des. 12 (1) (2012) 508–517.
[49] Y. Diao, M.E. Helgeson, A.S. Myerson, T.A. Hatton, P.S. Doyle, B.L. Trout, Controlled nucleation from solution using polymer microgels, J. Am. Chem. Soc. 133 (11) (2011) 3756–3759.
[50] Y. Diao, T. Harada, A.S. Myerson, T. Alan Hatton, B.L. Trout, The role of nanopore shape in surface-induced crystallization, Nat. Mater. 10 (11) (2011) 867–871.
[51] F.I. Rusinga, D.D. Weis, Soft interactions and volume exclusion by polymeric crowders can stabilize or destabilize transient structure in disordered proteins depending on polymer concentration, Proteins: Struct. Funct. Bioinform. 85 (8) (2017) 1468–1479.
[52] Y.-T. Lai, E. Reading, G.L. Hura, K.-L. Tsai, A. Laganowsky, F.J. Asturias, J. A. Tainer, C.V. Robinson, T.O. Yeates, Structure of a designed protein cage that self-assembles into a highly porous cube, Nat. Chem. 6 (12) (2014) 1065–1071.
[53] C.C. Jolley, M. Uchida, C. Reichhardt, R. Harrington, S. Kang, M.T. Klem, J. B. Parise, T. Douglas, Size and crystallinity in protein-templated inorganic nanoparticles, Chem. Mater. 22 (16) (2010) 4612–4618.
[54] K.W. Lei, T. West, X.Y. Zhu, Template-assembly of quantum dot molecules, J. Phys. Chem. B 117 (16) (2013) 4582–4586.
[55] L.H. Tan, S. Xing, T. Chen, G. Chen, X. Huang, H. Zhang, H. Chen, Fabrication of polymer nanocavities with tailored openings, ACS Nano 3 (11) (2009) 3469–3474.
[56] M. Plomp, A. McPherson, S.B. Larson, A.J. Malkin, Growth mechanisms and kinetics of trypsin crystallization, J. Phys. Chem. B 105 (2) (2001) 542–551.
[57] N.S. Trasi, L.S. Taylor, Effect of polymers on nucleation and crystal growth of amorphous acetaminophen, CrystEngComm 14 (16) (2012) 5188–5197.
[58] P.J.M. Smeets, K.R. Cho, R.G.E. Kempen, N.A.J.M. Sommerdijk, J.J. De Yoreo, Calcium carbonate nucleation driven by ion binding in a biomimetic matrix revealed by in situ electron microscopy, Nat. Mater. 14 (4) (2015) 394–399.
[59] D. Suryanarayana, P.A. Narayana, L. Kevan, Effect of molecular cage size on the motion and coordination of copper(2+) in crosslinked poly(vinyl alcohol) and poly(ethylene oxide) gels: electron spin echo and electron spin resonance studies, Inorg. Chem. 22 (3) (1983) 474–478.
[60] R.P. Padbury, J.S. Jur, Effect of polymer microstructure on the nucleation behavior of alumina via atomic layer deposition, J. Phys. Chem. C 118 (32) (2014) 18805–18813.
[61] G. Carotenuto, A. Longo, P. Repetto, P. Perlo, L. Ambrosio, New polymer additives for photoelectric sensing, Sensors Actuators B Chem. 125 (1) (2007) 202–206.
[62] G. Distefano, H. Suzuki, M. Tsujimoto, S. Isoda, S. Bracco, A. Comotti, P. Sozzani, T. Uemura, S. Kitagawa, Highly ordered alignment of a vinyl polymer by host–guest cross-polymerization, Nat. Chem. 5 (4) (2013) 335–341.
[63] F. Mallamace, C. Corsaro, D. Mallamace, S. Vasi, C. Vasi, P. Baglioni, S.V. Buldyrev, S.-H. Chen, H.E. Stanley, Energy landscape in protein folding and unfolding, Proc. Natl. Acad. Sci. 113 (12) (2016) 3159–3163.
[64] G.A. Papoian, J. Ulander, P.G. Wolynes, Role of water mediated interactions in protein−protein recognition landscapes, J. Am. Chem. Soc. 125 (30) (2003) 9170–9178.

CHAPTER 6

The Development of High-Temperature Superconductors and 2D Iron-Based Superconductors

Xun Geng*, Jiabao Yi†
*School of Materials Science and Engineering, UNSW, Sydney, NSW, Australia
†Global Innovative Centre for Advanced Nanomaterials, School of Engineering, The University of Newcastle, Callaghan, NSW, Australia

1 INTRODUCTION

After superconductivity [1] was first discovered, it has been of interest for research into condensed matter physics. T_c, has been the center of the research of superconductors since its discovery, including how to increase the critical temperature of superconductors. In 1987, the T_c of a new compound ($La_{2-x}Ba_xCuO_{4-y}$), a kind of Cu-based superconductor (Cuprates) [2], was found to be as high as 57 K, which is believed as the first real high-temperature superconductor [3]. High-temperature superconductors (HTS) [4] mostly belong to Type II superconductors, which have two critical external fields, H_{1c} and H_{2c}, compared with Type I superconductors (traditional superconductors) that have only one critical magnetic field [5]. Those high-temperature superconductor materials have demonstrated diverse potential applications such as cables for power transmission, magnetic resonance imaging (MRI) in medical service [6], nanometer-sized SQUIDs sensors for brain diagnostics [7], and maglev trains [8,9].

Ferro-pnictides (iron-based superconductors), another kind of high-temperature superconductor, were first found in 2008 using F doping in $AeAFe_4As_4$ (Ae contains Ca, Sr, Ba, and A presents I group elements) [10] and they have shown marvelous potential properties from bulk iron chalcogenide to single-layer FeSe [11]. These new materials not only offer researchers a new horizon of high-temperature superconductors, but they also provide diverse new theories for investigating the superconductivity

Nano-sized Multifunctional Materials
https://doi.org/10.1016/B978-0-12-813934-9.00006-2

mechanisms, including antiferromagnetic fluctuations [12] termed as a spin density wave [13], pseudo gap [14], and three-band theory [15]. Because these "Iron-based superconductors" cannot be explained by BCS theory, they are thus called unconventional superconductors [16,17].

Monolayer FeSe is the latest discovered superconductor based on iron-based superconductors. It has drawn significant attention in superconducting research due to its very simple two-dimensional (2D) structure and high critical temperature, up to 100 K [18], above the boiling point of liquid nitrogen. Currently, single-layer FeSe can be fabricated on $SrTiO_3$ substrate and has similar characteristics of doping-induced insulator-superconductor transition to cuprates [19]. Its properties have been investigated extensively by both theoretical calculations and experimental measurements with advanced facilities, such as angle-resolved photoemission spectroscopy (ARPES) [20].

It was discovered that electron doping can enhance the critical temperature of FeSe thin films from 8 to more than 80 K. Therefore, introducing electrons into FeSe thin films by doping and coating techniques may pave ways to achieve high performance 2D superconductors [21] toward high T_c and a strong tolerance to external magnetic fields. Experimentally, the T_c of an n-type heavy doped FeSe layer has been observed to increase significantly at normal pressure with a thickness of 1 ML [22]. A critical temperature up to 100 K has been achieved [23].

Hence, in this chapter, we mainly introduce the current research on iron-based superconductors and the prospects of 2D superconductors.

2 IRON-BASED SUPERCONDUCTOR

2.1 Fe-As System

2.1.1 Introduction

Since the discovery of $LaFeAsO_{1-x}F_x$ with a T_c of 26 K [24], diverse types of iron-based superconductors were reported, such as Co-doped $BaFe_2As_2$ [25] or $FeTe_xSe_{1-x}$. The properties of these new compounds are all investigated systemically in terms of transport measurements, the gap structure, and electronic behavior as well as their thermal conductivities [26–28].

In the research of $RFeAsO_{1-x}F_x$ ($R=1/4$ rare earth elements) iron pnictides, many new materials have been found, such as so-called 1111-type superconductors ($RFeAsO_{1-x}F_x$, R = La), oxygen-free pnictide superconductor-$Ba_{1-x}K_xFe_2As_2$ subsequently or 122-type superconductor (BFe_2As_2, B = Ca, Ba, Sr) [29,30] as well as the latest material class (FeSe 11-system and LiFeAs 111-system superconductors). Especially the quaternary iron pnictides (FeSe) have demonstrated high performance on superconducting transition temperatures (≈65 K), high upper critical fields, and multiband character [31]. The first superconducting $LaFeAsO_{1-x}F_x$ thin film was fabricated using a pulsed laser deposition (PLD) method, which is believed to be the most versatile path for the growth of 2D materials because of its capability for atomic growth as well as controllable and adjustable deposition parameters, especially $LaFeAsO_{1-x}F_x$ thin film growth on an LAO substrate [32]. However, it is still a difficult task to achieve high-performance superconductivity for oxypnictide-doped iron-based fluorine thin films, though some studies did succeed in the fabrication of $LaFeAsO_{1-x}F_x$ thin films having high T_c and a detailed superconducting transition phase diagram has also been provided [33–35]. One of the most difficult challenges is to control the stoichiometric level of the fluorine content [36]. Based on a series of experiments, it was found that how to control the oxidation during the deposition process is one of the key techniques to achieve high quality films because rare earth elements are usually highly reactive and sensitive to air.

2.1.2 Structure of Fe-As System Compounds

So far, in literature, LaOFeAs has been systematically studied by high-resolution neutron diffraction and XRD methods [37]. The basic structure of $LaFeAsO_{1-x}F_x$ is demonstrated in Fig. 1.

Typically, $LaO_{1-x}F_xFeAs$ demonstrates a tetragonal structure. However, in agreement with most previous studies, it has been found that the low-temperature phase of $LaO_{1-x}F_xFeAs$ is orthorhombic (Cmme).

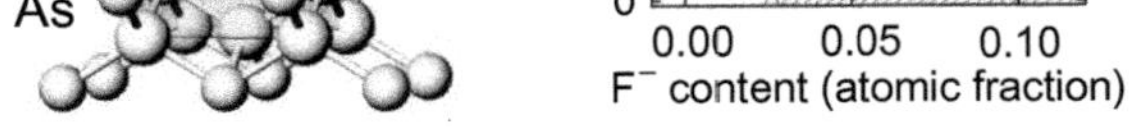

Fig. 1 The schematic diagram of the tetragonal structure of $LaO_{1-x}F_xFeAs$ (left) and the functional temperature change with the F^- content increasing (right). T_{anom} and T_{min} are used for indicating the temperature when the electrical resistivity of undoped LaOFeAs suddenly decreases and continues decreasing to the minimum value, respectively. The T_{onset} is the temperature where the superconducting transition just started. T_c is the critical temperature for F-doped LaOFeAs [38].

This structural phase transition (from tetragonal to orthorhombic) results in the most prominent feature of $LaO_{1-x}F_xFeAs$. Because of this long-range orthorhombic distortion, doping is needed to induce superconductivity. Although the structural features of the monoclinic space model are not supported by reasonable and reliable data, the phase alternation from P4-nmm to Cmme is unique. This change can be considered as one of the characteristics of iron elasticity and is closely related to magnetism. Due to the alignment and accumulation patterns of the ferromagnetic lines, the magnetic order at low temperature shows a kind of extremely high symmetry. Therefore, the crystal structure of $LaO_{1-x}F_xFeAs$ in the magnetic phase must be orthorhombic, but this distortion also increases its magnetic frustration. Moreover, the stability of the magnetic structure is determined by its large adjacent iron-iron network along the angular direction of the iron lattice. In this case, the interaction of adjacent square lattices offsets its magnetic moment and produces two completely decoupled antiferromagnetic lattices. This research has demonstrated the vital role of crystalline transformation in magnetic degeneration. This frustration is very similar to the recent magnetic elastic coupling phenomenon found in vanadium oxide halides [39–41].

2.1.3 Microstructure of 1111-Type Compounds and Their Superconductivity

Theoretically, the occurrence of magnetism in $LaO_{1-x}F_xFeAs$ can be believed to be an electronic nematic phase, but the magnetic anisotropy is due to the breaking of the tetragonal rotation axis by the magnetic structure. Besides, the similarity of the liquid crystal phase of $LaO_{1-x}F_iFeAs$ is further restricted, and it only shows four times and discontinuous rotational symmetry at the difference temperatures [42–44]. However, a measurement on the resistivity of $Ba(Fe/Co)_2As_2$ crystal has revealed obvious electronic anisotropy for La-iron-based superconductors. In some reports, La-Fe superconductors phase diagrams have been successfully mapped through detailed magnetic neutron diffraction measurements [45]. It was found that the orthorhombic structural precursors are still in the LaO ($LaO_{1-x}F_xFeAs$) and other F-doped REO-Fe series and are very difficult to remove. For the pure LaO series, the abnormal structure was easily found at the appearance of magnetism, and the variation trend of the structure is also consistent with the change of the thickness of iron-arsenic layers and the thermal expansion coefficient [46]. According to the results of band structure calculation results of the LaO series, the tetrahedral shape of iron-arsenic layers and the size of the magnetic moment are influenced by a strong ferromagnetic coupling effect, which suggests that the antiferromagnetism enhanced by orthorhombic distortion has a close connection with the change of the magnetic moment [47].

Near the Fermi surface, the electronic state is made of iron 3d orbits and arsenic p states. This kind of hybrid orbit produces a massive three-dimensional network packet, including two holes and another two multi-electronic areas [48–50]. In early stages, there were many attempts for electron doping at oxygen sites and for hole doping using La-group elements. Among them, the studies on the trend of T_c have shown that the substance has high dependence on fluorine ($T_c \approx 26\,K$ with the doping ratio 5–11 at.%) [51]. Moreover, the analysis results from many experiments have shown that the fluorine doping not only can improve the electron density of the conductive layer without any geometrical shape transition, but they also have illustrated a clear crystallization shift at low-temperature environments. According to the latest research, the phase diagram of pnictides by hole doping and electron doping is almost symmetrical compared with cuprates, as shown in Fig. 2.

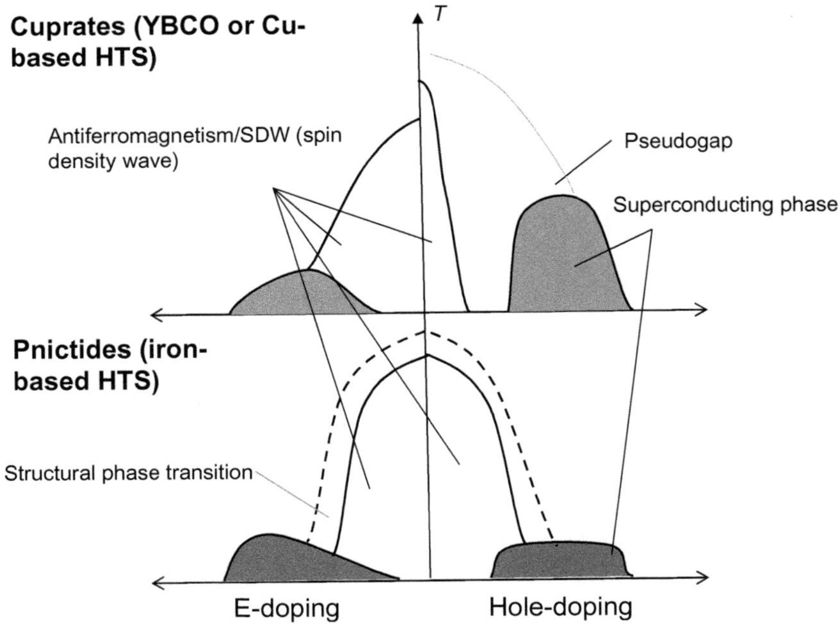

Fig. 2 Phases diagram of cuprates and pnictides [48–50,52].

The lattice constant decreases systematically with the change of doping concentration, and the distance between the (LaO) δ+ and (FeAs) δ− layers is also significantly reduced, which indicates that the electrodoping of iron-based superconductors not only enhances the polarization of the layers and the interaction of coulomb, but also suppresses both structural phase change and antiferromagnetism. Mössbauer spectrum analysis, muon spin relaxation, and neutron scattering experiments all support this conclusion [53–55].

Besides structure transition, the resistance anomaly can also be observed. An abrupt decrease of resistivity has been observed in resistivity versus temperature, which is proposed to be induced by spin density waves (SDW). In the case of fluoride doping, the abnormity peak of SDW was suppressed, which the T_c can be easily reduced to a lower value [56]. The data of the phase diagram comes from a wide variety of 111 and 1111 compounds. All these samples are synthesized under environmental pressures in solid state so that any external effects, such as external pressure, can be excluded to analyze the influence of the ion radius of the La-group elements on T_c [57,58].

2.2 Fe-Se System

2.2.1 Introduction

According to traditional BCS theory, superconductors are catalogued into two groups, type-I superconductors and type-II superconductors. FeSe, as the simplest type-II superconductor, has drawn significant attention. Theoretically, the superconducting transition temperature of FeSe in the form of a single crystal can be varied from about 9 to 38 K under normal pressure [59,60], but a higher critical temperature up to 100 K was observed in the FeSe monolayer by the optimized doping when the FeSe was grown on a $SrTiO_3$ substrate [18,61].

2.2.2 Structure of FeSe

FeSe is known to exist in two structures, one of which is the superconducting tetragonal phase of the PbO (β-FeSe) structural type with T_c=8–10 K [62] and the other is a nonsuperconducting hexagonal FeSe phase with a NiAs (δ-FeSe) structure. The nonsuperconducting δ-FeSe phase crystallizes at a temperature of 1075°C and the superconducting β-FeSe phase is formed below 450°C. Generally, the crystals are grown from a flux or a gas phase. There are some variations on composition, homogeneity region, and superconducting properties of $FeSe_{1-x}$ obtained in different groups [63,64]. Basically, three different types of FeSe structures have been reported and are illustrated in Fig. 3.

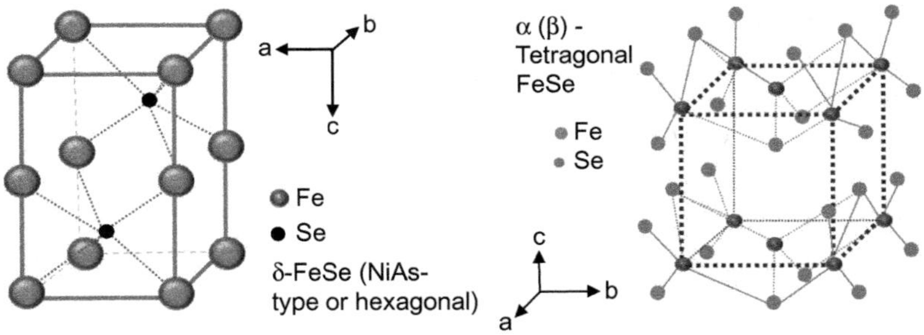

Fig. 3 The structure of FeSe [65,66].

FeSe has the simplest microstructure and the fewest toxic elements in the giant group of iron-based superconductors. The microstructure of FeSe is composed of two two-dimensional Fe_2Se_2 layers and forms the basic unit of Fe-based superconductors. Therefore, the simple structure of FeSe makes it an ideal example for understanding the mechanism of iron-based superconductors. Theoretically, FeSe can be crystallized into two polymorphs. One is a hexagonal NiAs-type structure (δ-FeSe) and the other is β-FeSe, and only the latter has advanced superconductivity. Fig. 4 shows a fragment of the combined phase diagram of the Fe-Se system: The composition range near 50 at.% shows that β-FeSe is implemented in a narrow homogeneity region near the stoichiometric composition Fe:Se = 1:1 from the Fe side.

The diagram further indicates that FeSe crystal material mainly contains two different structures, including β-iron-based selenium stabilized at room conditions (P4/nmm) and δ-iron based selenium stabilized at high

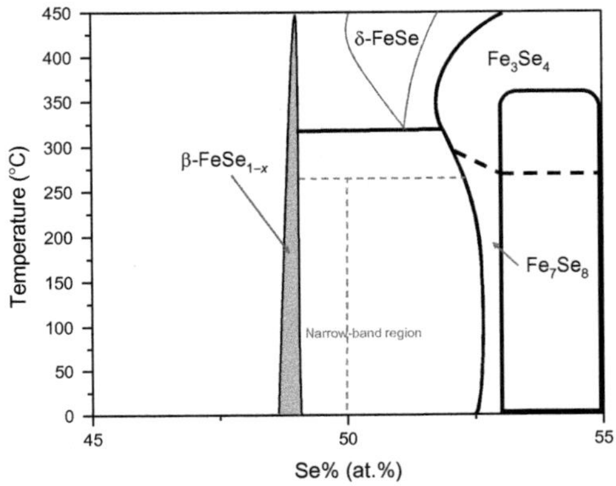

Fig. 4 Fragment of the combined Fe-Se phase diagram. The diagram shows that the narrow band near 50 at.% Se corresponds to the region where the β phase exists [67].

temperature (P63/mmc). Hexagonal Fe_7Se_8, monoclinic Fe_3Se_4, Fe_3O_4, and even pure iron phases also can be observed in the Fe-Se system under different heating conditions. The latest findings suggest that the low-temperature β-Fe-based selenium phase is occasionally surrounded by δ-FeSe during the cooling process and prevents the degradation of the β-phase [62]. The crystal structure of FeSe is regarded as the representative of the entire family of Fe-SC. In addition, the superconducting transition temperature $T_c \approx 8$ K [62] in bulk FeSe can be enhanced up to 37 K [68] by applying pressure and even to 50–100 K by growing a monolayer on a $SrTiO_3$ substrate, as previously discussed [69].

Although bulk FeSe only has a T_c of 8 K, it can increase to 15 K with half the Se substituted by Te [70]. Applying external pressure on $FeSe_{0.5}Te_{0.5}$ can enhance the T_c up to 23 K, being considerably higher than that of chemical doping, while the maximum T_c and superconducting properties vary significantly, depending on the systems for applying the pressure and the samples fabricated with different methods or facilities [70]. In general, bulk FeSe does not show very promising superconducting properties, including those with chemical doping with other elements, such as $(Na_{0.16}K_{0.70})Fe_{1.72}Se_2$ with $T_c = 29$ K [71], $Li_{1-x}FeSe$ at $T_c = 13$ K [21], or $Fe_{0.95}TM_{0.05}$-$Se_{0.5}Te_{0.5}$ (TM represents transition metals, such as manganese, iron, cobalt, nickel, copper, and zinc) with critical temperatures, T_c, around 15 K [72].

2.2.3 Microstructure of FeSe and Its Superconductivity

FeSe is the simplest system with an intricate evolution of the electronic band structure in iron-based superconductors. In FeSe, there is a nematic transition that is associated with spontaneous symmetry breaking between the x and y axis in the Fe plane, reducing the group symmetry of the lattice from tetragonal to orthorhombic, called nematic. It is believed to be a result of intrinsic electronic instability because the effect on electronic properties is much larger than ever expected based on the structural distortion observed. Moreover, in all other iron-based superconductors, the nematic transition is closely followed by the antiferromagnetism (AFM) that provides a natural reason in the mutual relation of these two phases. The nematic transition can be traced back to spin fluctuations that offer a solid support for the s six-pairing model. FeSe is quite different from all other Fe-SCs and is a sort of unique example for spin-fluctuation theories [73,74]. First, the nematic transition, which happens for FeSe crystals at about 90 K [63], is not accompanied by the AFM at all. Second, Fermi surface topology for some FeSe incarnations, such as the mentioned intercalates and single-layer films, can hardly support the s six-pairing. Now more and more evidence is shown in favor of charge-induced nematicity in FeSe [75].

It is hard to find all characteristics of copper-based superconductors in iron-based superconductors because the structure of iron-based superconductors is more complex in its three-dimensional structure compared with copper-based compounds after the electron coupling process [76]. At present, the parent state of the iron-based superconductor is classified as a conductor while the state of cuprates is believed to be an antiferromagnetic insulator. In addition, the hole-doped cuprates exhibit a strange "pseudo gap" state at critical temperature, which does not exist in the iron-based superconductor system. Commonly, Fermi surface criticality can be easily revealed by the angle-resolved photoemission (ARPES) for almost all the Fe-SCs. But for some other cases, such as the hole-doped cuprates, the Fermi surface criticality is hard to resolve in experiments directly, which can also be the case for some of the FeSe compounds [77,78].

Evidently, pure FeSe crystal is not optimal for superconductivity because the transition temperature increases with isovalent doping, surface doping, and increasing pressure. At the same time, the nematic phase is suppressed by doping and pressure. In addition, it seems to compete with superconductivity. There are many possible selenium-tellurium-sulfur combinations to study the isovalent doping in Fe(Se, Te, S) [79]. The most studied ternary system is $FeSe_{1-x}Te_x$. Even though its phase diagram is still not fully known for the whole doping range, the quality of the crystals is constantly improving. Besides AFM and superconducting regions, there is also a region in between AFM and the superconducting phases that has been considered as a weak superconductivity [80]. The transition temperature is slightly increased with sulfur doping (Fe(Se, S)). Doping-induced compounds, such as Fe(Te, S) can induce superconductivity, though FeTe and FeS are non-superconducting individually, which suggests either two different mechanisms of pairing or that two different peculiarities of the electronic band structure may be responsible for superconductivity [81,82].

Unlike other Fe-pnictides, the superconductivity of FeSe crystals cannot be the result of spin-driven nematic transition based on thermal expansion and verified by neutron magnetic resonance (NMR) results [83]. Ab initio calculations indicated that FeSe is close to magnetic instability, but magnetic order is rarely observed in FeSe, which could be explained by strong fluctuated-magnetism frustration or by the formation of a quantum paramagnet [84,85]. One nonelastic scattering neutron experiments indicats that the spin density waves (SDW) around the antimagnetism wave vector is in the FeSe system. Other neutron-scattering experiments showed that these spin fluctuations are coupled with orthorhombicity [68]. On the other hand, it has been suggested that because of very small Fermi energy in FeSe, there is a near degeneracy between magnetic fluctuations and the fluctuations in the current density waves channel. In addition, both an SDW channel and a charge-current density-wave (CDW) channel are comparable and strongly fluctuating at the ordering vector (π, 0) seen by APRES measurement [86]. Therefore, it may be concluded that in order to resolve the spin versus orbital fluctuations dilemma, the exact knowledge of the electronic structure with its orbital origin is required [87].

3 2D MATERIALS AND FeSe THIN FILMS

3.1 2D Materials

In 2004, two-dimensional material graphene revealed many advantages of the 2D structure compared with the 3D bulk system [88]. Since then, diverse 2D compounds have been figured out, including transition metal dichalcogenides (TMDs), black phosphorus, silicine, and hexagonal boron nitride [89–91]. These materials possess diverse electronic behaviors and have shown potential applications in a variety of devices, such as field effect transistors (FET), light-emitting devices (LED), and energy-harvesting devices [92,93]. 2D materials demonstrate their low-dimensional and periodical structures, which leads to sensitive electron transmission and high specific surface area. These characteristics ensure 2D materials advanced adsorption feature and large elasticity [94,95]. Therefore, 2D materials also can be used in many areas, as previously discussed, including batteries, catalysts, spintronics, and semiconductor devices. 2D materials can be synthesized by a variety of methods, such as chemical vapor deposition (CVD), microwave-assisted chemistry approaches, cathodic magnetron sputtering, and vacuum arc deposition [96–100]. Especially, the CVD and microwave chemistry approaches were mainly used for 2D materials in recent studies.

3.2 FeSe Thin Films

With the development of 2D materials, the excellent performance of 2D iron-based superconducting materials has been gradually noticed by scientists. The single layer of FeSe on $SrTiO_3$ (STO) substrate was observed to have a T_c of about 65 K [18]. An even higher critical temperature is expected, which opens a new frontier of science for superconductivity.

The first discovered Fe-based superconductors (LaFeAsO) have a T_c of 26 K. However, soon after, SmFeAsO showed a critical temperature as high as 56 K synthesized under ambient pressure [33,101,102]. More recently, the highest T_c record of iron-based materials has reached 65 K, but it is still hard to surpass the record of the copper-based superconductors. In recent years, using interface effects to improve the superconductivity of FeSe has become a major interest in the research community. The superconducting energy gap ($\Delta \approx 19.5$ meV) of FeSe has been directly observed by scanning tunnel microscopy (STM), which was deposited by molecular beam epitaxy (MBE) on an $SrTiO_3$ substrate. Based on that, the highest T_c of 100 K predicted by calculations can be achieved, given that the BCS constant ($2\Delta k_B/T_c$) can be applied to the FeSe system. It should be noted that $2\Delta/k_BT_c$ is an important constant for measuring superconducting critical temperature. Δ is the energy gap that decreases with the increase in temperature; k_B is the Boltzmann constant; and T_c is the critical temperature [23]. Experimentally, a 109 K critical temperature has been achieved in a single-layer FeSe film grown on a doped $SrTiO_3$ substrate [103]. Interface-enhanced electron-phonon coupling may have played an important role in the enhanced critical temperature. Further works are needed to clarify the superconducting mechanisms.

3.2.1 Microstructure of FeSe Thin Films

The structure of one unit-cell thick and multi-UC thick FeSe thin films has been demonstrated in Fig. 5. One unit-cell (1UC) thick iron-based selenium film grown on Nb-etched STO (001) substrate has shown the superconducting gap of about 20 meV in tunneling spectra [107]. Based on this value, it was concluded that T_c could be about 80 K, assuming the same superconducting mechanism as for the bulk FeSe with $2\Delta/k_BT_c$.

By reducing the spatial dimension and increasing the interface effect, the performance of iron-based selenium has been improved significantly compared with the bulk samples. The electronic structure of the superconducting single-layer FeSe film is unique, and the trace of electrons is only detected at the corner of the Brillouin zone and does not cross the center area. In particular, the annealing process of the FeSe membrane in a vacuum chamber can easily change the concentration of the carriers foran iron-based Se membrane in order to provide diverse selections for the study of the dependence on carriers [108].

The electronic structure demonstrates an electron-like pocket pattern near the zone of the corner without any Fermi surface overlapping around

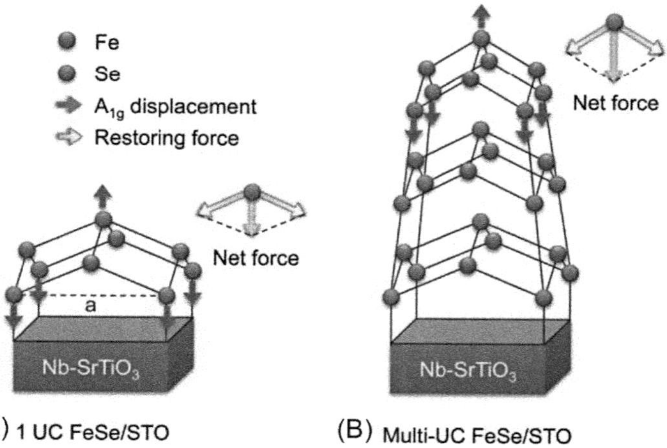

Fig. 5 The schematic diagram of structural models of 1UC FeSe (A) and multi-UC on an Nb-$SrTiO_3$ (001) surface (B), which is terminated by a smaller net restoring force and on O-deficient surface [104–106].

the zone center, as shown in Fig. 6. In the diagram of Fig. 6A, the high-symmetry cuts are measured at 20 K along the Γ-M direction and centered at Γ and M: the hole band is located 80 meV below E_F [49,110]. Comparing the intercalates with respect to single crystals, it is natural to expect that their electronic band structures are more two-dimensional, and shift below the Fermi level due to electron doping. One can see that the electronic structure of 1UC on STO is remarkably like that of intercalates, and the superconducting transition temperature is even higher (above 65 K). Because it is the single layer, the band structure is 2D in nature. However, the mechanism for electron doping is still not clear, though the formation of the electron gas at the interface with the STO has been widely known and studied for a long time [22,105,107,111–113].

According to reliable data, the oxygen vacancies between the monolayer FeSe and the substrate interface can not only provide diverse opportunities for electronic doping, but also can significantly synchronize the width of the iron 3^d orbit with the Fermi level to adjust the magnetic state to be antiferromagnetic. The LDA calculations of 1UC FeSe on the STO substrate revealed the appearance of an additional band of O 2_p surface states near the Fermi level with good nesting-like matching of the hole-Fe 3d band. In addition, the 1UC-FeSe-on-STO calculations show a rather small splitting of electron bands at the M-point [110,114]. Under critical temperatures, FeSe/STO thin films will step into the superconducting phase

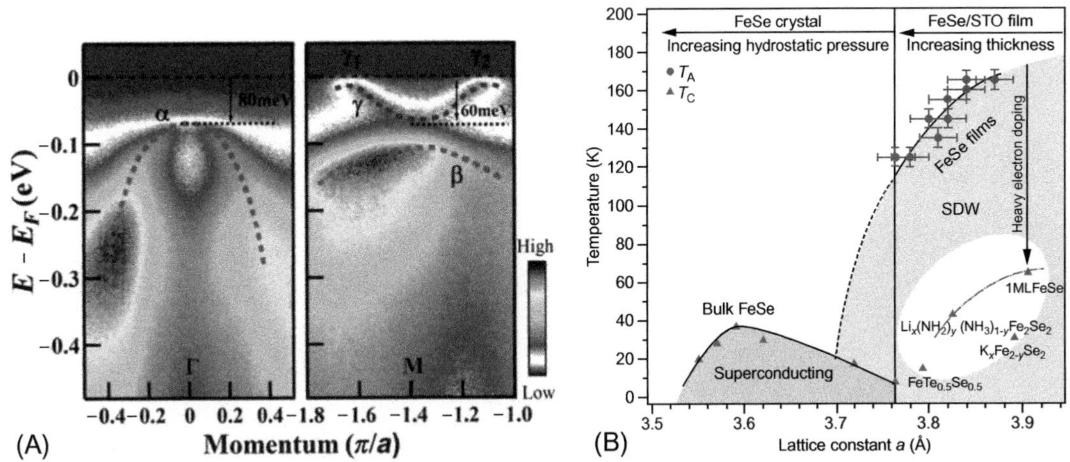

Fig. 6 (A) Electronic band diagram of Fe-SC's electron band in M-point and hole and electron bands in Γ point, respectively, correspond to proximity to the Lifshitz transition. (B) The universal phase diagram for the superconducting and structural transitions versus the lattice constant for all FeSe families [22,49,109].

(S phase) from the insulator phase (N phase). When applying low concentration carriers, the s-phase of a single-layer iron-based selenium also demonstrates insulating behavior. But when the gap width decreases with the carrier concentration ascent, going up to approximately 0.09, FeSe thin films demonstrate superconducting behavior [115].

Particularly, a similar enhancement of superconductivity related with similar electronic structure has been found in the topmost layer in a potassium-coated FeSe single crystal [116]. The superconductivity emerges when the interpocket scattering between two electron pockets is turned on by a Lifshitz transition of the Fermi surface, which suggests an underlying correlation among superconductivity, interpocket scattering, and nematic fluctuation in electron-doped FeSe superconductors. The results of this surface doping also confirm the recent observation of the two-dome phase diagram of the K-doped ultrathin FeSe films [117].

3.2.2 Latest Progress on FeSe Thin Films

First, although the resistivity of 1UC-thick films is tricky to measure and only the resistivity of 5UC-thick FeSe films has been presented in the main papers [118–120], 1UC-FeSe could be crucial for the superconductivity of FeSe that the superconducting transition occurs above 30 K, and it could be even above 50 K [121].

The statistic lines of the 1UC FeSe in Fig. 7 also have shown that T_c is really high and that today's best record is slightly above 100 K by the 4pp. method (in situ four-point probe electrical transport measurements) done

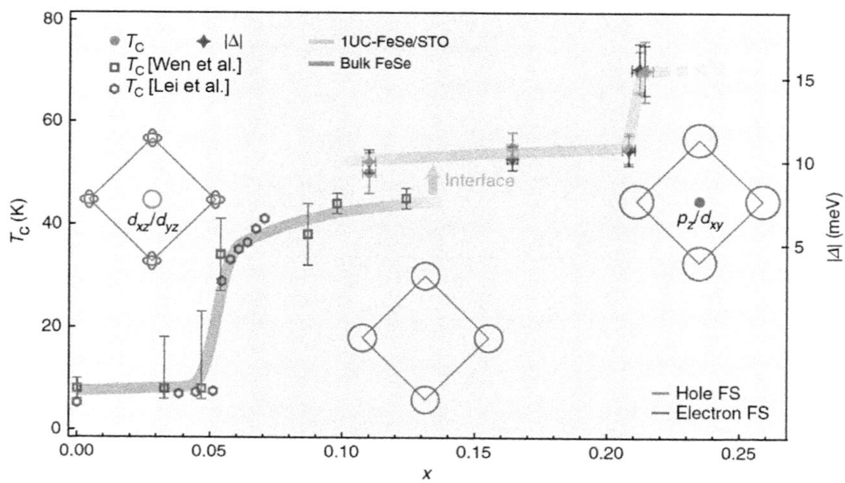

Fig. 7 Schematic phase diagram of FeSe [122].

by Jian-Feng Ge et al. [23], who used the 4pp. technique in the study of the FeSe monolayer deposited by an MBE system. The T_c of the FeSe single layer can reach up to 100 K in a wide range of magnetic strength. The ARPES spectra from the 1UC films on the $SrTiO_3$ (STO) substrate also have been measured following the discovery of high critical-temperature superconducting behaviors in such films. There is also a large amount of ARPES data that show the critical temperature of 60 K based on the value of the superconducting gap [75,123,124].

Second, the differences between the single-layer iron-based selenium system and the traditional La-Bi system (2201 type system) have been studied extensively. First, in the 2201 system, the parent compound is believed to be an antiferromagnetic insulator, and the gap between each node decreases from 0.10 to zero, accompanying the 3D antiferromagnetism disappearance when the doping increases. Moreover, according to the existing studies of the s-phase carrier evolution of iron-based selenium and the comparison data from the n-phase and s-phase analysis, scientists have indicated that the n phase may be a magnetic phase, but it is still not clear about the magnetism of the s phase when the carrier concentration is low. Second, the La-Bi 2201 system is a kind of quasi two-dimensional system, and it will become three-dimensional when it goes into the superconducting state, but the single-layer FeSe is believed to be a genuine and pure two-dimensional system, even in the superconducting state.

In any case, it is tempting to suppose that the band structure coincides with the result of the LDA band structure calculations at a high enough temperature or at least the Fe-SC topology, but there is an interaction called hopping selective renormalization, which means that different electrons affect hopping integrals dramatically. That interaction develops with lowering temperature and with the enhancement of the fluctuations of a certain order [125]. While the temperature-dependent ARPES is rather complicated, one may consider the temperature-dependent Hall measurements as a complementary tool that is especially sensitive when the Fe-SC goes through the topological Lifshitz transitions, such as from electron-like barrels to mixed electron hole-like propellers in $Ba_{1-x}K_xFe_2As_2$ (BKFA) [77]. Then, careful temperature-dependent ARPES measurements would be a key tool to identify the microscopic interaction that is responsible for both the T_c improvement and the nematic instability.

3.2.3 Synthesis of 2D FeSe

As the latest superconductor materials, iron-based superconductors have drawn sufficient attention in the research field of superconductivity.

There are a wide variety of reports that demonstrate how to synthesize bulk and thin film samples. Different techniques such as sputtering [126], evaporation [127], MBE, pulsed laser ablation (PLA) [128,129], chemical vapor deposition [130], and the PLD method [131] are all available to successfully prepare the thin films of high-performance superconductors. Most of these techniques work in a vacuum chamber and the oxygen partial pressures need to be controlled for obtaining good superconducting films. The temperature for the substrate during the deposition is a crucial parameter that determines a sample's microstructure, texture, or its epitaxy degree. Substrate-film interaction such as interdiffusion can affect the quality of films. Thus, it is advisable to develop new processes that are low temperature friendly to fabricate thin films [132,133].

Up to now, epitaxial β-FeSe thin films were deposited by several techniques: low-pressure metal organic chemical vapor deposition (LP-MOCVD) was used by Liu et al. and Wu et al. to grow FeSe films on GaAs (001) and SiO_2 substrates, respectively [134–136]. Particularly, Wu et al. presented a strong dependence of morphology on substrate temperature during deposition. However, there was less specific information related to the superconducting properties of these thin films. Tkachenko et al. obtained very rough superconducting thin films of β-FeSe on $LaAlO_3$ and $SrTiO_3$ by using the high gas pressure trap system (HGPTS). Pulsed laser deposition (PLD) is also a popular method for depositing β-FeSe on various substrates with transition temperatures of $T_{onset,c} = 11.8\,K$ [134,137,138]. The information about the morphology of PLD-grown thin films was published by Han et al. [135]. Their experiment indicated that the morphology and properties of the film are strongly dependent on the substrate temperature. Generally, a rough surface unexpectedly leads to good superconducting properties. Other results also indicate that FeSe can be nonsuperconducting. For example, Takemura et al. used the selenization technique by using an Se beam to irradiate the Fe target surface, thus achieving the FeSe layer on GaAs. However, the deposited film did not show superconducting behavior [139]. However, Qing et al., using a novel selenization technique, successfully produced superconducting β-FeSe thin films on $LaAlO_3$ [140].

Another available and widespread method for synthesizing FeSe thin films is the MBE technique. Within the MBE growth process, Fe from an electron-beam evaporator and Se from an effusion cell are coevaporated. Under a growth rate of approximately 0.1 nm/s, the thickness of samples can reach 500 nm. High-quality films in terms of crystal structure and superconductivity performance are obtained when the film is deposited on $YAlO_3$ (110) substrates, whereas less superconducting samples are obtained when the film was deposited on MgO (100) substrates [141]. The substrate temperature during deposition is 350°C. Alternatively, β-FeSe thin films were deposited using sputtering. The optimized composition of the FeSe film is composed of 7–8 wt% selenium and 13–14 wt% iron to achieve high-quality samples in terms of crystal structure and superconducting properties. Typically, the samples were prepared with thicknesses of 20–100 nm. The ideal substrate temperature for deposition is around 350°C, which is the same value as that for growing this film by MBE. In contrast to MBE growth, the epitaxial growth of superconducting thin films by sputtering was not only obtained on the $YAlO_3$ (010) substrate, despite the lattice mismatch of 11.8% for this substrate, but also achieved on STO (001) and MgO (100) substrates [142–144]. By comparing two methods of different sedimentary sources, multiconcentration gradient deposition can be performed in a single deposition process. This gradient is much stronger in the sputtering deposited samples than that in the MBE-grown thin films.

The results of energy dispersive X-ray spectroscopy (EDX), in some studies, indicate that there is some difference (about 20%) in the Fe/Se-content ratio (around 8 mm distance in the inspecting area). Due to the nonuniformity of composition in different areas, no absolute values for the sample stoichiometry could be determined. This makes the study of stoichiometry-dependent properties focusing on single thin film samples [145]. For thin films, structural investigations by X-ray diffraction (XRD) of samples grown by MBE demonstrated the β-FeSe is in the (001) direction. Fe_7Se_8 will still be in the system, but it will be less intensive. When using the MgO (100) substrate to fabricate FeSe (001) thin films, due to its lattice sequence and epitaxial relationship, only a (101) peak of FeSe can be observed in the XRD measurement, which is the same value of the lattice space as that of the substrate in the (220) direction. Therefore, the FeSe (100) plane and that of the MgO (010) overlap, and the weak FeSe (101) peak is visible on the relative plane of 45 degrees. In addition, the lattice constants can be calculated from the position of FeSe (00L) and the FeSe (101) surface, a (001) = 3.765 Å and c (101) = 5.53 Å. The lattice parameters are consistent with the report by Hsu et al. and the distance of the c-axis is slightly longer than 5.4847 Å [146–149].

The CVD method is a gradual film growth process that puts a particular substrate (depending on the proposed products) into a constant volume chamber, such as the Si substrate in synthesizing rhombic and triangular cross-sectional AlN nanorods [150], Ni substrate in NiAl coatings [151], diamond systems [152], and even an STO substrate in producing ceramic superconductors [153]. Raw materials are evaporated into the atmosphere and transited by vacuum or inert gases. When the raw elements are deposited on the substrate, several reactions will occur, such as thermal decomposition and redox. Other techniques such as plasma assistance can dramatically reduce the temperature of the reaction. According to different mechanisms of reaction, the CVD method has been developed into many branches. For example, YBCO ($YBa_2Cu_3O_{7-x}$) thin films were synthesized by MOCVD that has exhibited advanced superconducting properties [154].

Fig. 8 shows the schematic of a basic CVD system. The liquid source is vapor mixed with argon gas that is used for raw materials. Then, the precursor is sent into the reactor to shower the substrate with oxygen feeding. Finally, the wasted gas is pumped away. The main advantage of this method is that the synthesis is performed at nonmoral pressure and an easily controlled temperature, which is beneficial for the fabrication of high-quality, iron-based superconductors [156,157] and the cost can be reduced.

Recently, FeSe thin films were deposited on the GaAs (100) substrate by low-pressure metalorganic chemical vapor deposition (LP-MOCVD) reported by Feng et al. [158]. Hydrogen selenium (H_2Se) and iron-$(CO)_5$ gas were chosen as precursor materials. High-purity hydrogen was used as the carrier gas to push precursors into the reaction chamber. After 10 min of the annealing process in a hydrogen atmosphere at 600°C, the substrate could be ensured as free from contaminants. After 30 min of reaction, an FeSe film with a thickness of around 200 nm was obtained [158].

PLD is an efficient method to deposit thin films in a vacuum chamber by a high-power pulsed laser beam. It is a kind of physical vapor deposition (PVD) technique for depositing target materials on substrate. This process also needs an ultrahigh vacuum system or a specific background atmosphere,

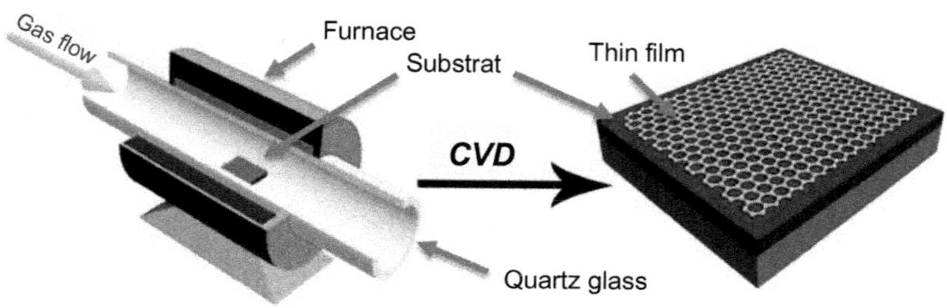

Fig. 8 The diagram of conception of chemical vapor deposition [155].

which is commonly chosen for depositing smooth thin films. The whole PLD method can be described as three processes, including the laser irradiation ablation process, high energetic ions plasma plume development, and self-growth on heated materials [159,160], and six subprocesses, including plasma dynamics, laser absorption, plasma creation, substrate deposition, film growth, and film nucleation [161–163]. The main structure of the PLD equipment is illustrated in Fig. 9.

Many studies have shown that LAO and STO are optimal substrates for the growth of FeSe thin films by PLD. For example, Nie et al. [165] deposited FeSe films on an STO substrate by a PLD system. The deposition power is approximately $1.5\,J/cm^2$ and the growth rate is approximately $0.12\,Å/s$. The thickness of the films produced by the PLD method can be thinner than 20 nm with better performance compared to the films of more than 100 nm in superconductivity, like 1 UC-FeSe thin films [166].

The solve-thermal method for the synthesis of the $Fe_3Se_4(en)_2$ tetrahedral chain was recently reported by Chongin Pak et al. [167]. This synthesis is carried out at 473 K under aerobic conditions with the necessary glove box. NH_4Cl is crucial for improving the solubility and productivity of reactions. The NH_4Cl sediment can be effectively washed off by water or ethanol, and $Fe_3Se_4(en)_2$ can remain stable in them, but it will be decomposed by dilute acid solutions. In addition, $AFeSe_2$ (A indicates Group I elements) also contains this structure. Because of its simplicity and the high activity of

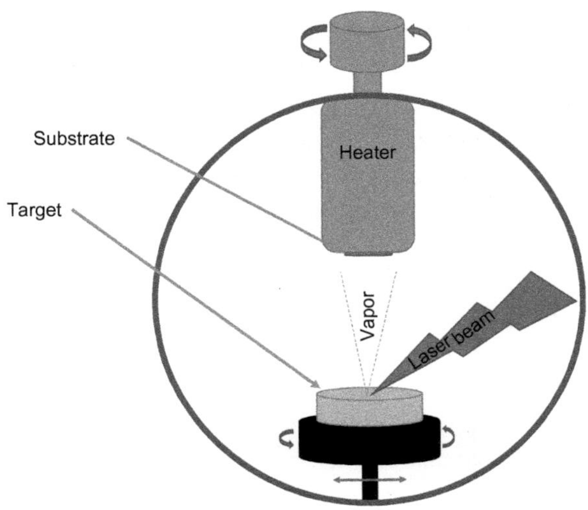

Fig. 9 Structure of PLD equipment [164].

the amino-metal complex, the synthesis method is more attractive. In addition, a new crystal structure (a tetrahedral $FeSe_2$ chain) is synthesized based on the simplest elements, Fe and Se. The antiferromagnetic order of $Fe_3Se_4(en)_2$ can be found in its chain structure. Theoretically, $FeSe_2$ has a larger chain than $AFeSe_2$. However, in $Fe_3Se_4(en)_2$, there is an Fe_2 structure that covers each Se atom a short distance outside the chain, from 2.3 to 2.8 Å. This compound provides a unique way to study the interaction of magnetism and electrons in isolated fragments of superconductors.

4 PROSPECTIVE

Superconductors have evolved from pure metals, metal alloys, complicated oxides, and iron-based superconductors. In addition, high pressure is beneficial to achieve high critical temperature, such as H_2S. Despite the significant progress of superconductors, metal and metal alloy superconductors are still the main materials for practical applications. For the high-temperature superconductors, problems for practical applications exist in terms of high cost, product manufacturing techniques, and application environments such as high pressure. In addition, the mechanism of high-temperature superconductor is still not clear.

The key point of the study of superconductors is to achieve high critical temperature and high current density. For metal and metal alloys, the critical temperature is very low, as predicated by BCS theory. Hence, the running cost is high due to cooling by liquid helium. Complicated oxides have shown a high critical temperature. However, due to the brittleness of oxides, fabricating superconductor wires/cables and devices is very difficult. In addition, the fabrication cost is very high. Other superconductors such as MgB_2 are also difficult for fabrication and the critical temperature is lower than the liquid nitrogen boiling point, suggesting a high running cost.

Iron-based superconductors were recently discovered. Though the bulk form does not show very high critical temperature, two-dimensional thin films show very promising properties. An FeSe monolayer has shown a critical temperature higher than 100 K. However, the mechanism is not clear and there is still no convincing explanation why FeSe has a high critical temperature and whether some more advanced materials with a high critical temperature can be found in the iron-based superconductors' family.

Based on the above analysis, searching for new materials for superconductors is still one of the major challenges. Though 2D-based FeSe has shown promising superconductivity, the mechanism is still not conclusive.

The understanding of superconductivity in such a simple compound may help in the search for high-quality superconductors with a high critical temperature.

Extreme conditions such as high pressure can induce high-temperature superconductivity. The mechanism must be further investigated. The understanding of the meca hanism may provide guidance for searching for other superconductor materials with high critical temperature. It is known that metal and metal alloys cannot achieve high critical temperature, as previously discussed. New technology or new techniques, such as interface techniques, nanostructure with quantum effects, or applying extreme conditions is required to achieve high-temperature superconductors.

Furthermore, for future research of superconductors, one cannot ignore the research of organic materials or polymers. Though this chapter does not cover this content, the research of organic superconductors has been extensively investigated. Until now, the critical temperature at normal pressure has been achieved up to 33K in alkali-doped fullerene $RbCs_2C_{60}$. Due to the complexity and the millions of kinds of forms in organic materials, high-performance superconductors with high critical temperature may be achieved. In addition, the understanding of the mechanisms should be useful for basic chemistry and physics science.

REFERENCES

[1] D.J. Scalapino, A common thread, Phys. C-Supercond. Appl. 470 (2010) S1–S3.

[2] G.E.A. Dubuis, On field effect studies and superconductor-insulator transition in high-Tc cuprates, Eur. Phys. J. Spec. Topics 222 (5) (2013) 1217–1221.

[3] N. Khare (Ed.), Introduction to high-temperature superconductors, In: Handbook of High-Temperature SuperconductorCRC Press, Boca Raton, FL, 2003.

[4] T. Shoji, High-temperature superconductivity, Jpn. J. Appl. Phys. **45** (12R) (2006) 9011.

[5] R.R. Hake, Thermodynamics of type-I and type-II superconductors, J. Appl. Phys. 40 (13) (1969) 5148–5160.

[6] R. Schwall, MRI in medical service, Trans. Magn. 23 (2003) 1287–1293.

[7] M. Schmelz, et al., Highly sensitive miniature SQUID magnetometer fabricated with cross-type Josephson tunnel junctions, Phys. C: Supercond. 476 (2012) 77–80.

[8] N. Shen, et al., The synthesis and performance of Zr-doped and W–Zr-codoped VO_2 nanoparticles and derived flexible foils, J. Mater. Chem. A 2 (36) (2014) 15087–15093.

[9] M. Tomita, et al., Resin-impregnated bulk RE–Ba–Cu–O current leads for the superconducting magnet on maglev train, Phys. C: Supercond. 372–376 (2002) 1216–1220.

[10] R.T. Gordon, London penetration depth measurements in Barium(Iron$_{1-x}$T$_x$)2Arsenic2 (T = cobalt, nickel,ruthenium, rhodium, palladium, platinum, cobalt + copper) superconductors, in: APA-6th, 2011, pp. 17–18.

[11] D. Mou, et al., Enhancement of the superconducting gap by nesting in $CaKFe_4As_4$: a new high temperature superconductor, Phys. Rev. Lett. 117 (27) (2016) 277001.

[12] K. Kobayashi, H. Yokoyama, Superconductivity and antiferromagnetism in the phase diagram of the frustrated Hubbard model within a variational study, Phys. C: Supercond. Appl. 470 (20) (2010) 1081–1084.

[13] A.M. Mounce, et al., Spin-density wave near the vortex cores in the high-temperature superconductor $Bi_2Sr_2CaCu_2O_{8+y}$, Phys. Rev. Lett. 106 (5) (2011). 057003.

[14] L. Craco, M.S. Laad, Normal state incoherent pseudogap in FeSe superconductor, Eur. Phys. J. B 89 (5) (2016).

[15] K. Yamaji, T. Yanagisawa, I. Hase, 3-Band theory of Fe pnictide superconductors, Phys. C: Supercond. Appl. 470 (20) (2010) 1060–1062.

[16] D.C. Johnston, Elaboration of the α-model derived from the BCS theory of superconductivity, Supercond. Sci. Technol. 26 (11) (2013) 115011.

[17] A.S. Alexandrov, Theory of high temperature superconductivity beyond BCS with realistic coulomb and Fröhlich interactions, J. Supercond. Nov. Magn. 26 (4) (2013) 1313–1317.

[18] S. He, et al., Phase diagram and electronic indication of high-temperature superconductivity at 65 K in single-layer FeSe films, Nat. Mater. 12 (7) (2013) 605–610.

[19] J. He, et al., Electronic evidence of an insulator-superconductor crossover in single-layer FeSe/$SrTiO_3$ films, Proc. Natl. Acad. Sci. U. S. A. 111 (52) (2014) 18501–18506.

[20] K. Nakayama, et al., Reconstruction of band structure induced by electronic nematicity in an FeSe superconductor, Phys. Rev. Lett. 113 (23) (2014) 237001.

[21] C.-M. Yang, et al., Enhancement of superconductivity in $FeSe_{1-x}$Texby Li doping, Supercond. Sci. Technol. 25 (9) (2012). 095010.

[22] S. Tan, et al., Interface-induced superconductivity and strain-dependent spin density waves in FeSe/$SrTiO_3$ thin films, Nat. Mater. 12 (7) (2013) 634–640.

[23] J.F. Ge, et al., Superconductivity above 100 K in single-layer FeSe films on doped $SrTiO_3$, Nat. Mater. 14 (3) (2015) 285–289.

[24] P.J. Baker, et al., Heat capacity measurements on FeAs-based compounds: a thermodynamic probe of electronic and magnetic states, New J. Phys. 11 (2) (2009). 025010.

[25] V. Mishev, et al., One-dimensional pinning behavior in co-doped $BaFe_2As_2$ thin films, Appl. Phys. Lett. 103 (23) (2013) 232601.

[26] Z.-H. Liu, et al., Photoemission study of iron-based superconductor, Chin. Phys. B 22 (8) (2013). 087406.

[27] X.-J. Zhou, et al., Band structure, Fermi surface, and superconducting gap in FeAs-based superconductors revealed by angle-resolved photoemission spectroscopy, Front. Phys. China 4 (4) (2009) 427–432.

[28] H. Luetkens, et al., The electronic phase diagram of the LaO$(1-x)$F(x)FeAs superconductor, Nat. Mater. 8 (4) (2009) 305–309.

[29] J.S. Kim, et al., The specific heat of the electron-doped La-1038 compound $(Ca_{0.85}La_{0.15})_{10}(FeAs)_{10}(Pt_3As_8)$, J. Phys. Condens. Matter 25 (13) (2013) 135701.

[30] Z. Yamani, et al., Phonon density of states and the search for a resonance mode in $LaFeAsO_{0.85}F_{0.15}$ (Tc = 26 K), J. Phys. Conf. Ser. 340 (2012). 012074.

[31] T. Schickling, F. Gebhard, J. Bunemann, Antiferromagnetic order in multiband Hubbard models for iron pnictides, Phys. Rev. Lett. 106 (14) (2011) 146402.

[32] P. Mele, et al., In-field characterization of $FeTe_{0.8}S_{0.2}$ epitaxial thin films with enhanced superconducting properties, Supercond. Sci. Technol. 23 (5) (2010) 052001.

[33] S. Haindl, et al., High upper critical fields and evidence of weak-link behavior in superconducting $LaFeAsO_{1-x}F_x$ thin films, Phys. Rev. Lett. 104 (7) (2010). 077001.

[34] E. Backen, et al., Growth and anisotropy of La(O, F)FeAs thin films deposited by pulsed laser deposition, Supercond. Sci. Technol. 21 (12) (2008) 122001.

[35] N. Kang, et al., Pressure dependence of the thermoelectric power of the iron-based high-Tcsuperconductor $SmFeAsO_{0.85}$, New J. Phys. 11 (2) (2009) 025006.
[36] J. Prakash, et al., Potassium fluoride doped LaOFeAs multi-band superconductor: evidence of extremely high upper critical field, Europhys. Lett. 84 (5) (2008) 57003.
[37] Z. Wang, et al., Observations of temperature stability of γ-zirconium hydride by high-resolution neutron powder diffraction, J. Alloys Compd. 661 (2016) 55–61.
[38] Y. Kamihara, et al., Iron-based layered superconductor $La[O_{1-x}F_x]FeAs$ ($x=0.05$–0.12) with Tc=26 K, J. Am. Chem. Soc. 130 (11) (2008) 3296–3297.
[39] T. Yildirim, Origin of the 150-K anomaly in LaFeAsO: competing antiferromagnetic interactions, frustration, and a structural phase transition, Phys. Rev. Lett. 101 (5) (2008). 057010.
[40] J.C. Wojdeł, I.P.R. Moreira, F. Illas, Chemical bonding and electronic and magnetic structure in LaOFeAs, J. Am. Chem. Soc. 131 (3) (2009) 906–907.
[41] R. Sen, et al., Oxo-vanadium (IV) dihydrogen phosphate: preparation, magnetic study, and heterogeneous catalytic epoxidation, Langmuir 24 (11) (2008) 5970–5975.
[42] H. Luetkens, et al., The electronic phase diagram of the $LaO_{1-x}F_xFeAs$ superconductor, Nat. Mater. 8 (2009) 305–309.
[43] N. Qureshi, et al., Crystal and magnetic structure of the oxypnictide superconductor $LaFeAsO_{1-x}F_x$: a neutron-diffraction study, Phys. Rev. B 82 (18) (2010) 184521.
[44] X. Luo, X. Chen, Crystal structure and phase diagrams of iron-based superconductors, Sci. China Mater. 58 (1) (2015) 77–89.
[45] F. Rullier-Albenque, et al., Hall effect and resistivity study of the magnetic transition, carrier content, and Fermi-liquid behavior in $Ba(Fe_{1-x}Co_x)_2As_2$, Phys. Rev. Lett. 103 (5) (2009). 057001.
[46] Z.-A. Ren, et al., Superconductivity and phase diagram in iron-based arsenic-oxides $ReFeAsO_{1-\delta}$ (Re=rare-earth metal) without fluorine doping, Europhys. Lett. 83 (1) (2008) 17002.
[47] J.H. Markna, Synthesis and Characterization of Perovskite Structured Mixed Oxide Compounds, Materials Chemistry and Physics, Saurashtra University, 2006.
[48] X. Tao, et al., Crystal growth and phase diagram of 112-type iron pnictide superconductor $Ca_{1-y}La_yFe_{1-x}Ni_xAs_2$, Supercond. Sci. Technol. 30 (9) (2017). 095002.
[49] L. Xu, et al., Electronic structure and superconductivity of FeSe-related superconductors, J. Phys. Condens. Matter 27 (18) (2015) 183201.
[50] D.N. Basov, A.V. Chubukov, Manifesto for a higher Tc, Nat. Phys. 7 (2011) 272.
[51] Y. Kamihara, et al., Iron-based layered superconductor $La[O_{1-x}F_x]$ FeAs ($x=0.05$–0.12) with Tc=26 K, J. Am. Chem. Soc. 130 (11) (2008) 3296–3297.
[52] H. Luetkens, et al., The electronic phase diagram of the $LaO_{1-x}F_xFeAs$ superconductor, Nat. Mater. 8 (2009) 305.
[53] K. Kubo, P. Thalmeier, Antiferromagnetism with ferro-orbital order in Fe pnictides, J. Phys. Conf. Ser. 200 (1) (2010). 012097.
[54] Q. Huang, et al., Doping evolution of antiferromagnetic order and structural distortion in $LaFeAsO_{1-x}F_x$, Phys. Rev. B 78 (5) (2008). 054529.
[55] T. Nomura, et al., Crystallographic phase transition and high-Tc superconductivity in LaFeAsO:F, Supercond. Sci. Technol. 21 (12) (2008) 125028.
[56] H.S. Lee, et al., High-pressure growth of fluorine-free $SmFeAsO_{1-x}$ superconducting single crystals, Supercond. Sci. Technol. 22 (7) (2009). 075023.
[57] P.M. Aswathy, et al., An overview on iron based superconductors, Supercond. Sci. Technol. 23 (7) (2010). 073001.
[58] D.A. Zocco, et al., Effect of pressure on the superconducting critical temperature of La $[O_{0.89}F_{0.11}]FeAs$ and $Ce[O_{0.88}F_{0.12}]FeAs$, Phys. C: Supercond. 468 (21) (2008) 2229–2232.

[59] A.M. Zhang, et al., Superconductivity at 44 K in K intercalated FeSe system with excess Fe, Sci. Rep. 3 (2013) 1216.
[60] K. Liu, et al., Nematic antiferromagnetic states in bulk FeSe, Phys. Rev. B 93 (20) (2016).
[61] L. Sun, et al., Reemerging superconductivity at 48 K in iron chalcogenides, Nature 483 (7387) (2012) 67–69.
[62] K. Onar, et al., Solid state synthesis and characterization of bulk β-FeSe superconductors, J. Alloys Compd. 620 (2015) 210–216.
[63] M.K. Wu, et al., Recent advances in beta-$FeSe_{1-x}$ and related superconductors, Sci. Technol. Adv. Mater. 14 (1) (2013) 014402.
[64] M. de Souza, et al., Synthesis, structural and physical properties of δ′-$FeSe_{1-x}$, Eur. Phys. J. B 77 (1) (2010) 101–107.
[65] U. Pachmayr, N. Fehn, D. Johrendt, Structural transition and superconductivity in hydrothermally synthesized FeX (X=S, Se), Chem. Commun. 52 (1) (2016) 194–197.
[66] X. Zhou, et al., The preparation and phase diagrams of $(7Li_{1-x}Fe_xOD)FeSe$ and $(Li_{1-x}Fe_xOH)FeSe$ superconductors, J. Mater. Chem. C 4 (18) (2016) 3934–3941.
[67] F. Nitsche, et al., Room-temperature synthesis, hydrothermal recrystallization, and properties of metastable stoichiometric FeSe, Inorg. Chem. 51 (13) (2012) 7370–7376.
[68] Q. Wang, et al., Strong interplay between stripe spin fluctuations, nematicity and superconductivity in FeSe, Nat. Mater. 15 (2) (2016) 159–163.
[69] L. Jiao, et al., Superconducting gap structure of FeSe, Sci. Rep. 7 (2017) 44024.
[70] W. Wang, et al., Scotch tape induced strains for enhancing superconductivity of $FeSe_{0.5}Te_{0.5}$ single crystals, Appl. Phys. Lett. 105 (23) (2014) 232602.
[71] A. Krzton-Maziopa, et al., Superconductivity in alkali metal intercalated iron selenides, J. Phys. Condens. Matter 28 (29) (2016) 293002.
[72] A.M. Zhang, et al., Effects on superconductivity of transition-metal doping in FeSe (0.5)Te(0.5), J. Phys. Condens. Matter 22 (24) (2010) 245701.
[73] C. Platt, et al., Functional renormalization group for multi-orbital Fermi surface instabilities, Adv. Phys. 62 (4–6) (2013) 453–562.
[74] M.D. Lumsden, et al., Magnetism in Fe-based superconductors, J. Phys. Condens. Matter 22 (20) (2010) 203203.
[75] Y.V. Pustovit, A.A. Kordyuk, Metamorphoses of electronic structure of FeSe-based superconductors (review article), Low Temp. Phys. 42 (11) (2016) 995–1007.
[76] P. Dai, J. Hu, E. Dagotto, Magnetism and its microscopic origin in iron-based high-temperature superconductors, Nat. Phys. 8 (2012) 709.
[77] D.P. Chen, C.T. Lin, The growth of 122 and 11 iron-based superconductor single crystals and the influence of doping, Supercond. Sci. Technol. 27 (10) (2014) 103002.
[78] B. Zeng, et al., Anisotropic structure of the order parameter in $FeSe_{0.45}Te_{0.55}$ revealed by angle-resolved specific heat, Nat. Commun. 1 (2010) 112.
[79] S. Sahoo, S.M. Mobin, S. Ghosh, Direct insertion of sulfur, selenium and tellurium atoms into metallaborane cages using chalcogen powders, J. Organomet. Chem. 695 (7) (2010) 945–949.
[80] P. Mele, et al., Pulsed laser deposition and in-field characterization of $FeTe_{0.8}S_{0.2}$ epitaxial thin films with enhanced superconducting properties, Phys. C: Supercond. Appl. 470 (20) (2010) 1033–1037.
[81] Y.A. Izyumov, E.Z. Kurmaev, FeAs systems: a new class of high-temperature superconductors, Physics-Uspekhi 51 (12) (2008) 1261–1286.
[82] T. Taen, et al., Critical current densities and vortex dynamics in $FeTe_xSe_{1-x}$ single crystals, Phys. C: Supercond. Appl. 470 (20) (2010) 1106–1108.

[83] A.E. Bohmer, et al., Origin of the tetragonal-to-orthorhombic phase transition in FeSe: a combined thermodynamic and NMR study of nematicity, Phys. Rev. Lett. 114 (2) (2015). 027001.
[84] J. Kang, et al., Superconductivity in FeSe thin films driven by the interplay between nematic fluctuations and spin-orbit coupling, Phys. Rev. Lett. 117 (21) (2016) 217003.
[85] R.S. Kumar, et al., Pressure induced high spin-low spin transition in FeSe superconductor studied by X-ray emission spectroscopy and ab initio calculations, Appl. Phys. Lett. 99 (6) (2011). 061913.
[86] A. Kumar, et al., First-principles analysis of electron correlation, spin ordering and phonons in the normal state of $FeSe_{1-x}$, J. Phys. Condens. Matter 22 (38) (2010) 385701.
[87] A.V. Chubukov, I. Eremin, D.V. Efremov, Superconductivity versus bound-state formation in a two-band superconductor with small Fermi energy: applications to Fe pnictides/chalcogenides and doped $SrTiO_3$, Phys. Rev. B 93 (17) (2016) 174516.
[88] R. Mas-Balleste, et al., 2D materials: to graphene and beyond, Nanoscale 3 (1) (2011) 20–30.
[89] U.K. Sen, et al., High-rate and high-energy-density lithium-ion battery anode containing 2D MoS_2 nanowall and cellulose binder, ACS Appl. Mater. Interfaces 5 (4) (2013) 1240–1247.
[90] N. Huo, Y. Yang, J. Li, Optoelectronics based on 2D TMDs and heterostructures, J. Semicond. 38 (3) (2017). 031002.
[91] A.F. Khan, et al., Surfactant-exfoliated 2D hexagonal boron nitride (2D-hBN): role of surfactant upon the electrochemical reduction of oxygen and capacitance applications, J. Mater. Chem. A 5 (8) (2017) 4103–4113.
[92] M.R. Neupane, Electronic and vibrational properties of 2D materials from monolayer to bulk, 2015 International Workshop on Computational Electronics (IWCE), 2015.
[93] A. Bedia, F.Z. Bedia, B. Benyoucef, 2D device modeling and simulation/FET for biomedical applications, Phys. Proc. 21 (2011) 35–41.
[94] X. Wang, et al., Atomic-scale recognition of surface structure and intercalation mechanism of Ti3C2X, J. Am. Chem. Soc. 137 (7) (2015) 2715–2721.
[95] T. Suzuki, et al., All-optical single-shot ultrafast 2D-burst imaging using a linearly frequency chirped pulse, 11th Conference on Lasers and Electro-Optics Pacific Rim (CLEO-PR), IEEE, 2015.
[96] A. Shashurin, et al., Synthesis of 2D materials in arc plasmas, J. Phys. D: Appl. Phys. 48 (31) (2015) 314007.
[97] M.I.B. Utama, et al., Detrimental influence of catalyst seeding on the device properties of CVD-grown 2D layered materials: a case study on $MoSe_2$, Appl. Phys. Lett. 105 (25) (2014) 253102.
[98] A. Habibi, et al., Fast synthesis of turbostratic carbon thin coating by cathodic plasma electrolysis, Thin Solid Films 621 (2017) 253–258.
[99] H. Samassekou, et al., Viable route towards large-area 2D MoS_2 using magnetron sputtering, 2D Materials 4 (2) (2017) 021002.
[100] M. Medhisuwakul, et al., Development and application of cathodic vacuum arc plasma for nanostructured and nanocomposite film deposition, Surf. Coat. Technol. 229 (2013) 36–41.
[101] Y.Q. Cai, et al., The phenomenon of the superheating of YBCO thin film and its application in processing REBCO bulk, Supercond. Sci. Technol. 19 (7) (2006) S423–S428.
[102] Y. Zhang, et al., Compare of the electronic structures of F- and Ir-doped SmFeAsO, Phys. C: Supercond. Appl. 470 (20) (2010) 1073–1076.

[103] J.-F. Ge, et al., Superconductivity above 100 K in single-layer FeSe films on doped $SrTiO_3$, Nat. Mater. 14 (3) (2015) 285–289.
[104] D. Huang, J.E. Hoffman, Monolayer FeSe on $SrTiO_3$, Ann. Rev. Condens. Matter Phys. 8 (1) (2017) 311–336.
[105] M. Xu, X. Song, H. Wang, Substrate and band bending effects on monolayer FeSe on $SrTiO_3$(001), Phys. Chem. Chem. Phys. 19 (11) (2017) 7964–7970.
[106] S. Yang, et al., Thickness-dependent coherent phonon frequency in ultrathin FeSe/$SrTiO_3$ films, Nano Lett. 15 (6) (2015) 4150–4154.
[107] Y. Zhang, et al., Superconducting gap anisotropy in monolayer FeSe thin film, Phys. Rev. Lett. 117 (11) (2016) 117001.
[108] M. Ohno, et al., Thermal conductivity and annealing effects in the iron-based superconductor $FeSe_{0.3}Te_{0.7}$, J. Phys. Soc. Jpn. 83 (4) (2014) 044704.
[109] Y. Wang, Theory of Gap Symmetry and Structure in Fe-Based Superconductors, in: ProQuest Dissertations and Theses, 2014 3647874.
[110] Q. Wang, Angle-Resolved Photoemission Spectroscopy Studies on Cuprate and Iron-Pnictide High-Tc Superconductors, in: Physics Graduate Theses and Dissertations, 49, University of Colorado at Boulder, 2011.
[111] J.J. Lee, et al., Interfacial mode coupling as the origin of the enhancement of T(c) in FeSe films on $SrTiO_3$, Nature 515 (7526) (2014) 245–248.
[112] I.A. Nekrasov, et al., Electronic structure of NaFeAs superconductor: LDA+DMFT calculations compared to the ARPES experiment, JETP Lett. 102 (1) (2015) 26–31.
[113] A.A. Kordyuk, et al., Electronic band structure of ferro-pnictide superconductors from ARPES experiment, J. Supercond. Nov. Magn. 26 (9) (2013) 2837–2841.
[114] H. Lohani, et al., Investigation of correlation effects in FeSe and FeTe by LDA+U method, Phys. C: Supercond. Appl. 512 (2015) 54–60.
[115] R. Peng, et al., Tuning the band structure and superconductivity in single-layer FeSe by interface engineering, Nat. Commun. 5 (2014) 5044.
[116] J.J. Seo, et al., Superconductivity below 20 K in heavily electron-doped surface layer of FeSe bulk crystal, Nat. Commun. 7 (2016) 11116.
[117] Y. Miyata, et al., High-temperature superconductivity in potassium-coated multilayer FeSe thin films, Nat. Mater. 14 (8) (2015) 775–779.
[118] Q.-Y. Wang, et al., Interface-induced high-temperature superconductivity in single unit-cell FeSe films on $SrTiO_3$, Chin. Phys. Lett. 29 (3) (2012). 037402.
[119] Q. Wang, et al., Thickness dependence of superconductivity and superconductor-insulator transition in ultrathin FeSe films on $SrTiO_3$(001) substrate, 2D Materials 2 (4) (2015) 044012.
[120] Z.C. Huang, et al., Electronic structure and superconductivity of single-layer FeSe on Nb:$SrTiO_3$/$LaAlO_3$ with varied tensile strain, 2D Materials 3 (1) (2016) 014005.
[121] K. Hanzawa, et al., Electric field-induced superconducting transition of insulating FeSe thin film at 35 K, Proc. Natl. Acad. Sci. 113 (15) (2016) 3986–3990.
[122] X. Shi, et al., Enhanced superconductivity accompanying a Lifshitz transition in electron-doped FeSe monolayer, Nat. Commun. 8 (2017) 14988.
[123] M. Abdel-Hafiez, et al., Impurity scattering effects on the superconducting properties and the tetragonal-to-orthorhombic phase transition in FeSe, Phys. Rev. B 93 (22) (2016).
[124] I. Nekrasov, et al., Electronic structure of FeSe monolayer superconductors, Low Temp. Phys. 42 (10) (2016) 891–899.
[125] A. Linscheid, et al., High T_c via spin fluctuations from incipient bands: application to monolayers and intercalates of FeSe, Phys. Rev. Lett. 117 (7) (2016). 077003.
[126] Z.Q. Li, et al., Epitaxial growth $La_{0.67}Ca_{0.33}MnO_3$ thin film by radio frequency sputtering method, Appl. Surf. Sci 254 (21) (2008) 6959–6961.

[127] N.T. Khoa, et al., Photodecomposition effects of graphene oxide coated on TiO_2 thin film prepared by electron-beam evaporation method, Thin Solid Films 520 (16) (2012) 5417–5420.
[128] I. Yagi, et al., Direct observation of contact and channel resistance in pentacene four-terminal thin-film transistor patterned by laser ablation method, Appl. Phys. Lett. 84 (5) (2004) 813–815.
[129] E. Venzmer, et al., Morphology of Superconducting FeSe Thin Films Deposited by Co-sputtering and MBE, arXiv preprint arXiv: 1505.02630, 2015.
[130] S.-M. Zhou, et al., Synthesis of ordered α-FeSe nanorod array by CVD and its high Tc, Mater. Lett. 65 (11) (2011) 1741–1743.
[131] A. Tsukada, et al., Pulsed laser deposition conditions and superconductivity of FeSe thin films, Appl. Phys. A 104 (1) (2011) 311–318.
[132] N. Khare, Handbook of High Temperature Superconductors, first ed., CRC Press, Boca Raton, FL, 2003.
[133] R.A. Roy, et al., Substrate and temperature effects in lead zirconate titanate films produced by facing targets sputtering, J. Mater. Res. 7 (06) (2011) 1455–1464.
[134] L. Li, et al., Superconductivity and magnetism in FeSe thin films grown by metal–organic chemical vapor deposition, Supercond. Sci. Technol. 24 (1) (2011). 015010.
[135] Y. Han, et al., Preparation and superconductivity of iron selenide thin films, J. Phys. Condens. Matter 21 (23) (2009) 235702.
[136] C.-J. Liu, et al., Transport properties of Bi-doped FeSe superconductor up to 700 K, Appl. Phys. Lett. 104 (25) (2014) 252602.
[137] K. Iida, et al., Generic Fe buffer layers for Fe-based superconductors: epitaxial $FeSe_{1-x}Te_x$ thin films, Appl. Phys. Lett. 99 (20) (2011) 202503.
[138] O. Tkachenko, et al., Synthesis, Crystal Growth and Epitaxial Layer Deposition of FeSe0. 88 Superconductor and Other Poison Materials by Use of High Gas Pressure Trap System, arXiv preprint arXiv: 1108.5069, 2011.
[139] Y. Takemura, et al., Characterization of FeSe thin films prepared on GaAs substrate by selenization technique, J. Appl. Phys. 81 (8) (1997) 5177–5179.
[140] C. Qing-Lin, et al., A novel selenization technique for fabrication of superconducting $FeSe_x$ thin film, Phys. C: Supercond. 471 (13–14) (2011) 428–431.
[141] R. Schneider, et al., Excess conductivity and Berezinskii-Kosterlitz-Thouless transition in superconducting FeSe thin films, J. Phys. Condens. Matter 26 (45) (2014) 455701.
[142] D.G. Schlom, et al., A thin film approach to engineering functionality into oxides, J. Am. Ceram. Soc. 91 (8) (2008) 2429–2454.
[143] M.A. Rahman, J.K. Park, On the growth and coalescence of the epitaxial islands of L10-FePt (001) thin-film, J. Appl. Phys. 120 (16) (2016) 165302.
[144] E.A. Stepantsov, et al., Investigation into the growth and structure of thin-film solid solutions of iron-based superconductors in the $FeSe_{0.92}$-$FeSe_{0.5}Te_{0.5}$ system, Crystallogr. Rep. 58 (5) (2013) 735–738.
[145] M.-L. Li, et al., Microwave-assisted synthesis of flower-like β-FeSe microstructures, CrystEngComm 12 (10) (2010) 3138.
[146] J. Zhuang, et al., Unabridged phase diagram for single-phased FeSe(x)Te($1-x$) thin films, Sci. Rep. 4 (2014) 7273.
[147] B. Yuan, et al., Facile synthesis of flake-like $FeSe_2$ particles in open-air conditions, New J. Chem. 36 (10) (2012) 2101.
[148] B. Yuan, et al., One-step synthesis of cubic FeS_2 and flower-like $FeSe_2$ particles by a solvothermal reduction process, Dalton Trans. 41 (3) (2012) 772–776.
[149] M.L. Amigó, et al., Influence of the Fe concentration on the superconducting properties of $Fe_{1-y}Se$, J. Low Temp. Phys. 179 (1–2) (2014) 15–20.

[150] L. Yu, et al., Synthesis of rhombic and triangular cross-sectional AlN nanorods on Si substrate via thermal CVD, Mater. Lett. 65 (10) (2011) 1499–1502.
[151] J. Romanowska, et al., Zirconium influence on microstructure of aluminide coatings deposited on nickel substrate by CVD method, Bull. Mater. Sci. 36 (6) (2013) 1043–1048.
[152] M.-S. Chun, T. Teraji, T. Ito, Interfacial charge transfer and hole injection in α-NPD organic overlayer–CVD diamond substrate system, Appl. Surf. Sci. 237 (1–4) (2004) 470–477.
[153] N.M. Hwang, Non-Classical Crystallization of Thin Films and Nanostructures in CVD and PVD Processes, Springer Netherlands, Dordrecht, Netherlands, 2016.
[154] H. Kono, et al., Degradation characteristics of YBCO coated conductors due to fault-current in power cable applications, Phys. C: Supercond. Appl. 470 (20) (2010) 1334–1337.
[155] J. Chen, et al., Oxygen-aided synthesis of polycrystalline graphene on silicon dioxide substrates, J. Am. Chem. Soc. 133 (44) (2011) 17548–17551.
[156] R. Muydinov, et al., High throughput of reel-to-reel MOCVD-YBCO on different CSD-and MOCVD-buffered cube textured Ni-substrates, Phys. Proc. 36 (Suppl. C) (2012) 1468–1474.
[157] O. Stadel, et al., MOCVD and MOD of YBCO and buffer layers on textured metal tapes, IEEE Trans. Appl. Supercond. 19 (2009) 3160–3163.
[158] Q.J. Feng, et al., High oriented FeSe thin film on GaAs(100) substrate prepared by low-pressure metalorganic chemical vapor deposition, J. Magn. Magn. Mater. 279 (2–3) (2004) 435–439.
[159] M.R.R. Vaziri, et al., Microscopic description of the thermalization process during pulsed laser deposition of aluminium in the presence of argon background gas, J. Phys. D: Appl. Phys. 43 (42) (2010) 425205.
[160] H. Kim, Interface Studies on the Organic-Inorganic Hybrid Heterojunctions and Device Applications of Nanowire Materials, in: ProQuest Dissertations and Theses, 143, Brown University, 2010, p. 3430189.
[161] L.A. Gautier, et al., Field electron emission enhancement of graphenated MWCNTs emitters following their decoration with Au nanoparticles by a pulsed laser ablation process, Nanotechnology 26 (4) (2015). 045706.
[162] J. Brcka, et al., Investigation of plasma expansion in the PLD process by means of an electrical probe, Plasma Sources Sci. Technol. 3 (2) (1994) 128.
[163] R. Castro-Rodriguez, et al., Nucleation of strontium titanate films grown by PLD on silicon: a kinetic model, Thin Solid Films 307 (1–2) (1997) 306–310.
[164] U. Bansode, et al., Hybrid perovskite films by a new variant of pulsed excimer laser deposition: a room-temperature dry process, J. Phys. Chem. C 119 (17) (2015) 9177–9185.
[165] Y. Nie, A Study of Iron Chalcogenide Superconductivity and Cobaltite Magnetism, University of Connecticut, 2011.
[166] S. Tan, et al., Interface-induced superconductivity and strain-dependent spin density waves in $FeSe/SrTiO_3$ thin films, Nat. Mater. 12 (7) (2013) 634–640.
[167] J.T. Greenfield, et al., Control over connectivity and magnetism of tetrahedral $FeSe_2$ chains through coordination Fe-amine complexes, Chem Commun. (Camb.) 51 (25) (2015) 5355–5358.

FURTHER READING

[168] C. Pak, et al., Chemical excision of tetrahedral FeSe(2) chains from the superconductor FeSe: synthesis, crystal structure, and magnetism of Fe(3)Se(4)(en)(2), J. Am. Chem. Soc. 135 (51) (2013) 19111–19114.

CHAPTER 7

Nanoscaled Magnetic Oxide: Remarkable Properties and Potentials for Applications

Nguyen Hoa Hong
Department of Physics and Astronomy, Seoul National University, Seoul, Republic of Korea

1 INTRODUCTION

The main idea of using semiconducting materials for devices is to exploit the charge of electrons in those compounds. In a more advanced way, if semiconductors are simultaneously ferromagnetic, then one can use both charge and spin of electrons in the same material. A combined material in this case can indeed become multifunctional: the spin of electrons that carries information can actually serve as an added degree of freedom in novel electronic devices, along with its previously mentioned function of having charge. One can see that, by embedding ferromagnetism (FM) into semiconductors, we can expect to have a new family of promising materials for future devices.

In early 2000, some simulation work suggested that FM could be obtained at high temperature in *p*-type diluted magnetic semiconductors (DMS) if Mn could be doped below 5 at.% into ZnO host lattices [1]. Their density functional theory (DFT) calculations showed that FM should be observed in a DMS if the host material has a wide band gap. Very soon after, Yoshida-Katayama and Sato importantly pointed out in their simulation work that room temperature FM could be induced into ZnO if a transition metal (TM) other than Mn could be a dopant [2].

Because the theoretical predictions were seriously significant, many experimental groups started to study along this direction. In the first experimental report, room temperature FM was obtained in TiO_2 thin films doped with cobalt. However, the magnetization was only 0.3 μ_B per Co atom [3]. In the following years, many investigations were done, either looking for a new semiconducting oxide that might be room temperature

Nano-sized Multifunctional Materials
https://doi.org/10.1016/B978-0-12-813934-9.00007-4

ferromagnetic, or trying to obtain materials whose magnetic moments would be greater [4, 5]. The key issue for industrial applications is that such FM must be intrinsic, that is, it must really stem from the doped matrices but not from TM clusters. Because the mechanism that is responsible for the magnetic interactions in this kind of oxide is still not clear, and indeed controversial, many laboratories had been trying to elucidate standing issues. It had been suggested by experiments that the induced magnetic properties in DMS systems cannot simply result from the Rudeman-Kittel-Kasuya-Yoshida (RKKY) interaction, but most probably relate to defects and/or oxygen vacancies that might be very favored at the surface and interface between semiconductor and insulator [6, 7]. Particularly, the properties of semiconducting oxides such as ZnO, TiO_2, SnO_2, In_2O_3, etc., are quite different from other traditional semiconductors such as GaAs or GaN. Therefore, it appears to be more complicated to cope with due to the existence of oxygen in those materials. Over the last few years, people have become more careful and critical when analyzing and reporting the magnetic properties of semiconducting oxides. It is the reason why this family has been called "diluted magnetic semiconducting oxides" (DMSO) in order not to mix it up with other classical semiconductors.

After the discovery of FM in undoped HfO_2, TiO_2, In_2O_3, SnO_2, CeO_2, and MgO [6, 8–13], people started giving special attention to magnetism that would be due to defects and oxygen vacancies. It is certain that we should be prudent when judging the real role of TM doping in introducing FM in a semiconducting oxide: Does the FM indeed comes from the doping? Or does it just add some supplementary moment to what already exists in the oxide host due to the confinement effects of nanometer-sized materials? Indeed, it was proven that the observed FM was seen only in ultrathin films and nanopowders, suggesting that one cannot ignore the importance of confinement effects in low dimensional systems.

In this chapter, I will review the results of this research domain, based on its theoretical models reported and experimental results shown so far in order to better understand the mechanism of these materials. I expect that this review could give the readers an overall picture about this family of materials, and eventually, it may give some new guides for future spintronic devices.

I do not plan to make the chapter a complete report about DMSO. However, I wish to express my own viewpoint about some standing issues, based on my long working experiences in this domain.

2 THEORETICAL BACKGROUND

In this session, I would like to summarize previous theoretical works on DMS systems that actually have guided experimental studies over the last two decades. For a while, researchers have been questioning whether all are really correct because, in fact, completed experiments have not really supported theoretical predictions, even though at first they appeared "very reasonable."

The group of Dietl et al. first suggested this direction [1], stating that by doping few percent of Mn into ZnO, one should be able to induce room temperature FM if the doped compound is *p*-type. Following this idea, Sato et al. studied more deeply for the whole range of doping of TM [2, 14]. Their main points could be summarized as follows.

The Fermi level (E_F) of ZnO lies in the majority impurity band. The important energy gain arises from the broadening of the impurity band when concentration c increases. When c is increased, the density of state (DOS)-weight is transferred from a value that is around E_F to lower energies, leading to an energy gain. That would stabilize the ferromagnetic state. This energy gain is proportional to the bandwidth W of the impurity band, which is proportional to the square root of the concentration. This is indeed already known as Zener's double exchange [15, 16].

According to the electronic structure of TM impurity in semiconductors [17], fivefold degenerated *d*-states of TM impurity are split into doubly degenerated d_g-states and threefold degenerated d_e-states in the tetrahedral coordination. Two d_g-states have the symmetry of $3z^2-r^2$ and x^2-y^2. Three d_e-states have the symmetry of xy, yz, and zx. The wave functions of these d_e-states are extended to anions, thus, d_e-states can hybridize with O-2*p* states that make the host valence band. Therefore, the bonding states (t^b) and the antibonding counterparts (t^a) are created. Moreover, the wave functions of d_g-states are extended to the interstitial region, thus the hybridization of d_g-states with the host valence band is weak and the d_g-states actually remain as nonbonding states.

In ZnO-based DMS, the magnetic properties could be decided by the competition between ferromagnetic double exchange and antiferromagnetic superexchange. In order to have the FM, there must be itinerant electrons. It was discovered that the ferromagnetic state can be stable when delocalized t^a-states are partially occupied. According to this rule, the induced magnetism in DMS could be explained. Remembering that TM in general may have either a 2^+ or a 3^+ charge state in II–VI or III–V [14], respectively, then their 3*d*-electron configurations could be expected.

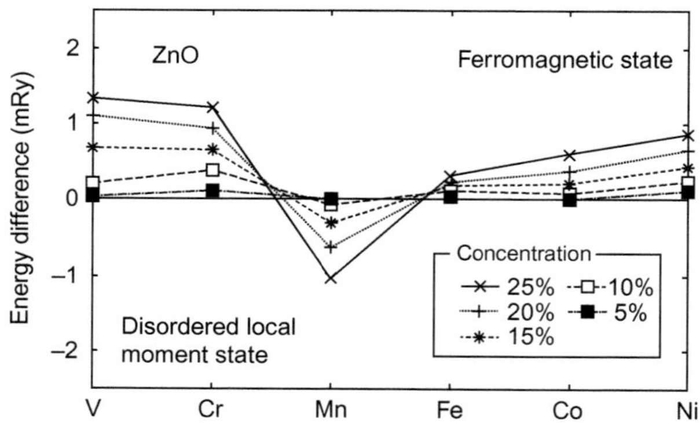

Fig. 1 Simulated values of energy difference versus the TM element dopant for ZnO crystals *(Reproduced from reference K. Sato, et al., Computational materials design of ZnO-based semiconductor spintronics, in: N.H. Hong (Ed.), Magnetism in Semiconducting Oxides, Transworld Research Network, 2007 with permission).*

The computed total energy difference per one formula unit between the ferromagnetic state and the spin-glass state for V-, Cr-, Mn-, Fe-, Co-, and Ni-doped ZnO can be seen in Fig. 1 [14]. One can see that the energy difference is positive, indicating that the ferromagnetic state is more stable than the spin-glass state. For the Mn-doped ZnO, the disordered local moment state is most stable while for V- and Cr-doped ZnO, the ferromagnetic states are more favored than the disordered local moment states.

Indeed, for DMSO, TM dopants (few at.%) are indeed quite isolated with no magnetic nearest neighbors. Thus, the usual picture of magnetism in insulators and magnetic semiconductors cannot be used to explain what really happens in DMSO systems that we will mention later in the following sections.

3 IMPORTANT FEATURES

3.1 Semiconducting oxides doped with transition metals

After the theoretical predictions in 2000, many experimental studies were done in the field of diluted magnetic semiconducting oxides (DMSO). People have made a lot of attempts to find new compounds that can be ferromagnetic at room temperature, or even better, compounds that have a large magnetic moment [4,5]. It is quite essential for devices that such room temperature ferromagnetism (FM) is indeed intrinsic, but does not originate from any dopant clusters. Because the mechanism of the induced magnetism in this system is still unclear, many research laboratories have been trying to clarify this standing issue. It seems that the magnetism in DMSO compounds does not simply result from the RKKY interaction, but originates from defects and/or oxygen vacancies that might be favored during the growth of thin films [6, 7]. Many systematic studies on TM-doped oxide thin films were done in order to justify the FM mechanism of these types of materials.

All TM-doped TiO_2, ZnO, SnO_2, and In_2O_3 films fabricated by optimal conditions could show ferromagnetic behavior at room temperature [18]. A summary of magnetization values versus element of dopant is replotted for TiO_2 and In_2O_3 systems in Fig. 2. Room temperature FM was detected by the [$M(H)$] curves taken at 300 K that demonstrate very clear hysteresis loops, and T_C is above room temperature [9, 18]. The observed magnetization is quite large in some case, for example, the greatest value as of 4.2 μ_B per

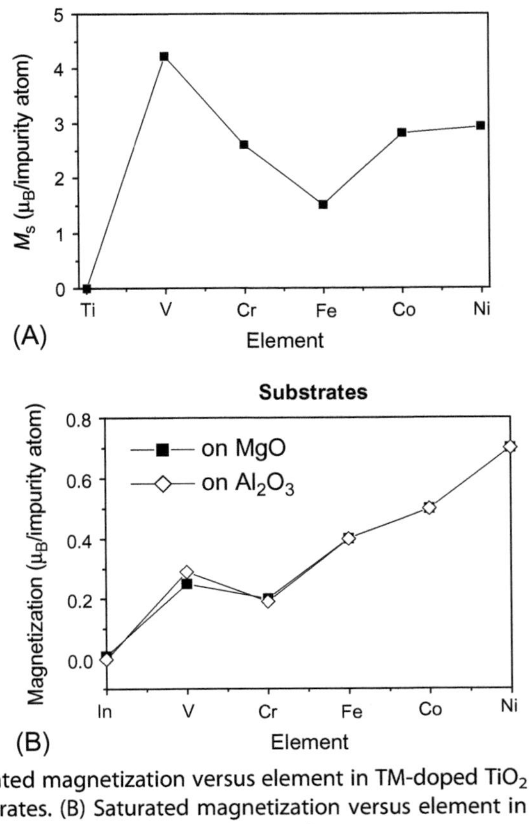

Fig. 2 (A) Saturated magnetization versus element in TM-doped TiO_2 thin films grown on $LaAlO_3$ substrates. (B) Saturated magnetization versus element in TM-doped In_2O_3 thin films grown on MgO and Al_2O_3 substrated.

V atom was found in V-doped TiO_2 films [5]. This result is worth noting because we all know that vanadium itself is paramagnetic. The observed FM, therefore, is supposed to come from the doped matrix but not vanadium clusters. A similar trend was observed for TM-doped In_2O_3 films deposited in both MgO and Al_2O_3 substrates (Fig. 2B).

This assumption has been confirmed by the magnetic force microscopy (MFM) measurements that were performed at room temperature (see atomic force microscopy (AFM) and magnetic force microscopy (MFM) data collected for Ni-doped In_2O_3 film shown in Fig. 3). The Ni-doped In_2O_3 film shows a uniform MFM image (Fig. 3B) that was significantly different from the AFM image suggesting a ferromagnetic effect that rules out the thought about clusters (Fig. 3A).

Fig. 3C shows the profiles of the topography and MFM signals recorded on the onward and backward sweepings. The solid line reflects the height of

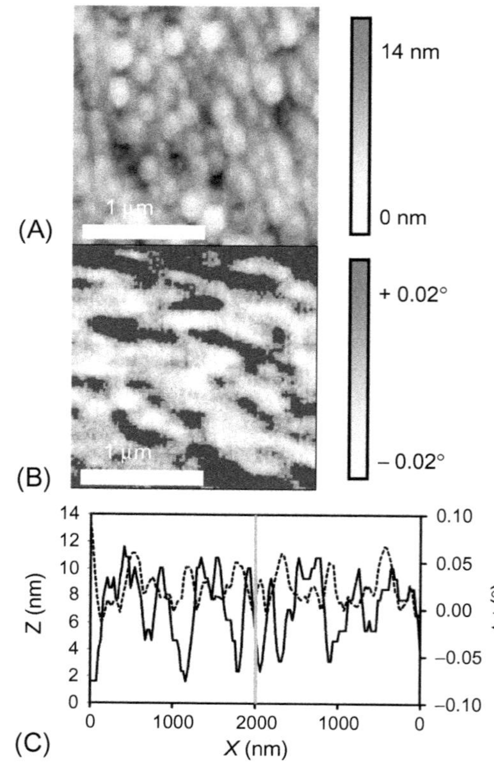

Fig. 3 AFM-MFM of Ni-doped In_2O_3 films on Al_2O_3.

the topography image while the dashed line reflects the phase of the MFM image. It is clearly seen that there is a perfect symmetry for all the peaks or, in other words, sweepings back and forth give exactly the same signals, indicating firmly that the detected signals are real magnetic signals [9].

As seen from Fig. 2, the chemical trends for TM-doped TiO_2 and In_2O_3 films look quite analogic to the theoretical prediction for TM-doped ZnO single crystals (refer to Fig. 1). However, except one point is that we have obtained FM in Mn-doped TiO_2 while, according to the latter, antiferromagnetic should be the ground state of Mn-doped ZnO [2]. This could be explained by the difference of the matrices (TiO_2 and In_2O_3 in our case) and the texture of films, which are in general different in terms of oxygen and defect formations.

Much less popular than TiO_2 and ZnO, TM-doped SnO_2 has been a research subject of fewer groups. While some groups reported about nonmagnetic Mn-doped SnO_2 [19], other teams observed room temperature FM in SnO_2 thin films doped with Co ($M_s = 7.5\,\mu_B$/Co) and with Fe ($1.8\,\mu_B$/Fe) [20, 21]. It has been interpreted that such a large magnetic moment was the consequence of unquenched orbital contributions [20].

TM-doped SnO_2 films were deposited on (001) $LaAlO_3$ (LAO), (001) $SrTiO_3$ (STO), and *R*-cut Al_2O_3 substrates with appropriate growth conditions. When only the V-doped SnO_2 films grown on Al_2O_3 substrates are paramagnetic, Cr-doped SnO_2 and Ni-doped SnO_2 films grown on three types of substrates as well as V-doped SnO_2 films on both LAO and STO substrates are well ferromagnetic at room temperature. All films have T_C above 400 K. A great magnetization of 6 μ_B per atom was obtained in Cr-doped SnO_2 films [22–24]. It is impossible to attribute this large value to Cr metal clusters because, in fact, Cr metal is paramagnetic at high temperatures and antiferromagnetic below 308 K [25]. Because the LAO substrate is diamagnetic, it is also not possible to attribute the large ferromagnetic signals of the Cr-doped SnO_2 films on LAO to the substrates. Moreover, the magnetic moment of 2–6 μ_B per impurity atom in TM-doped SnO_2 films on LAO is too large to be attributed to any precipitation. In fact, one can explain only that in these cases, the orbital quenching effects are likely absent.

Another important feature to be noticed about TM-doped SnO_2 films is their substrate effect. Films that were fabricated with the very same conditions on $SrTiO_3$ (STO) and Al_2O_3 substrates have a magnetic moment of one order smaller than that of films grown on LAO substrates. The [*M*-(*T*)] curves of TM-doped SnO_2 films on STO and Al_2O_3 showed a rise at very low temperatures, indicating a sure existence of clusters. This is because, according to Ref. [22], if there are some antiferromagnetic precipitations, their contributions could cause a reduction of the magnetic moment. The reason should be the big difference in the substrate's morphology that must strongly influence the interface between the film and its substrate. The LAO substrate has a very smooth and flat surface while the STO substrate is very rough with lots of steps on it. This can be the reason to make the film's morphology as well as the structural nature of the compound to be modified at the interface [22]. As a result, some clusters could be better formed when films were deposited on STO substrates (also similarly on Al_2O_3 substrates), in comparison with films grown on LAO substrates.

The main goal of our study on the ZnO-based system is actually not to find the observed FM but to clarify the mechanism of its FM. If compared to the TiO_2 or SnO_2 hosts that we discussed earlier, doping TMs in ZnO does not result in any huge magnetic moment (in fact, it is about one order smaller than that of TM-doped TiO_2 or TM-doped SnO_2 films). However, these compounds are seen to be very sensitive to defects and/or oxygen vacancies. Therefore, studying the ZnO case indeed might help a lot to elucidate the nature of magnetism in DMSOs.

For V-doped ZnO films, by changing the substrate temperature of only 50°C, the magnetization could be changed by one order of magnitude [26]. This then suggests that growth conditions can play a key role in tailoring the magnetic properties of TM-doped ZnO. We then have taken this advantage to deal with the most controversial case, Mn-doped ZnO. As mentioned earlier, a theoretical work claimed that antiferromagnetic should be the ground state of Mn-doped ZnO [27], and Mn doping alone cannot produce FM in the ZnO system. In order to achieve FM for ZnO, one must codope Mn with Cu [28]. Our work on Mn-doped ZnO showed that oxygen vacancies could play a more crucial role than that of additional carriers [29]. It was found that doping Mn alone does not result in room temperature FM if inappropriate conditions were applied. However, by applying appropriate ones, it should turn to become possible. The substrate temperature and oxygen pressure during the growth process indeed could create necessary defects and/or oxygen vacancies that act similarly to an *n*-type doping. This hypothesis is in good agreement with the explanations for the magnetism in HfO_2 films [6], which was later confirmed by a theoretical work supposing that vacancies can be necessary ingredients to create additional bands inside the semiconducting gap that is responsible for such FM [30]. Our work on Cr-doped ZnO films has shown that an oxygen annealing could certainly improve the crystallinity of films, but at the same time, it might destroy their ferromagnetic ordering [7]. In this system, a perfect crystallinity does not go along with FM, and filling oxygen vacancies enormously would degrade ferromagnetic ordering. Indeed, defects and oxygen vacancies must certainly play a key role in tuning FM in DMSO [7].

The question about an intrinsic nature of FM in DMSO thin films discussed above has suggested that we reexamine the role that a 3*d* element dopant indeed plays. X-ray absorption spectroscopy (XAS) and X-ray magnetic circular dichroism (XMCD) measurements on Cr-, Mn-, and Co-doped TiO_2 films were carried out at the Cr, Mn, and Co $L_{2,\ 3}$ edges. Typical normalized XMCD data for the Mn L_3 edges of a TiO_2 film doped with 2% of Mn are shown in Fig. 4.

It is obviously seen that the field dependence of the XMCD is certainly paramagnetic. A very similar feature was also observed for Co- and Cr-doped TiO_2 films [31]. Our results have revealed that the main contribution to the ferromagnetic signal basically comes from the TiO_2 host matrix but not the TM doping. Very similar results were also reported for Co-doped ZnO [32]. The Co contribution to the magnetic properties of Co-doped ZnO is indeed paramagnetic. Or, one can say that indeed the observed FM in ZnO does not come from the RKKY interaction. These facts explain very well why the Curie temperature in DMSO, for example, does not depend much on the type and concentration of dopant. This

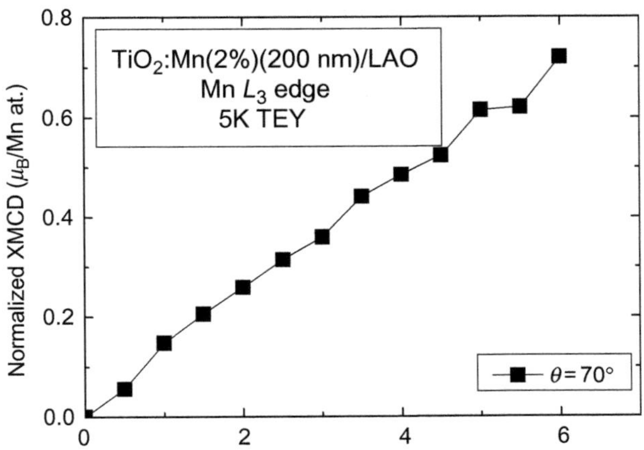

Fig. 4 Normalized XMCD data for a 200 nm-thick $Mn_{0.02}Ti_{0.98}O_2$ film grown on $LaAlO_3$ substrate.

finding has suggested that new theories are needed to rightly explain the mechanism of the DMSO family.

TM-doped TiO_2, SnO_2, In_2O_3, and ZnO thin film fabricated with appropriate growth conditions could be ferromagnetic at room temperature, and in many cases, even a very large magnetic moment could be obtained. This finding could introduce many potential candidates for spintronic applications, and in a more fundamental aspect, they can help us to understand the mechanism of these materials. One should note that for nanostructured materials such as thin films and nanoparticles, besides the RKKY interaction that might play some certain role in inducing magnetism in these compounds, there must be another important source for magnetism such as defects and/or oxygen vacancies. We must answer whether the TM doping is indeed important for introducing FM into nonmagnetic semiconducting oxides, or TM doping in fact just acts as some catalyst to add some paramagnetic contribution to the magnetic ordering already existing in the host.

3.2 Ferromagnetism Due to Defects and Oxygen Vacancies in Nanometer-Sized Pristine Semiconducting Oxides

The observed FM in HfO_2 thin films grown on sapphire or silicon substrates was first reported in 2004 [6]. This unexpected finding attracted a lot of attention in the materials community about a new phenomenon; later on, it was usually called "d^0 magnetism."

Indeed, the thin film structure should be a key point here because defects and/or oxygen vacancies that are favored during the film deposition can actually become a magnetic source. Oxygen vacancies can indeed play a role that is similar to an *n*-type doping. Following this suggested angle, many experiments have shown that defects can tailor the magnetic properties of DMSO thin films. It was shown that defects could intentionally introduce FM into the ZnO system [33]. In some other cases, it was also discovered that a good crystallinity could significantly degrade the ferromagnetic ordering [7]. Moreover, it was also found that filling oxygen vacancies might enormously reduce the magnetic moment in DMSO films [7]. Some simulation on HfO_2 showed that isolated cation vacancies in HfO_2 could form high-spin defect states, and as a result, they could be ferromagnetically coupled with a rather short-range magnetic interaction leading to a ferromagnetic ground state [30]. We showed that FM could be indeed obtained in HfO_2 thin films grown on yttrium stabilized zirconia (YSZ) substrates [8]. But we also noted that the XMCD experiments performed on HfO_2 films do not show any magnetic signal on the Hf site. All these controversial issues have suggested that we experimentally investigate the magnetic properties of several types of undoped semiconducting oxides. We will surely confirm that pristine oxides could be room temperature ferromagnetic in thin films and nanoparticles. And more importantly, TM doping in fact does not play any crucial role in inducing FM in those semiconducting oxides.

Pulsed-laser ablated TiO_2, HfO_2, ZnO, In_2O_3, and SnO_2 films were deposited from sol-gel made-ceramic targets on (100) LAO, (100) YSZ, *C*-cut Al_2O_3, (001) MgO, and (100) LAO substrates, respectively. The typical thickness is 220 nm for TiO_2, HfO_2, and SnO_2 films, 600 nm for In_2O_3 films on MgO, and 375 nm for ZnO films on Al_2O_3. Magnetic moment data were basically collected when the magnetic field was applied parallel to the film plane. In Fig. 5, it is shown that pristine oxide films can be ferromagnetic at 300 K. For HfO_2 and TiO_2 films, magnetization is rather large (M_s is about 30 emu/cm^3 for HfO_2 and 20 emu/cm^3 for TiO_2), but for In_2O_3 films on MgO and ZnO films on Al_2O_3, it is about one order less (i.e., only about a few emu/cm^3) [8]. Harris' group also published similar results for TiO_2 films [34]. Our TiO_2 films made by another method (spin coating) also showed FM at room temperature [35]. Certainly we can ensure that the observed phenomenon is not at all mistaken. It is impossible to attribute such a huge value of the magnetic moment in the case of HfO_2 and TiO_2 films to any type of impurity. For the TiO_2 case, neither Ti^{4+} nor O^{-2} is magnetic. Also for the HfO_2 case, neither Hf^{4+} nor O^{-2} is magnetic. At first glance, one may think that it should be due to impurities. Concerning the purity of the targets, we can say confidently that such a possibility is very small because impurities of less than 10^{-2} wt.% could not result in such great magnetic

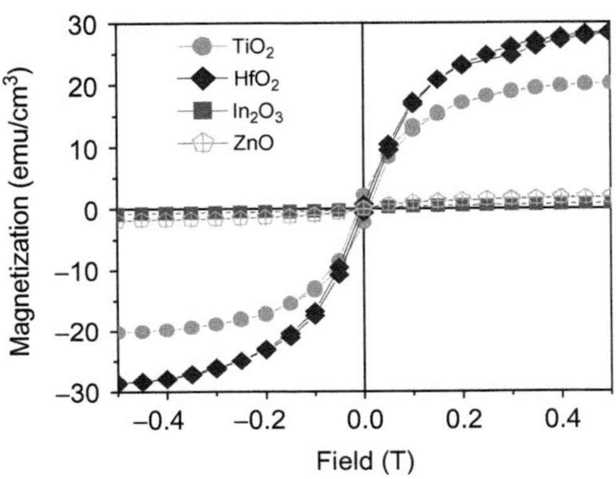

Fig. 5 *M-H* curves for undoped oxide films.

moments. From structural properties of the grown films, it is seen that there is no trace of any alien phase that could be seen from X-ray diffraction patterns [23].

One should notice that In_2O_3 films grown on MgO are room temperature ferromagnetic but with a quite modest magnetic moment. There is another important feature here to be noticed, and that is that In_2O_3 films grown on Al_2O_3 substrates with the same sets of conditions are diamagnetic [8]. There has been no report so far about magnetic In_2O_3 because In^{3+} could not be the source of magnetism. Even though In_2O_3 has a tendency to be associated with oxygen vacancies [36], the fact that FM is observed on only one type of substrate strongly implies some kind of defects that might induce magnetism. Similarly, in the ZnO case, neither Zn^{2+} nor O^{2-} is magnetic; therefore, there should be no magnetic source in pure ZnO, if one just thinks based on conventional concepts.

Concerning SnO_2 films, one cannot attribute the observed FM to any dopant, and additionally, there is no 3*d* electron involved. Therefore, there is no source for any magnetic interaction to exist, in principle. SnO_2 films of some other group are diamagnetic while ours are well ferromagnetic. As for SnO_2 nanoparticles, the results are different as well: Hays et al. reported that their nanoparticles of SnO_2 are nonferromagnetic [37] while some of Rao's group have confirmed that their SnO_2 nanoparticles were a mixture of ferromagnetic and paramagnetic phases [12]. Our explanation for the observed FM in our undoped semiconducting oxide films is that, most likely, oxygen vacancies formed during the growth are an essential factor, along with the confinement effects in nanosized structures. However, growth conditions and how to make them precisely controllable must be a real issue to solve. Ref. [12] stated that thermal treatments could drastically influence the magnetic properties of SnO_2 nanoparticles. This can explain why, in one case, the SnO_2 films could be strongly ferromagnetic, but in the other case [12], the paramagnetic phase is more dominant.

The possible formation of oxygen vacancies near the surface of SnO_2 thin films from the oxygen *K*-edge was investigated by X-ray emission and absorption. It was found that the distribution of O-2*p* unoccupied states for ferromagnetic SnO_2 thin films is completely different from that of the SnO_2 films postannealed in oxygen that are diamagnetic. This observation suggests that oxygen vacancies should be the main source of magnetism in SnO_2 thin films. This possibility was then confirmed by obtaining the lowest energy levels of the structural defects (impurities or neutral vacancies) with two localized carriers near the surface of the SnO_2 film using a quantum-mechanical approach combined with the image charge method. It was found that a magnetic triplet state is the ground state of those defects around the SnO_2 surface, whereas the nonmagnetic singlet is the ground state of the SnO_2 bulk [11].

Similar results were reported for FM in CeO_2 and MgO [11, 13]. We then can conclude that pristine semiconducting oxide materials can be ferromagnetic if:

(1) They are nanostructured thin films or nanoparticles.
(2) Defects and/or oxygen vacancies exist in the systems.

In Fig. 6, it is shown that bulks of TiO_2, HfO_2, ZnO, and In_2O_3 (pieces cut from ceramic targets from which films were made) are diamagnetic. It implies that indeed FM was induced only in the thin film form, or in other words, it is unique for low-dimension structures (also seen in Ref. [12] for nanoparticles and Ref. [38] for films made by spin coating). One should assume that confinement effects should play some role here. We see from Fig. 7 that the ultrathin film such as 5 nm thick has a much larger saturated magnetization than the 200 nm thick one. It suggests that besides the reason of confinement effects in low-dimension systems, the observed FM must be induced by defects and/or oxygen vacancies that are mostly formed at and near to the surface and/or interface (between the film and insulating substrate).

The assumption about FM due to defects and oxygen vacancies is reinsured by the data shown in Fig. 8. Fig. 8 shows *M-T* curves for the as-grown HfO_2 film as well as for the HfO_2 postannealed films in an oxygen atmosphere, and also of those films after being vacuum heated. While heating in vacuum likely does not significantly change the magnetic moment of

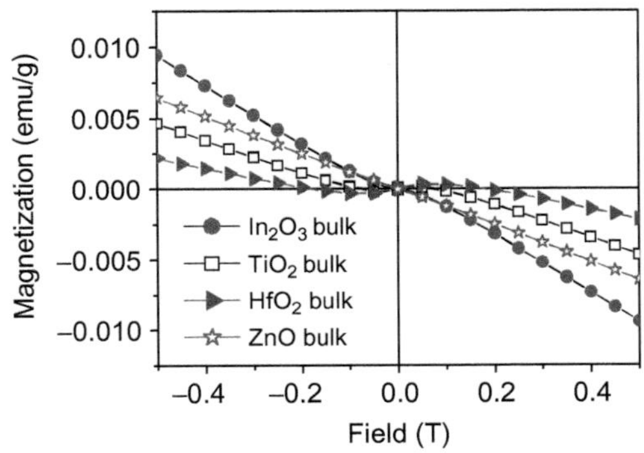

Fig. 6 *M-H* of pristine bulk oxides.

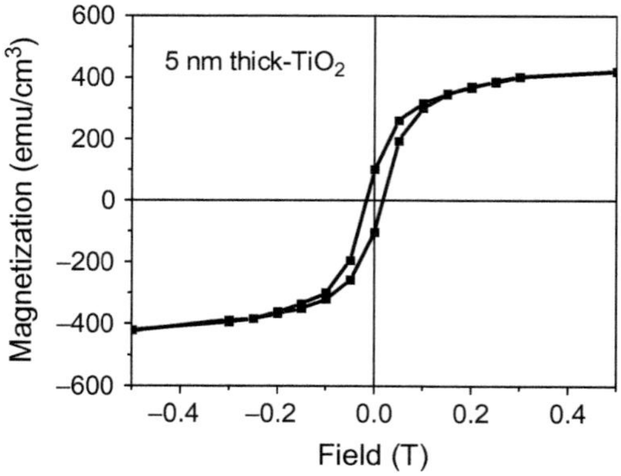

Fig. 7 *M-H* curve for a 5 nm thick TiO_2 film taken at 300 K.

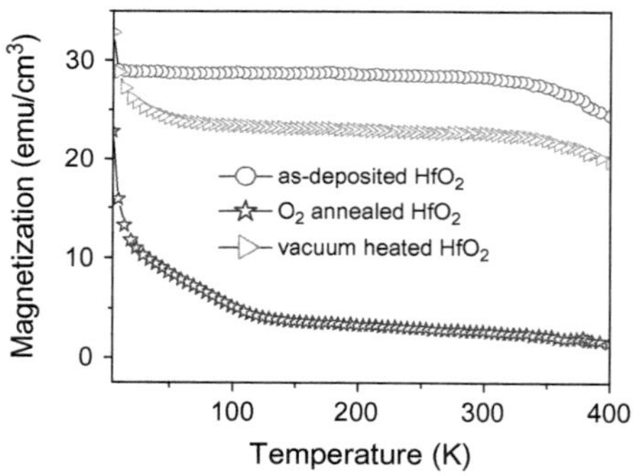

Fig. 8 *M*(*T*) data taken at 0.5 T of as-grown, oxygen-annealed, and vacuum-heated HfO_2.

the deposited films (i.e., it does not easily remove oxygen from the system), the annealing vacuum degrades the magnetism of the undoped film drastically. Filling up oxygen vacancies could degrade the ferromagnetic ordering of the films enormously. It seems that the observed FM is directly related to oxygen vacancies because when the oxygen vacancies could be filled, the magnetic moment is obviously reduced [8].

For a deeper study, the chemical and orbital selectivity of XAS and XMCD measurements were carried out to prove that the observed FM in laser-ablated TiO_2 films is intrinsic. It was found that the ferromagnetic signals directly originate from the O-2*p* and, to a lesser extent, from the Ti-3*d* electrons. This FM stems from a surface region of a thickness of several nanometers, which is known to be rich in oxygen vacancies [38].

In order to explain our experimental results on nanoscaled pristine oxides, Huong proposed a model of FM for HfO_2 and TiO_2 thin films to investigate the possibility of magnetism due to oxygen vacancies and confinement effects [36]. Due to the special structure of HfO_2 and TiO_2, each oxygen atom is surrounded by three or four Hf (or Ti) atoms and the local symmetry is D_{3h} and D_{4h}, accordingly. An oxygen vacancy would result in the losing of bonding of two *d*-electrons in the outer shell of the Hf (Ti) atom. It was assumed that in the local D_{3h} or D_{4h} molecular orbital high symmetry, the two electrons do not really leave their own shell and remain as the *d*-electrons, then becoming a d^2 impurity center. The result of the exchange interaction of these two *d*-electrons with each other, and with the molecular orbital field, will be a splitting of the energy level of the impurity band around the vacancy, and as a consequence, a large magnetic moment could be achieved. Besides the orbital local symmetry, the electrons are confined in the two-dimensional confinements. The thin film configuration should have two effects: in one way, it enhances the formation of the oxygen vacancy and in another way, it enhances the coupling between the *d*-electrons and the interaction between those electrons and the local field. The boundary conditions at the surfaces of the thin film in the strong confinement approximation make the matrix elements for the exchange interaction much larger in comparison with the bulk case [36].

The tight binding calculation was applied for HfO_2, TiO_2, and In_2O_3 thin films to consider the effect of the molecular orbital field and the impurity band created by the exchange interaction of the electrons created by the oxygen vacancy and trapped around the vacancy (the 2D confinement effects of thin films were well taken into account). The splitting of energy levels of this impurity band has been obtained and the 2D confinement results in a large exchange interaction matrix element, which is very different in comparison with the case of bulks. A high spin state with a magnetic moment per vacancy of $3.18\,\mu_B$ for TiO_2, $3.05\,\mu_B$ for HfO_2, and $0.16\,\mu_B$ for In_2O_3 had been found. If supposing the vacancy concentration is about 3%, the theoretical magnetic moment will be completely comparable to the experimental values. According to this model, FM is very possible for thin films of pristine semiconducting oxides, especially for the configurations that would favor oxygen vacancies (thin films and nanoparticles) [36].

In many oxides, if the observed FM is due to defects, then it would be very difficult to make them controllable for applications. Therefore, the search for a more suitable candidate is always continued. So far, there is only one type of oxide that shows an intrinsic induced FM: cubic Mn-doped ZrO_2 films.

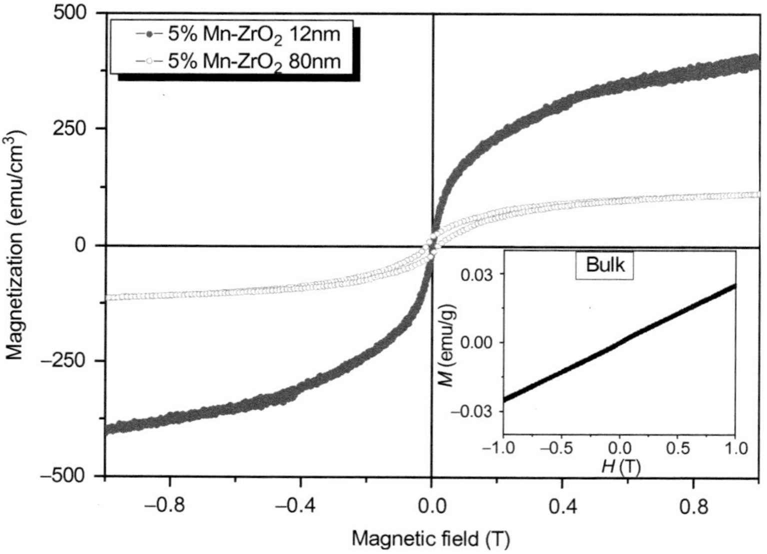

Fig. 9 *M-H* curves for Mn-doped ZrO_2 films with different thicknesses. The inset shows *M-H* indicating paramagnetic behavior of the corresponding target.

Undoped Mn-doped ZrO_2 film is paramagnetic [39], and doping Mn really introduces FM into the ZrO_2 host (see Fig. 9). One cannot blame that FM in Mn-doped ZrO_2 is due to defects because it shows clearly that it comes only after Mn doping [39]. One can see from the inset of Fig. 9 also that the Mn-doped ZrO_2 bulk is not ferromagnetic, and the FM exists only in thin film form.

To be able to generalize this case, we have investigated the magnetic properties of Fe/Co/Ni-doped ZrO_2 laser-ablated thin films in comparison with the known results of Mn-doped ZrO_2. It is found that doping with a TM can really induce room temperature ferromagnetism in ZrO_2. Density functional theory (DFT) simulations have confirmed the experimental data by showing that the magnetic moments of Mn- and Ni-doped ZrO_2 thin films are much larger than that of Fe- or Co-doped ZrO_2 thin films. Most importantly, our theoretical results have confirmed that Mn- and Ni-doped ZrO_2 show a ferromagnetic ground state in comparison to Co- and Fe-doped ZrO_2, which favor an antiferromagnetic ground state (see a summary in Table 1). This might give an important guide for material research [40].

Experimental results have shown that FM could certainly be obtained in pristine low-dimensional semiconducting oxides. The observed FM is likely

Table 1 Total magnetic moment, magnetic moment per atom, energy, and coupling state for 6.25% Mn/Fe/Co/Ni-doped ZrO_2

(6.25%)	μ_{tot} (μ_B)	μ_{Mn} (μ_B)	μ_{Zr} (μ_B)	μ_O (μ_B)	ΔE (eV)	Coupling
$ZrMnO_2$	3.010	3.620	0.011	−0.090	0.141	FM
$ZrFeO_2$	4.010	3.940	0.010	−0.004	−0.162	AFM
$ZrCoO_2$	3.262	3.042	0.007	0.028	−0.232	AFM
$ZrNiO_2$	3.236	1.861	−0.004	0.101	0.034	FM

due to oxygen vacancies and/or defects. This assumption is strongly confirmed by XMCD measurements. Our theoretical model also suggests that confinement effects should play a crucial role in tailoring the magnetic properties of low-dimensional materials.

This important finding suggests a new direction to search for potential materials for spintronic applications. By downscaling semiconducting oxides to nanosize and by applying suitable conditions that may favor oxygen vacancies/defects, room temperature FM can be achieved. And as a result, spin and charge could be exploited simultaneously in one device. Making these properties controllable is still very challenging, but it should be a must if we want use them for spintronic applications.

4 CONCLUSION AND PERSPECTIVES

Investigations on DMSO systems have been performed by many research groups around the world, showing many interesting results but also raising many difficult questions. Standing issues include:

(1) Different laboratories have applied different growth conditions and techniques that result in samples with different qualities such as different concentrations of defects and oxygen vacancies, different conductivities (*n*-type or *p*-type, insulating or conductive), different homogeneity (clusters or not), etc.

(2) Role of dimensionality (e.g., bulks, films, and particles behave differently).

It is sure that that RKKY interaction cannot be the main cause for the induced FM in DMSO systems, and defects should play a key role here. However, even though it is very interesting to study, in reality, in order to bring DMSO into industries, we need to clarify whether those defects could be stabilized and kept in control because otherwise, they would be meaningless for applications. Interestingly, research on very thin films and nanoparticles of DMSO has shown that downscaling these materials to the nanometer scale can be a right method in order to turn DMSO from being diamagnetic (as in bulks) to being ferromagnetic (as in 2D, 1D, or 0D structures). It wisely suggests an easy way to exploit the wonderful facets of the nanoworld for spintronics. The DMSO research still requires a lot of effort from both experimentalists and theorists to work together toward a higher level in order to be able to apply this for spintronic devices.

ACKNOWLEDGMENTS

The author would like to thank her ex-research team in Tours (France) and her team at Nanomag Lab at Seoul National University as well as other collaborators for their cowork that led to our results in the DMSO field. The work on ZrO_2 was supported by project 3348-20100041 of the National Research Foundation of Korea. The chapter is supported financially by grant 2017060271 of the National Research Foundation of Korea.

REFERENCES

[1] T. Dietl, H. Ohno, F. Matsukura, J. Cibert, D. Ferrand, Science 287 (2000) 12019.
[2] K. Sato, H. Katayama-Yoshida, Jpn. J. Appl. Phys. Part 2 39 (2000) L555.
[3] Y. Matsumoto, M. Murakami, T. Shono, T. Hasegawa, T. Fukumura, M. Kawasaki, P. Ahmet, T. Chikyow, S. Koshihara, H. Koinuma, Science 291 (2001) 854.
[4] N.H. Hong, J. Mag. Mag. Mater. 303 (2006) 338.
[5] N.H. Hong, S. Sakai, A. Hassini, Appl. Phys. Lett. 84 (2004) 2602.
[6] M. Venkatesan, C.B. Fitzgerald, J.M.D. Coey, Nature 430 (2004) 630.
[7] N.H. Hong, J. Sakai, N.T. Huong, N. Poirot, A. Ruyter, Phys. Rev. B 72 (2005) 45336.
[8] N.H. Hong, J. Sakai, N. Poirot, V. Brizé, Phys. Rev. B 73 (2006) 132404.
[9] N.H. Hong, J. Sakai, N.T. Huong, A. Ruyter, V. Brizé, J. Phys. Condens. Matter 18 (2006) 6897.
[10] N.H. Hong, N. Poirot, J. Sakai, Phys. Rev. B 77 (2008) 33205.
[11] G.S. Chang, J. Forrest, E.Z. Kurmaev, A.N. Morozovska, M.D. Glinchuk, J. A. McLeod, A. Moewes, T.P. Surkova, N.H. Hong, Phys. Rev. B 85 (2012) 165319.
[12] A. Sundaresan, B. Bhagravi, N. Rangarajan, U. Siddesh, C.N.R. Rao, Phys. Rev. B 74 (2006) 161306(R).
[13] C. Martínez-Boubeta, J.I. Beltrán, L. Balcells, Z. Konstantinović, S. Valencia, D. Schmitz, J. Arbiol, S. Estrade, J. Cornil, B. Martínez, Phys. Rev. B 82 (2010) 024405.
[14] K. Sato, et al., Computational materials design of ZnO-based semiconductor spintronics, in: N.H. Hong (Ed.), Magnetism in Semiconducting Oxides, Transworld Research Network, India, 2007.
[15] K. Sato, P.H. Dederichs, H. Katayama-Yoshida, J. Kudrnovsky, J. Phys. Condens. Matter 16 (2004) S5491.
[16] K. Sato, P.H. Dederichs, H. Katayama-Yoshida, Europhys. Lett. 61 (403) (2003) 2003.
[17] A. Zunger, Solid State Phys. 39 (1986) 276.
[18] N.H. Hong, J. Sakai, W. Prellier, A. Hassini, A. Ruyter, F. Gervais, Phys. Rev. B 70 (2004) 195204.
[19] H. Kimura, T. Fukumura, M. Kawasaki, K. Inaba, T. Hasegawa, H. Koinuma, Appl. Phys. Lett. 80 (2002) 94.

[20] S.B. Ogale, R.J. Choudhary, J.P. Buban, S.E. Lofland, S.R. Shinde, S.N. Kale, V. N. Kulkarni, J. Higgins, C. Lanci, J.R. Simpson, N.D. Browning, S. Das Sarma, H. D. Drew, R.L. Greene, T. Venkatesan, Phys. Rev. Lett. 91 (2003) 77205.
[21] J.M.D. Coey, A.P. Douvalis, C.B. Fitzgerald, M. Venkatesan, Appl. Phys. Lett. 84 (2004) 1332.
[22] N.H. Hong, J. Sakai, W. Prellier, A. Hassini, J. Phys. Condens. Matter 17 (2005) 1697.
[23] N.H. Hong, A. Ruyter, W. Prellier, J. Sakai, N.T. Huong, J. Phys. Condens. Matter 17 (2005) 6533.
[24] N.H. Hong, J. Sakai, Physica B 358 (2005) 265.
[25] C. Kittel, Introduction to Solid State Physics, seventh ed., John Wiley, New York, 1996.
[26] N.H. Hong, J. Sakai, A. Hassini, J. Phys. Condens. Matter 17 (2005) 199.
[27] Q. Wang, Q. Sun, B.K. Rao, P. Sena, Phys. Rev. B 69 (2004) 233310.
[28] N.A. Spaldin, Phys. Rev. B 69 (2004) 125201.
[29] N.H. Hong, V. Brizé, J. Sakai, Appl. Phys. Lett. 86 (2005) 82505.
[30] D.P. Pemmaraju, S. Sanvito, Phys. Rev. Lett. 94 (2005) 217205.
[31] N.H. Hong, A. Barla, J. Sakai, N.Q. Huong, Phys. Status Solidi C 4 (2007) 4461.
[32] A. Barla, G. Schmerber, E. Beaurepaire, A. Dinia, H. Bieber, S. Colis, F. Scheurer, J.-P. Kappler, P. Imperia, F. Nolting, F. Wilhelm, A. Rogalev, D. Muller, J.-J. Grob, Phys. Rev. B 76 (2007) 125201.
[33] D.A. Schwartz, D.R. Gamelin, Adv. Mater. 16 (2004) 2115.
[34] S.D. Yoon, Y. Chen, A. Yang, T.L. Goodrich, X. Zuo, D.A. Arena, K. Ziemer, C. Vittoria, V.G. Harris, J. Phys. Condens. Matter. 18 (2006) L355–L361.
[35] A. Hassini, J. Sakai, J.S. Lopez, N.H. Hong, Phys. Lett. A 372 (3299) (2008) 2008.
[36] N.Q. Huong, Magnetism due to oxygen vacancies in undoped oxide thin films, in: N. H. Hong (Ed.), Magnetism in Semiconducting Oxides, Transworld Research Network, India, 2007.
[37] J. Hays, A. Punnoose, R. Baldner, M.H. Engelhard, J. Peloquin, K.M. Reddy, Phys. Rev. B 72 (2005) 075203.
[38] Barla A., Hong N.H. et al., Unpublished XMCD Results on TiO_2 System (2006–2007).
[39] N.H. Hong, C.-K. Park, A.T. Raghavender, A. Ruyter, E. Chikoidze, Y. Dumont, J. Mag. Mag. Mater. 324 (2012) 3013.
[40] N.H. Hong, M.B. Kanoun, S. Goumri-Said, J.-H. Song, E. Chikoidze, Y. Dumont, A. Ruyter, M. Kurisu, J. Phys. Phys. Condens. 25 (2013) 436003.

CHAPTER 8

How to Make Zinc Ferrites Become Ferromagnetic?

A.T. Raghavender[*,†,‡]
[*]Department of Physics, Electronics, and Communication Engineering, Nishitha College of Engineering and Technology, Hyderabad, India
[†]Abhigyaan Labs Private Limited, Hyderabad, India
[‡]Aksharaarka Enterprise Private Limited, Hyderabad, India

1 INTRODUCTION

Nanotechnology and nanoscience have been considered as rapidly emerging fields for the past two decades. They have unique, unusual, unexpected, unimaginable, and superior properties compared to their bulk counterparts. Due to their unique properties, the physical and chemical properties were extensively investigated both experimentally and theoretically [1–7]. At nanoscale, the quantum size effects are strongly influenced by core and surface area [7, 8]. Due to new and highly sophisticated measurement techniques, growth conditions, and tools, it has become easy to understand the peculiar behavior of nanomaterials. These materials have potential applications for creating and fabricating new nanoscale devices. Also, it was easy to grow artificial materials with the controlled properties using different chemical compositions.

The unexpected magnetism in a few materials is due to nanosize of particles, thin films, and single crystals that induces changes in the material. When the dimension of the material is reduced, it results in the reduction of a coordinating number of atoms, which in turn reduces the electrons hopping from one site to another. Therefore, the bandwidth reduces and increases the coulomb interaction. This results in the tendency of electrons to enhance in making the magnetism to appear at the nanoscale [9–19]. The surface interface layer reduction is due to reduced symmetry and boundary conditions. They play important roles in observing the magnetic properties in the materials whose dimensions are reduced. The defects due to additives may create vacancies in the magnetic material. These vacancies play important roles in observing collective magnetic properties at the nanoscale [20–22].

Nano-sized Multifunctional Materials
https://doi.org/10.1016/B978-0-12-813934-9.00008-6

Among all the spinel ferrites, $ZnFe_2O_4$ was observed to have interesting properties. Spinel $ZnFe_2O_4$ crystallizes into two main cations with 8 tetrahedral (A) sites and 16 octahedral [B] sites. The general cation distribution can be written as $(Zn)[Fe_2]O_4$. Among the nanostructured $ZnFe_2O_4$ spinel ferrites, several researchers have reported a peculiar nature in this material [23–27]. The stable phase of $ZnFe_2O_4$ has the normal spinel structure with cation distribution $(Zn^{2+})_A[Fe_2^{3+}]_BO_4$ for the FCC structure. In the stable phase, Zn^{2+} has strong tendency to occupy only (A) sites. It is very well known that $ZnFe_2O_4$ will have a mostly normal spinel structure. This ferrite behaves like an antiferromagnetic material with a Néel temperature around 10 K. Therefore, $ZnFe_2O_4$ was considered to have short-range magnetic ordering. The long-range magnetic ordering in $ZnFe_2O_4$ can be achieved only around 1.5 K. The magnetic transitions at a low temperature for $ZnFe_2O_4$ are due to the weak superexchange interactions between Fe^{3+} ions at [B] sites. To induce magnetism into nonmagnetic $ZnFe_2O_4$ or to change its magnetic structure, several researchers have adopted nonequilibrium techniques/methods such as coprecipitation, ball milling, sputtering, rapid quenching, ion irradiation, the pulsed laser deposition technique (PLD), molecular beam epitaxy (MBE), hydrothermal, glycene as fuel, a self-propellant method, forced hydrolysis using a polyol medium, a self propagating low-temperature combustion method, an oxalic acid route, a nonaqueous solution synthesis, a liquid phase precipitation method, electrospinning, the lacto ferrin-assisted method, sol-gel, etc. All these methods/techniques induce the metastable phase in $ZnFe_2O_4$. Thus, the disorder arrangements of Zn^{2+} and Fe^{3+} at the (A) and [B] sites can make $ZnFe_2O_4$ to become ferromagnetic at room temperature.

The main objective of this work is to explore the different magnetic properties of $ZnFe_2O_4$ spinel ferrites so that we can understand how to make this nonmagnetic material become ferromagnetic at room temperature.

2 ZNFE$_2$O$_4$ NANOMATERIALS: MAGNETIC PROPERTIES

The $ZnFe_2O_4$ nanoparticles were synthesized using the sol-gel method [28–30]. The XRD patterns of the $ZnFe_2O_4$ annealed samples from 500°C to 1000°C are presented in Fig. 1. The XRD patterns indicate the formation of a single-phase spinel structure without any secondary phases. It is observed that with increasing annealing temperature, the XRD peaks became sharp and the full width half maximum (FWHM) decreased. The decrease in the FWHM is the evidence of enhanced particle size with

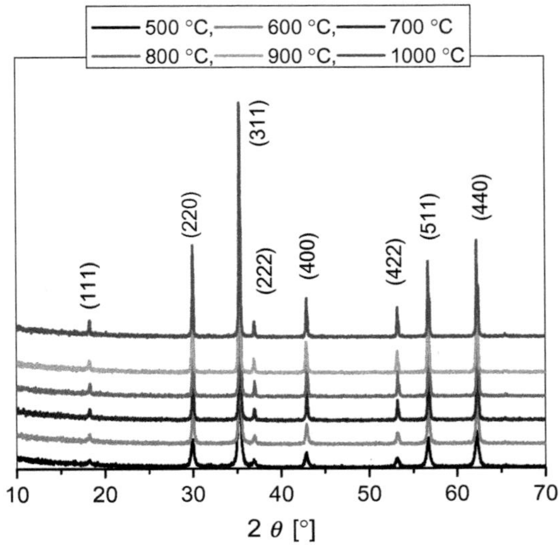

Fig. 1 X-ray diffraction patterns of sol-gel synthesized $ZnFe_2O_4$ nanoparticles annealed at different temperatures.

the annealing temperature. The lattice parameters were slightly smaller than the standard value of 8.44 Å. A close examination of Fig. 1 reveals that the intensity of the diffraction peaks increases and the peaks shift toward the bigger angles. To find any inversion of cations taking place in $ZnFe_2O_4$, the study used the intensity ratios of 220/400 lines for calculations. The intensities ratio increased with increasing annealing temperature. This signifies that there is no inversion of cations. Therefore, these samples correspond to normal spinel ferrites. This kind of behavior was observed for nanosize $ZnFe_2O_4$ with an increasing milling time [31]. When the inversion parameter increases in $ZnFe_2O_4$ nanoparticles, a transition from nonmagnetic to ferromagnetic will take place. Due to inversion, the redistribution of Fe^{3+} and Fe^{2+} ions and oxygen vacancies take place, therefore, creating superexchange interactions and the spin disorder.

The annealing temperature has a significant effect on the $ZnFe_2O_4$ nanoparticles. This is directly related to the crystallization of the nanoparticles. A straight line of ln (D) against $1/T$ (Fig. 2) is plotted according to the Scott equation (1) for homogeneous growth rate conditions for nanocrystallites [32, 33]. The Scott equation approximately describes the growth rate of nanocrystallites from the thermal treatment of the amorphous compound.

$$D = C\exp(-E/RT) \tag{1}$$

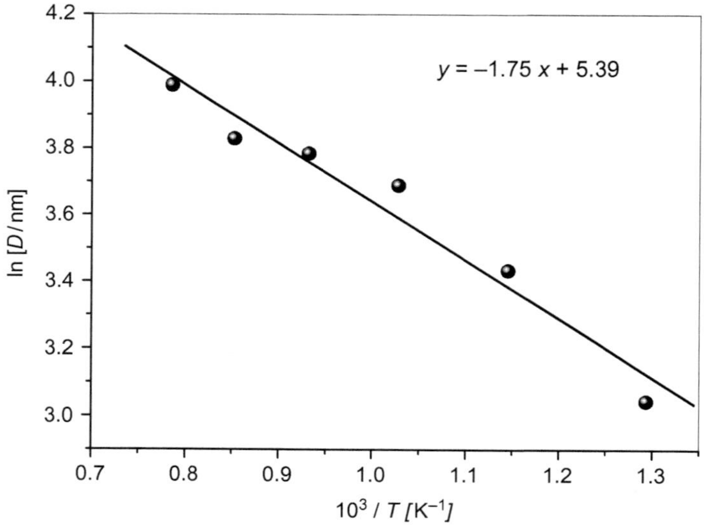

Fig. 2 Plot of ln (*D*) against 1/*T*. Line is fit to equation $D = C\exp(-E/RT)$ for $ZnFe_2O_4$ nanoparticles annealed at different temperatures.

where D is the XRD crystallite size, C is a constant, E is the activation energy per mole for crystal growth, R is the ideal gas constant, and T is the absolute temperature.

There is a good linear relationship between ln (D) and $1/T$. The E values could be calculated from the slope of the ln (D) vs. $1/T$ line as $E = 19.9$ kJ/mol. It can be considered that the crystallite grows primarily by means of an interfacial reaction. Linear dependence also shows that the growth of $ZnFe_2O_4$ nanocrystallites is easily affected by the annealing temperature. The activation energy values for these samples were observed to be $E = 19.9$ kJ/mol. This value is nearly equal to 18.5 kJ/mol for $ZnFe_2O_4$ prepared by ball milling [33].

The *M-H* curves for $ZnFe_2O_4$ nanoparticles were measured at room temperature with a maximum applied field of 10 kOe, as shown in Fig. 3. The 500°C annealed $ZnFe_2O_4$ nanoparticles showed superparamagnetic behavior at room temperature. The influence of the annealing temperature on the magnetic properties of $ZnFe_2O_4$ is clearly observed in Fig. 3. Except the 500°C annealed $ZnFe_2O_4$ sample, all other samples showed a clear paramagnetic behavior with a linear curve [34, 35]. It was reported that $ZnFe_2O_4$ is antiferromagnetic at room temperature and the spin disorder is not homogeneous. Therefore, there is no additive magnetic contribution at room temperature [36]. Fig. 4 shows the magnetic properties of $ZnFe_2O_4$

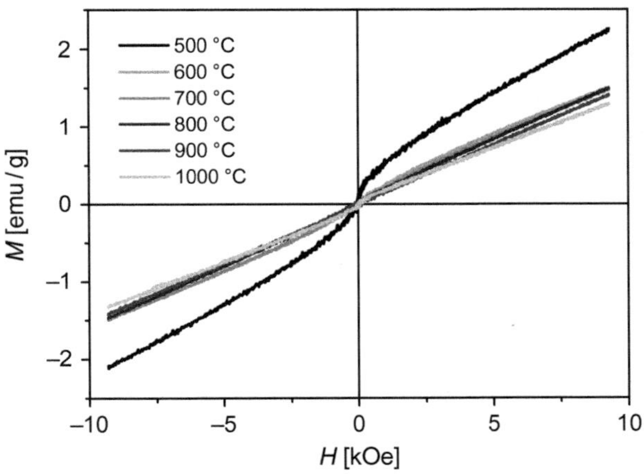

Fig. 3 *M-H* curves for $ZnFe_2O_4$ nanoparticles annealed at different temperatures.

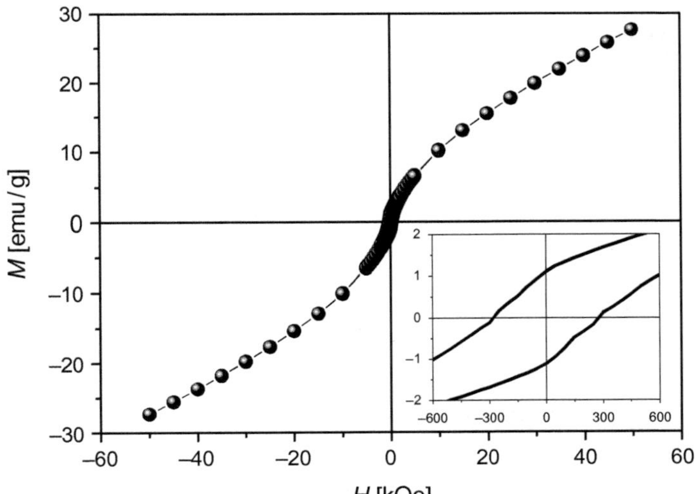

Fig. 4 *M-H* curves for $ZnFe_2O_4$ samples measured at 5 K with applied field of 50 kOe. The inset shows the expanded lower field curves.

nanoparticles annealed at 500°C. This sample is measured at 5 K by applying the magnetic field of 50 kOe. These *M-H* curves show a typical superparamagnetic behavior with no hysteresis while the remanence and coercivity are almost zero [34–38]. The maximum magnetization value of 1.3 emu/g was observed at room temperature and 28 emu/g at 5 K. These values are considerably in agreement with $ZnFe_2O_4$ samples synthesized by other methods. The transformation from the paramagnetic to the ferromagnetic state with temperature is generally observed only for $ZnFe_2O_4$ nanoparticles. It may be attributed to the fact that, in nanoparticles, the surface effects play an important role in observing such unusual magnetic properties. This in turn leads to noncollinear magnetic moments on the surfaces. This can be explained in terms of core shell morphology of the nanocrystals consisting of ferrimagnetically aligned core spins, and a spin glass-like surface layer [39]. If this surface layer is absent, the magnetization of the particles would saturate with an increase in the applied field up to a particular magnetic field. When the core magnetic moments align with the magnetic field, at some point of the magnetic field, the response of the *core mode* of the magnetization response is saturated, and the core magnetization of the system behaves in a usual Langevin-like way. Beyond this stage, any increase in the magnetic field on the particles has an effect only on the surface layer, thus the increase in the magnetization of the particles slows down [40]. The origin of surface spin disorder for ferrite nanocrystals may be due to broken exchange bonds, a high anisotropy layer on the surface, or a loss of the long-range order in the surface layer. These effects are particularly strong in the case of $ZnFe_2O_4$ because of the superexchange interactions through the oxygen ions [41]. The magnetization in this case is closely related to the existence of a magnetically disordered surface layer, in which the direct competition of exchange interactions between surface spins may take place [40].

Fig. 5 shows the XRD patterns of $ZnFe_2O_4$ nanoparticles synthesized using the oxalate-based precursor method [42–44]. Using TEM and XRD analysis, the particle size for $ZnFe_2O_4$ was observed to be 8 nm. Fig. 6 shows the *M-H* curves for $ZnFe_2O_4$ nanoparticles measured at 300 K and 5 K. A superparamagnetic behavior with maximum magnetization of 14.4 emu/g and with coercivity less than 50 Oe was observed for $ZnFe_2O_4$ nanoparticles at 300 K. The observed magnetization signifies the strong ferromagnetic behavior at 300 K. The saturation magnetization was observed to be 18.4 emu/g and coercivity was around 13.4 kOe at 5 K. This is the highest magnetization value ever reported for $ZnFe_2O_4$ nanoparticles having an 8 nm size. $ZnFe_2O_4$ nanoparticles synthesized using the oxalic acid route showed ferromagnetic properties at much higher temperatures. This was mainly due to confinement effects that are known to exist in nanometer-sized particles [44], and also due to the superexchange interactions of Fe^{3+} ions at (A) and [B] sites. Several studies have proved that the preparation, synthesis techniques, annealing process, etc., play important

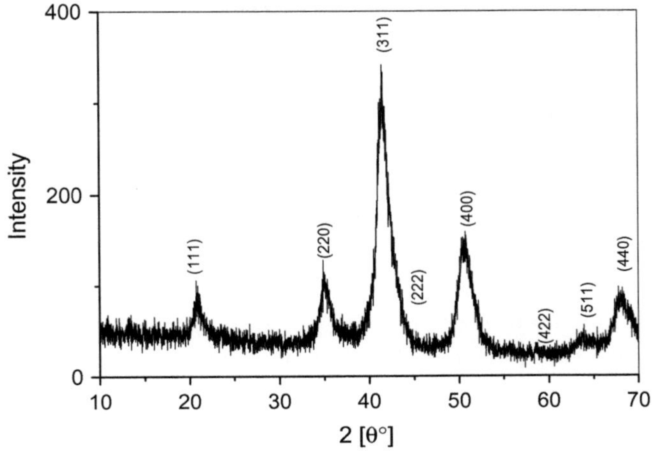

Fig. 5 X-ray diffraction patterns of oxalic acid-based synthesized $ZnFe_2O_4$ nanoparticles.

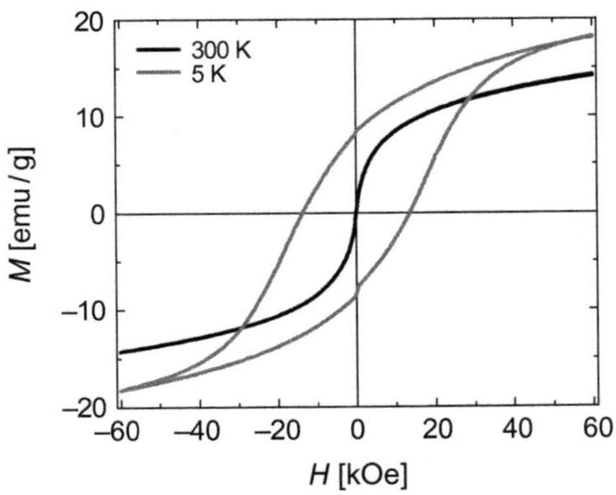

Fig. 6 *M-H* curves for $ZnFe_2O_4$ nanoparticles measured at 5 and 300 K.

roles in achieving the higher magnetization in $ZnFe_2O_4$ compared to the bulk counterparts [45–47].

Lotgering et al. [48] revealed that, if a small number of Fe^{3+} ions occupy (A) sites, these ions might form a magnetic cluster consisting of a Fe^{3+} ion at [B] sites through a coupling known as (A–B) interactions. By using neutron diffraction studies for $ZnFe_2O_4$ nanoparticles having 96 and 29 nm size, it was confirmed that Fe^{3+} ions occupying (A) sites were around 1.08 and 0.142 [49]. In bulk $ZnFe_2O_4$, it was observed that 4% of Fe^{3+} ions occupy (A) sites [50].

Chi et al. [27] measured the magnetization for 9.8 and 13.4 nm size $ZnFe_2O_4$ particles at 5 K with the maximum applied field of 5 T. The magnetization did not saturate even with an applied field of 5T. This indicates the coexistence of superparamagnetic clusters, and it was not possible to eliminate the canted spin structure in the $ZnFe_2O_4$ nanoparticles [27, 51–53]. The highest magnetization observed for the lowest particle size of 9.8 nm was about 52.9 emu/g. They observed that magnetization decreased with increasing particle size. The decreasing trend in magnetization with increasing particle size was due to cation distribution among the $ZnFe_2O_4$ nanoparticles. It is very well known that bulk $ZnFe_2O_4$ is a normal spinel ferrite in which the Zn^{2+} ions entirely occupy (A) sites and the Fe^{3+} ions occupy [B] sites. Bulk $ZnFe_2O_4$ has a Néel temperature around 10 K due to the (B-B) interactions. Below the Néel temperature, bulk $ZnFe_2O_4$ behaves like ferrimagnetic and the above paramagnetic. But in the case of nanostructured $ZnFe_2O_4$ the magnetization shows entirely different behavior [47, 48, 54, 55]. In $ZnFe_2O_4$ nanoparticles, the observed large magnetization may be due to the structural changes from normal to mixed spinel structure represented by Eq. (2).

$$(Zn_{1-x}Fe_x)_A[Zn_xFe_{1-x}]_BO_4 \tag{2}$$

where x is the inversion parameter

The inversion parameter x was calculated using the cation distribution formula (Eq. 3).

$$(Zn_{0.78}Fe_{0.22})_A[Zn_{0.22}Fe_{1.78}]_BO_4 \tag{3}$$

The inversion parameter was observed to be 0.22, indicating the probability of Fe^{3+} ions occupying (A) sites. This value also explicitly shows that, the $ZnFe_2O_4$ nanoparticles behave like inverted metallic ions, which contributes in observing magnetic properties.

The larger magnetization observed in $ZnFe_2O_4$ nanoparticles is due to the strong (A-B) interactions between Fe^{3+} ions at [B] sites, and forced Fe^{3+} ions at (A) sites, if the particles are under a certain limited size only [56]. It is observed that $ZnFe_2O_4$ nanoparticles having sizes of 10 nm showed magnetic behavior with coercivity less than 50 Oe [45, 57–59]. The Mössbauer investigations revealed that the quadrupole splitting is 0.40 mm/s for $ZnFe_2O_4$ nanoparticles. This value is larger than the bulk value of 0.35 mm/s. The large quadrupole splitting in $ZnFe_2O_4$ for 9 nm size particles is attributed to the cationic inversion in the synthesized samples [58–60].

$ZnFe_2O_4$ synthesized by the hydrothermal method with an average particle size of 10 nm showed superparamagnetic behavior [61]. Mössbauer measurements also confirmed the superparamagnetic behavior in $ZnFe_2O_4$ nanoparticles. Studies have been carried out on $ZnFe_2O_4$ by varying the amount of fuel (glycine) to tune the structural and magnetic properties. By varying the fuel content, the particle size was observed to change significantly. The magnetization in these samples showed superparamagnetic behavior. With the higher glycine-to-nitrate ratio, the magnetization increased drastically. The coercivity and remanence ratio was also observed to increase with the fuel ratio. These kinds of magnetic changes with the glycine/nitrate ratio may have happened due to the increase in the grain size, domain structure, and anisotropy in the crystals [62].

$ZnFe_2O_4$ synthesized by the nitrate method showed negligible remanence and coercivity as a typical behavior of superparamagnetism at room temperature [63]. The superparamagnetic behavior in $ZnFe_2O_4$ nanoparticles is due to structural changes from a normal to a mixed spinel structure. Due to the nano size, the degree of cation inversion is affected. Several authors investigated the size-dependent magnetic properties of $ZnFe_2O_4$. They observed that, as the particle size decreases, the cation inversion increases, thereby enhancing the magnetic properties [64, 65]. EPR studies revealed that, at low temperature, spin canting and cation inversions exist in the samples due to changes in the crystal size. Apart from the size of the particles, the magnetic properties are sensitive to synthesis and preparation conditions [31, 58, 66, 67].

The mechanism of magnetic properties in nanosize $ZnFe_2O_4$ is not only limited to superparamagnetism, but it may also behave like the ferrimagnetic or the coexistence of antiferromagnetic or ferromagnetism or the spin glass state [68–71]. $ZnFe_2O_4$ nanopowder synthesized by a self-sustainable method showed the saturation magnetization of 6.9 emu/g and coercivity around 260 Oe. In general, for magnetic nanoparticles below a certain critical size and temperature, they behave like the superparamagnetic [43]. It was observed that in nanostructured $ZnFe_2O_4$ material, the Zn^{2+} ions will have free 4s and 4p orbitals. They are ready for covalent bond formation with oxygen ions. The Fe^{3+} ions have the tendency to occupy the tetrahedral (A) sites. Therefore, in nano $ZnFe_2O_4$, Fe^{3+} ions are distributed between the (A) and [B] sites.

$ZnFe_2O_4$ is widely used as a catalyst and at the nanoscale, it was observed to modify the surface properties. Due to the unusual magnetic properties of

$ZnFe_2O_4$, it is expected that $ZnFe_2O_4$ will be used as a potential material in several industrial applications [31, 48–50, 54, 72–78]. Researchers have confirmed that when the coprecipitation technique is employed, the high inversion parameter can be achieved over the other methods [54, 72, 74]. Therefore, the magnetization was observed to be enhanced when compared to the bulk counterparts [77, 78].

It was reported by Chinnaswamy et al. [69, 77] that, with a small milling time, the magnetization increased enormously. But, as the milling time was increased to a very high level, the produced nanoferrites showed degraded magnetization due to surface disorder and spin glass-like behavior.

In a few $ZnFe_2O_4$ nanoparticles, the reduced magnetization compared to the bulk has been explained on the basis of the magnetic layer present on the surface of the nanoparticles. The magnetic dead layer is not commonly present in all the ferrites, and it is limited to only a few kinds of ferrites. It was also reported that, due to higher milling times, the $ZnFe_2O_4$ powder may also decompose to α-Fe_2O_3 and γ-$ZnFe_2O_4$. It is believed that, in nanostructured particles, a large number of atoms are located on the surface, meaning that the surface defects arise due to surface tension. Surface energy plays an important role in changing the positions of surface atoms, thereby enhancing the magnetic properties. Usually $ZnFe_2O_4$ shows long-range magnetic ordering only at the liquid helium temperature. But due to the nanosize, nowadays it is possible to observe magnetization even at room temperature. The ball-milled $ZnFe_2O_4$ samples showed $T_N \approx 115\,K$, which is 10 times higher than that of the bulk. At low temperatures, the nanoparticles start forming the clusters to have a superparamagnetic behavior. Researchers have observed that the preparation techniques such as coprecipitation may show huge structural and chemical disorders due to mechanical displacement and atomic disorder under high stress and strain. These chemical, structural, and atomic disorders in a nanostructured material such as $ZnFe_2O_4$ will contribute to achieving the magnetic properties. In the magnetic nanoparticles, when their sizes are below a particular dimension, there is a probability that thermally activated magnetization increases. In such cases, the surface effects also play important roles in observing the unusual magnetic properties. Therefore, the surface magnetic properties or surface effects play an important role in tuning the magnetic properties [79, 80].

Also, the cation distribution will have a large effect near the particle surface compared to the nonmagnetic/dead layer atoms [81, 82]. When the ferromagnetic (FM) coupling occurs with antiferromagnetic (AFM) material, it will result in FM-AFM exchange interactions. This will modify the magnetic properties in an unexpected way [83].

$ZnFe_2O_4$ nanoparticles of sizes 6.6 and 14.8 nm were synthesized by forced hydrolysis in a polyol medium [84]. The magnetic properties showed a strong dependence on the particle size due to unusual cation distribution. For these nanoparticles, a superparamagnetic behavior was observed at room temperature. At a very low temperatures, ferromagnetic or ferrimagnetic behavior was observed. As a general trend, magnetization was observed to increase with increasing particle size. From the Mössbauer analysis, they observed that the charge of Fe^{3+} remains unchanged, confirming a perfect stoichiometry [84].

Thermodynamically, the degree of structural inversion in $ZnFe_2O_4$ is defined by the amount of Fe^{3+} ions occupying (A) sites. The total Fe^{3+} ions in the system depend on the parameters such as ionic radii, cation distribution, and stabilization energies at the (A) and [B] sites. All the above parameters depend on the synthesis method, due to which a metastable local structure could be observed. $ZnFe_2O_4$ nanoparticles synthesized by a few other methods showed nonzero magnetization at a very low temperature [49, 85–89]. One has to remember that all the above conditions are extremely valid only for $ZnFe_2O_4$ [84].

When the concentration of Fe^{3+} ions in the (A) site is below a certain permissible limit of 0.33, this will correspond to structural inversion around 16.5%. If the inversion is around 25%, it will lead to a collinear Néel ferromagnetic structure and if the inversion is around 15%–18%, it may have a Yafet-Kittel-type canting arrangement. The magnetic frustration arises due to the competing interactions of (A-B) and (B-B), due to which the magnetic disorder on the [B] site gives rise to the spin glass-like behavior. The magnetic clusters are formed due to the short-range magnetic ordering. These clusters freeze in the random orientation below spin freezing temperatures when the magnetic concentration is too high, and this is based on the Yafet-Kittel model [90]. Canting in (A) and [B] sublattices is observed due to a long-range magnetic ordering. In such cases, the magnetic ordering temperature will be higher than the well-known Néel temperature for the normal $ZnFe_2O_4$ and could be as high as 460 K [91].

$ZnFe_2O_4$ nanoparticles with particle sizes as of 2–12 nm have been prepared by a soft chemical coprecipitation method [92]. The $ZnFe_2O_4$ with a 2 nm particle size showed a paramagnetic behavior. Those with sizes above 2 nm showed superparamagnetic behavior. With the increase in particle size, the magnetization also increased drastically. The degree of inversion in $ZnFe_2O_4$ was observed to depend on the particle size. This is confirmed from neutron powder diffraction analysis [49, 92].

For $ZnFe_2O_4$ at a low temperature, usually ferromagnetism is observed because, the interactions at [B] sites will be more dominant compared to (A) sites. Because of surface disorder, the antiferromagnetic and ferromagnetic nanoparticles appear to break down in several sublattices [93]. Due to this, a different spin configuration appears in the system having similar energies. The interactions between the core and surface spins result in large exchange bias and anisotropy. For a lower particle size, the low magnetization or nonmagnetic properties are observed due to the presence of a dead or inert layer. The smaller particles have a greater fraction of surface spins, which tend to be in a canted state or a spin glass state, resulting in a lower net magnetic moment. If the size of the particles increases, the surface effects become less significant, thereby showing enhanced magnetization.

$ZnFe_2O_4$ nanoparticles were synthesized by a new coprecipitation technique using area [94]. The obtained particle size was 13 nm. The magnetic properties for these nanoparticles were observed only at 10 K. $ZnFe_2O_4$ nanoparticles having 40 nm sizes were synthesized by a self-propagating, low-temperature combustion method. These samples showed a very high magnetization value of 78.54 emu/g, a remanence magnetization of 74.97 emu/g, and a coercivity of 1297.52 Oe. These kinds of enhanced magnetic properties at room temperature were attributed mainly to pronounced grain growth. The magnetic anisotropy inhibits the alignment of magnetic moment toward an applied magnetic field [95, 96].

$ZnFe_2O_4$ nanoparticles with sizes of 4 and 6 nm were synthesized using the citrate method [97]. The 6 nm particles did not show any magnetic anomaly but, when the particle size was reduced to 4 nm, the magnetic structure was changed by creating inversion. The Fe^{3+} ions at (A) sites were due to strong (A-B) interactions in the sample. This resulted in the ferromagnetic properties at a low temperature. $ZnFe_2O_4$ nanoparticles synthesized using multistep homogenous nonaqueous solutions [98] showed magnetic properties at room temperature. The annealing temperature showed a significant effect on the particle size. The magnetic properties were observed only at 5 K. For $ZnFe_2O_4$ nanoparticles, the magnetic anomalies at a higher field could be attributed to competition between different magnetic interactions in the ferrite systems. The EXAFS analysis revealed the cationic or spins disorder in the prepared $ZnFe_2O_4$ nanoparticles. The inversion was observed to decrease with increasing particle size. The 4 nm nanoparticles showed inversion around 80%, the 6 nm nanoparticles showed inversion around 40%, and the 7 nm particles showed negligible inversion. Theoretical studies on $ZnFe_2O_4$ suggested that the electrostatic interaction energy plays an important role in understanding the cation distribution in all the spinel ferrites.

Yu et al. [58] synthesized $ZnFe_2O_4$ nanoparticles of 300 nm by the hydrothermal technique, using zinc sheet and ferric chloride. The as-prepared $ZnFe_2O_4$ nanoparticles showed a room temperature magnetization value of 54.6 emu/g. The room temperature higher magnetization value for $ZnFe_2O_4$ compared to other reports might be due to the interfacial reaction associated with the microstructure of the particles [58, 88].

Raeisi Shahraki et al. [99] synthesized $ZnFe_2O_4$ nanoparticles at a very low temperature using the coprecipitation method. The particle sizes were observed to be less than 10 nm. Even though they have synthesized $ZnFe_2O_4$ nanoparticles using standard coprecipitation, maintaining a very low temperature during preparation, all their samples showed linear magnetization curves, whereas other coprecipitated $ZnFe_2O_4$ samples showed magnetization behavior. These kinds of discrepancies mainly depend on the preparation condition and the chemicals.

Another coprecipitation method for $ZnFe_2O_4$ nanoparticles showed that these ferrites change structure from normal to mixed spinel only at a particular annealing temperature. They concluded that Zn^{2+} ions shifts from (A) sites to [B] sites and the shifting ratio was the same. The smallest particle size of 16 nm showed the inversion parameter value of 0.207, and the inversion decreased with increasing particle size [100]. $ZnFe_2O_4$ nanoparticles prepared by the liquid phase precipitation method at low temperature showed the particle size around 20.35 nm. The maximum magnetization value was 80 emu/g at 300 K, having a superparamagnetic behavior [101].

Zinc ferrite nanofibers prepared using the electrospinning method showed magnetic properties at room temperature due to nanoparticles [102]. The *M-H* curves showed superparamagnetic behavior with coercivity less than 50 Oe. $ZnFe_2O_4$ nanoparticles synthesized by the thermal treatment method and calcined at a very high temperature from 723 to 870 K [103]. Poly vinyl pyrrolidone (PVP) was used as capping agent to stabilize the nanoparticles and prevent them from agglomeration. All the higher temperature annealed $ZnFe_2O_4$ nanoparticles showed negligible coercivity. They observed that as the particle size increases, the saturation magnetization increases linearly. The observed magnetic properties might be due to the effect of the inversion parameter which changes as a result of a change in the particle size. Usually, the degree of inversion is large for small particles compared to the large particles. The magnetic properties in bulk $ZnFe_2O_4$ are due to intrasublattice (B-B) exchange interactions and not due to (A-A) or (A-B) superexchange interactions. It is observed in general that, in ferrites where the magnetization is strong, the (A-B) superexchange interactions are much stronger than (A-A) and (B-B) [104]. The (A-B) superexchange interactions are usually observed for small particles under certain limitations for ferrites like $ZnFe_2O_4$. Therefore, in nanoparticles we could observe a much larger magnetization than in bulk particles [59, 65, 104]. It has been quite commonly observed that, having a similar particle size shows different magnetic properties due to preparation, the synthesis method, the annealing temperature, etc. Therefore, the structural, magnetic, and electrical properties of spinel ferrites mainly only depend on the preparation method.

$ZnFe_2O_4$ nanoparticles synthesized using the nitrate method were irradiated using an ion beam [105, 106]. After irradiation, these samples were annealed at 800°C and 1000°C. A nearly 200% increase in the magnetization value was observed for irradiated samples at room temperature. Due to radiation, the defects were induced in the samples, by which the magnetization was observed to be higher compared to other researchers. Also, the defects in the form of cracks might have contributed to the cation inversion in the $ZnFe_2O_4$ nanoparticles. When the swift heavy ion radiation is subjected to $ZnFe_2O_4$, it will align the randomly oriented ions toward the ion beam. Due to this, the spin of the particles is expected to align in the same direction. Besides this, the ion beam also induces the cation inversion due to local heat generated in the system. Therefore, the observed magnetization in the irradiated $ZnFe_2O_4$ samples is the combination effect of the above facts. $ZnFe_2O_4$ nanoparticles synthesized using the nitrate method showed interesting magnetic properties [71]. Due to the effect of particle size, they changed from the paramagnetic phase to the superparamagnetic phase. The paramagnetic nature in nanoparticles was associated due to Fe^{2+} ions and improved the crystal structure. The superparamagnetic nature was associated due to the dominant value of Fe^{3+} ions and the lower particle size. It was observed that isolated Fe^{3+} ions due to the Fe-O bonds at the surface of nanoparticles will contribute in observing the enhanced magnetic properties in $ZnFe_2O_4$ ferrites [71].

Zinc ferrite nanoparticles prepared using ultrasonic cavitation assisted solvothermal synthesis techniques showed the magnetization value of 24.32 emu/g mainly due to the highly disordered surface of the nanoparticles [107].

They proposed that, due to the large degree of inversion in their prepared $ZnFe_2O_4$ nanoparticles, they observed strong ferromagnetic coupling due to Fe^{3+} ions at [B] sites [108].

$ZnFe_2O_4$ nanoparticles prepared from the aerogel process showed interesting low temperature calorimetric and magnetic properties [72]. These samples showed the magnetic order at a higher temperature. Due to certain preparation methods at a low temperature, the magnetic ordering in $ZnFe_2O_4$ used to shift to high temperature or disappear. The observed magnetic properties are due to the Fe^{3+} distribution between the (A) and [B] sites. The as-prepared samples showed a large occupancy of Fe^{3+} ions at (A) sites, but usually these ions prefer to occupy [B] sites, as in the case of bulk $ZnFe_2O_4$. The heat capacity peaks near 10 K for $ZnFe_2O_4$ bulk particles were observed to be greatly suppressed due to the preparation method. The inversion parameter for bulk $ZnFe_2O_4$ was observed to be $C \approx 0$. This confirms the long-range antiferromagnetic ordering with Fe^{3+} ions residing only at [B] sites. As the inversion parameter starts increasing its value beyond "0," then slowly the $ZnFe_2O_4$ nanoparticles start showing variation in their structure. Due to this, the structural transformation takes place in the system with an increasing inversion parameter, and $ZnFe_2O_4$ cannot be totally inversion free. The beauty of $ZnFe_2O_4$ is, if you want to suppress the inversion, then higher annealing temperature, higher ball milling time, and higher particle size should be employed. But in such cases, the magnetization will be observed only at a lower temperature and they will behave simply like the bulk. These kinds of magnetic anomalies were confirmed by Westrum and Grimes [109].

Neutron diffraction measurements also revealed that the superexchange measurements are induced in antiferromagnetic $ZnFe_2O_4$ due to Fe^{3+} ions [110]. A large inversion parameter in $ZnFe_2O_4$ is observed only if the material has low temperature calorimetric properties. The highest inversion for $ZnFe_2O_4$ was observed for ball-milled samples with $x = 0.6$. Mössbauer measurements showed ferrimagnetic ordering at room temperature. The magnetic anisotropy linked to low-temperature antiferromagnetic ordering was observed to vanish completely [72]. Therefore, the synthesis technique plays a wide role in making the $ZnFe_2O_4$ material ferromagnetic. Neutron diffraction measurements on $ZnFe_2O_4$ confirmed long ago that (A-B) coupling is absent in $ZnFe_2O_4$, and this material is ideally suited for observing (B-B) interactions. Therefore, it is obvious to see only antiferromagnetic properties at a very low temperature [110].

$ZnFe_2O_4$ nanoparticles prepared by the supercritical sol-gel method showed a low-range magnetic order above 30 K. The inversion parameters in their samples were observed to be 0.21 for as-prepared samples. With an increase in the annealing temperature, the inversion parameter decreased to 0.05 [111]. Therefore, in bulk $ZnFe_2O_4$, the inversion parameter is zero. The magnetic ordering in $ZnFe_2O_4$ arises due to weak J_{BB} exchange interactions. This will yield a complex antiferromagnetic ordering at 10 K [112]. It is proved that nanostructured $ZnFe_2O_4$ will have a much larger inversion than bulk. $ZnFe_2O_4$ powders exhibit spontaneous magnetization near 30 K. The Mössbauer measurements confirm the collinear spin structure around this temperature [46, 49].

The coexistence of the ferromagnetic and antiferromagnetic properties of Fe^{3+} inverted $ZnFe_2O_4$ was synthesized by ball milling [70]. These samples were investigated by nuclear magnetic resonance (NMR) analysis under different temperatures and magnetic fields. The NMR measurements at 4.2 K revealed that the antiferromagnetic origin in $ZnFe_2O_4$ nanoparticles is observed to be the same as that of antiferromagnetic bulk $ZnFe_2O_4$, that is, the (B-B) interactions are observed to be dominant at 4.2 K. This argument is further supported by the Néel temperature for antiferromagnetic bulk at 10 K. The antiferromagnetic long-range order disappears at 10 K and is influenced by the ferromagnetic ordering. The overall spin structure of $ZnFe_2O_4$ is not as simple as compared to other ferrites. Its property changes over the temperature range and has a mixture of two types of magnetic orderings. The canting angle with the external applied magnetic field is estimated to be 54 degrees. The ferromagnetism observed in $ZnFe_2O_4$ is mainly due to the nanocrystalline nature of the samples. The nanosize influences the spin canting angle. NMR data further confirmed that the ferromagnetic nature exists with a transition temperature of 460 K, and the antiferromagnetic nature with a transition temperature of 10 K. It was clearly observed that (B-B) interactions induce antiferromagnetic interactions in [B] sites. The (A-B) interactions induce the ferrimagnetic coupling between (A) and [B] sites. Still, the spin structure in the nanocrystalline $ZnFe_2O_4$ was observed to be a controversial topic until now. The spins contributing both ferromagnetic and antiferromagnetic ordering at [B] sites are canted. They are consistent with the theory of spin canting when the two interactions compete. The spins of Fe^{3+} ions at (A) sites are ferromagnetically coupled and at [B] sites, they are antiferromagnetically coupled.

$ZnFe_2O_4$ nanoparticles having sizes 6–13 nm were investigated using X-ray absorption near edge spectroscopy (XANES) and extended X-ray absorption fine structure [113]. Also, they performed density functional theory (DFT) calculations to verify the experimental data. They found a good agreement between theoretical calculations and experimental work. From their XANES analysis, they confirmed that Zn^{2+} ions transfer from (A) to [B] sites whereas Fe^{3+} ions transfer from [B] to (A) sites without altering the long-range structural properties. The direct way to find the cationic inversion in $ZnFe_2O_4$ was confirmed from XANES. Also, the increase in the grain size contributes to observing the variation in the inversion parameter. The transfer of Zn^{2+} and Fe^{3+} ions in nanostructured $ZnFe_2O_4$ causes a considerable change in cation distribution. This influences the inversion parameter by modifying the electronic structure of the material. The inversion in the system is linked to the whole particle and not to any distorted surface layer. Therefore, XANES gives the direct proof of nonequilibrium cation distribution in materials such as $ZnFe_2O_4$. If a large number of Zn^{2+} ions occupy the [B] sites, they cause broken superexchange (A-B) interactions, thereby producing inhomogeneous cation inversion, in which case either ferromagnetic (A-B) or antiferromagnetic (B-B) interactions would exist [70, 113]. It is proved that the XANES approach used in the study will serve as a powerful tool in determining the structure of the materials in detail. The magnetic and Mössbauer results are in agreement with the observed ferromagnetic nature of the $ZnFe_2O_4$ nanoparticles.

$ZnFe_2O_4$ nanoparticles were synthesized by the Pechini method to study the effect of annealing temperature on the structural and magnetic properties [114]. All the annealed samples showed a superparamagnetic behavior at room temperature. The saturation magnetization increased with increasing annealing temperature and the coercivity decreased. The changes in the magnetic properties were mainly due to changes in the particle size [115, 116]. The increase in the saturation magnetization and the remnant magnetization was attributed to spin noncolinearity at the surface. The change in the particle size resulted in observing the large surface area and interparticle interactions. The highest magnetization was observed for bigger particles due to the fact that the external applied field is proportional to the size of the particles. The lower magnetization values are due to smaller particle size. These magnetic changes are attributed to surface distortion. The interactions between the oxygen and metallic ions reduce the net magnetic moment. This effect is especially observed for nanoparticles due to large surface-volume ratio [117]. Another factor as of magnetocrystalline anisotropy is known to depend on the crystalline nature of the samples. At low annealing temperatures, usually dislocations and point defects are present within the interfacial lattice sites of the samples. The higher the defects, the lower the magnetic properties or magnetization values.

Lactoferrin (LF)-assisted $ZnFe_2O_4$ nanoparticles were synthesized with sizes from 11 to 29 nm using the hydrothermal method [118]. The LF content in the $ZnFe_2O_4$ nanoparticles showed remarkable magnetic properties. All the samples showed a superparamagnetic behavior at room temperature. With the increase in LF content, the magnetization also increased as a consequence of particle size. The enhanced magnetization with increasing LF content may be due to the fact that the LF, Zn^{2+}, and Fe^{3+} ions disperse uniformly and need some time to diffuse. With the presence of LF, the Zn^{2+} and Fe^{3+} ions bind closely to open the interdomain cleft of LF. Therefore, the concentration of Zn^{2+} and Fe^{3+} ions in the open interdomain cleft will be higher than other areas. The particles with LF would result in better crystallinity. Also, the specific surface area of LF content in $ZnFe_2O_4$ is smaller than undoped $ZnFe_2O_4$. This surface area becomes much smaller with higher LF content. A paramagnetic to ferromagnetic transition was observed in $ZnFe_2O_4$ ferrites during the atomic hydrogenation process at room temperature. The magnetic properties were observed to enhance upon increasing the atomic hydrogenation time [119]. From the XRD analysis, there was no evidence of A-O-B superexchange phenomena. But the atomic distances between Fe-Fe, Zn-Zn, Fe-O, Zn-O, and Fe-Zn were observed to decrease. The observed paramagnetic to ferromagnetic transition and the higher magnetic properties in the samples were mainly due to the hydride and superexchange interactions of the Fe^{2+} and Fe^{3+} ions. Nonhydrogenated samples showed a paramagnetic nature while, with increasing atomic hydrogenation, the magnetization increased systematically and showed ferromagnetic-like behavior and the coercivity was observed to be more than 360 Oe. Therefore, the paramagnetic to ferromagnetic transition was observed mainly due to atomic hydrogenation. From the XANES and EXAFS spectral analysis, the transformation of Fe^{3+} and Fe^{2+} was observed during atomic hydrogenation.

3 $ZnFe_2O_4$ THIN FILMS: MAGNETIC PROPERTIES

$ZnFe_2O_4$ thin films were grown on *R*-cut Al_2O_3 substrates by using the pulsed-laser deposition (PLD) technique (KrF, $\lambda = 248$ nm, 13 Hz repetition rate, 2.1 J/cm^2) [120, 121]. The typical thickness of the films was about 130 nm. The substrate temperature was 650°C. During deposition, the oxygen partial pressure PO_2 was kept as 10^{-4} Torr. After deposition, films were

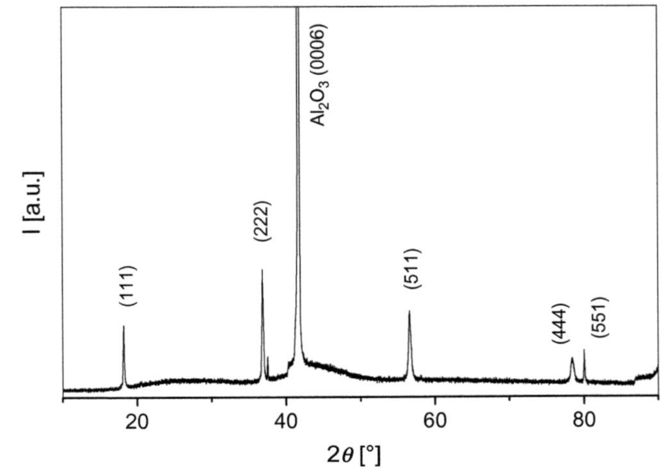

Fig. 7 X-ray diffraction patterns of $ZnFe_2O_4$ thin film.

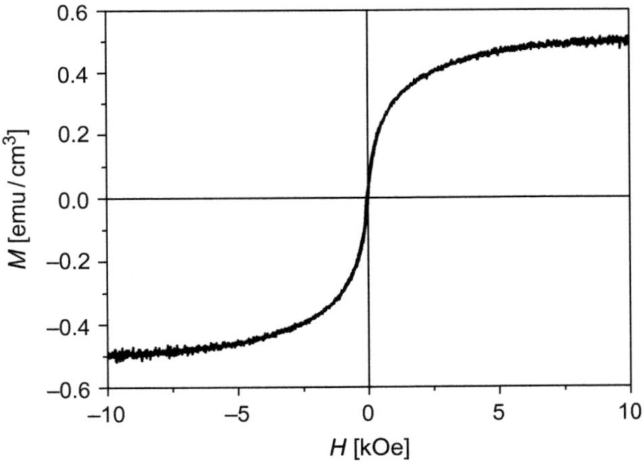

Fig. 8 *M-H* curve for $ZnFe_2O_4$ thin film measured at room temperature.

annealed at 500°C for 30 min under the same oxygen pressure as during deposition and then cooled down slowly to room temperature. Fig. 7 shows the XRD patterns of the $ZnFe_2O_4$ thin films corresponding to the (*Fd3m*) phase. Fig. 8 shows the *M-H* curves for $ZnFe_2O_4$ thin films measured at 300 K. The $ZnFe_2O_4$ thin films clearly exhibit superparamagnetic behavior at room temperature. $ZnFe_2O_4$ is normal spinel with Zn^{2+} ions in the (A) sites and Fe^{3+} in the [B] sites with an antiparallel arrangement of magnetic moments [122]. The superexchange interactions between the

(A) and [B] sites do not seem to be favorable at room temperature; hence the nanoparticles show paramagnetic behavior. Similar kinds of results were observed by other researchers [65, 122–124]. The *M-H* loops of $ZnFe_2O_4$ thin films measured at 5 K showed ferromagnetic behavior. The remarkable ferromagnetic behavior of $ZnFe_2O_4$ thin films compared to $ZnFe_2O_4$ nanoparticles may be due to nonequilibrium distribution of Fe^{3+} ions at (A) and [B] sites [123–126].

$ZnFe_2O_4$ thin films deposited using sputtering on glass substrate near room temperature showed the ferromagnetic behavior [127]. The particle size was observed to be 10 nm. A higher saturation magnetization of 32 emu/g was observed with an external magnetic field of 30 kOe. The ferromagnetic properties in these $ZnFe_2O_4$ thin films were due to the preparation fabrication technique. They argued that, due to the scattering method and the annealing process, the Zn^{2+} and Fe^{3+} ions on the thin films were distributed randomly. Therefore, the Fe^{3+} ions occupy both (A) and [B] sites. Because of this, a strong superexchange interaction takes place among them. This gives rise to a strong ferrimagnetic nature, thereby showing higher magnetization values. The spin freezing in sputtered $ZnFe_2O_4$ thin films was observed to be around 325 K. It is very well known that the $ZnFe_2O_4$ structure has diamagnetic Zn^{2+} atoms and magnetic Fe^{3+} atoms occupying both (A) and [B] sites, respectively. In the normal spinel structure, the negative superexchange interactions occur among Fe^{3+} in [B] sites, therefore it shows an antiferromagnetic nature. The observed ferrimagnetic nature in sputtered $ZnFe_2O_4$ thin films may be due to a change in the crystal structure, which is in the metastable phase and is quite different from the normal spinel structure. The possible mechanism for observing the ferromagnetic nature in $ZnFe_2O_4$ thin films might also be due to the random arrangement of Zn^{2+} and Fe^{3+} ions. In the sputtered $ZnFe_2O_4$ thin films, extremely high energies are involved so that solid material vaporizes to obtain highly excited vapor specimens on the substrate. In the sputtering process, the rapid vapor pressure is quenched on the substrate. Due to this, the random distribution of Zn^{2+} ions and Fe^{3+} ions is frozen in a metastable state. Therefore, in the sputtering process, the Fe^{3+} ions occupy both the (A) and [B] sites. The presence of strong superexchange interactions between J_{AB} gives rise to observing the high magnetization at room temperature. The cation distribution for the magnetic ordering $ZnFe_2O_4$ can be formulated using Eq. (4):

$$(Zn_{0.39}Fe_{0.61})_A[Zn_{0.61}Fe_{1.39}]_B O_4 \tag{4}$$

When a high magnetic field of 5T is applied to $ZnFe_2O_4$ thin films, due to this strong energy, the magnetic anisotropy leads to the nonalignment of the nanocrystals. This contributes in observing the collective magnetic moment. Therefore, the external magnetic field plays an important role in orienting the nanocrystals in the field direction. Also, the distribution of Zn^{2+} and Fe^{3+} ions in $ZnFe_2O_4$ is completely random, represented by the formula (Eq. 5):

$$(Zn_{0.33}Fe_{0.67})_A[Zn_{0.67}Fe_{1.33}]_BO_4 \tag{5}$$

In the sputtered $ZnFe_2O_4$ thin films, the spinel structure seems to be distorted even though Zn^{2+} ions prefer to occupy the (A) site [128]. During fabrication/deposition/preparation of thin films, due to the annealing process, there are chances of producing Fe^{2+} ions that slightly deviate from the normal spinel structure, thereby showing magnetic properties in those films [129, 130]. Generally, the ferromagnetic properties are observed for an Fe-rich $ZnFe_2O_4$ material. EXAFS studies have also revealed that Zn^{2+} ions occupy 60% at [B] sites, which might also contribute to the ferrimagnetic nature in $ZnFe_2O_4$ thin films [66]. Nakashima et al. [66, 124, 127] concluded in their studies that the rearrangement of Fe^{3+}-Fe^{2+} ions in $ZnFe_2O_4$ sputtered films is the main cause for exhibiting ferromagnetic properties. And the contribution from Fe^{3+} is much less pronounced. In the spinel ferrites structure, the J_{AB} interactions are higher than J_{BB} and J_{AA}, due to which the spinel ferrites show a long-range ferrimagnetic order [131, 132].

According to the proposed phase diagram of Hubsch et al. [133], it is believed that in spinel oxides, if the Fe^{3+} ion concentration at (A) sites is below 0.33, which is known as the percolation limit, the long-range magnetic ordering in spinels is broken, and at [B] sites the magnetic frustration starts appearing. Due to this, in [B] sites the clusters are then arranged in different sizes. They behave in an antiferromagnetic way with short-range magnetic ordering. The frustration due to the interactions between ferrimagnetic and antiferromagnetic interactions, and the magnetic disorder at the [B] site give rise to spin glass-like behavior [52, 132–136]. The superparamagnetic phase is one of the expected magnetic states in $ZnFe_2O_4$ thin films if they have nanoparticles on their surfaces, and due to dipole interactions and cluster growth [137, 138]. The unexpected magnetization in sputtered $ZnFe_2O_4$ thin films measured at 30 K is mainly due to the disorder arrangements of Zn^{2+} and Fe^{3+} ions in the spinel structure [139]. These films showed ferromagnetic properties due to A-O-B superexchange interactions. Some of the Fe^{3+} ions transfer to tetrahedral (A) sites and activate the superexchange A-O-B interactions [124].

$ZnFe_2O_4$ thin films were grown on different substrates such as yttria-stabilized zirconia (111), $SrTiO_3$ (STO) (100), and Si (100) substrates using RF magnetron sputtering. All these thin films showed excellent crystal structure [140]. The observed magnetic properties in $ZnFe_2O_4$ thin films grown on different substrates may be due to the random orientation of magnetic particles on the thin films.

$ZnFe_2O_4$ films deposited by magnetron sputtering on Si (111) substrates showed magnetic properties at room temperature. The saturation magnetization was observed to be 303 emu/cm^3, much higher than the reported value for $ZnFe_2O_4$ thin films [125, 126, 141, 142]. In the magnetron sputtering technique, the rapid cooling process leads to the formation of solid-state phases from the vapor phase. Because of these phase changes, the random distribution of Zn^{2+} and Fe^{3+} ions takes place in (A) and [B] sites [127, 143]. The ferromagnetic properties observed for $ZnFe_2O_4$ thin films at room temperature may be due to the occupancy of few Fe^{3+} ions at the (A) sites [124, 144].

Epitaxial $ZnFe_2O_4$ thin films were grown on a (001) $SrTiO_3$ substrate using the PLD technique [145]. The easy axis orientation was along (111) the crystallographic direction. These films showed magnetic properties and Curie temperature (T_C) greater than 350 K. $ZnFe_2O_4$ thin films prepared by the sputtering method and the first principle XANES and EXAFS studies were carried out to clearly understand cation distribution disorder [66]. Mössbauer spectroscopy is one of the most powerful tools to investigate the macroscopic analysis of magnetic materials. Usually, it gives information only for symmetry and bond covalence of Fe ions. In Mössbauer analysis, if the sample does not have any Fe ions, then we cannot expect magnetic properties in samples. Therefore, another technique such as extended X-ray absorption fine structure (EXAFS) is widely used for the analysis of magnetic materials. From EXAFS analysis, we can find interatomic distances and coordination numbers, which can be used to determine the accurate nature of cation disorder among spinel ferrites [146–148].

Techniques such as X-ray absorption near edge structure (XANES) analysis give information of the specific atoms, the chemical bond, and the electronic state. As the XANES and EXAFS give the complementary information, the combination of two these techniques is expected to give more accurate, effective information of the specific atom's local environment. Theoretically, researchers have tried to find the complete structural and magnetic properties of $ZnFe_2O_4$ by introducing a concept called *core hole*. In this process, one electron is removed from the core orbital and then one electron is placed in the unoccupied orbital.

From their XANES and EXAFS analysis, Nakashima et al. [66] concluded that more Zn^{2+} ions reside on [B] sites in spite of its full tendency to occupy only (A) sites. As the Zn^{2+} ions are occupying [B] sites, a few Fe^{3+} ions displace to (A) sites, thereby producing the cation disorder as evidenced from the Fe k-edge XANES data. For a small inversion, $ZnFe_2O_4$ remains antiferromagnetic, and if the inversion increases beyond a certain limit, the superexchange interactions between (A) and [B] sites occur. In this situation, the J_{AB} interactions become significant in observing the ferrimagnetic behavior. All the above discussions purely depend on the fabrication techniques of thin films, by which a nonequilibrium state is produced. Combining the first principle calculations using the core hole effect and XANES analysis, one can achieve important information on the structural and magnetic properties of metastable $ZnFe_2O_4$ and other spinel ferrites [66].

$ZnFe_2O_4$ thin films with a thickness of 57 nm were fabricated under low oxygen pressure to induce the oxygen vacancy defects for observing the magnetic properties [149]. Most of the room temperature ferromagnetic properties observed for $ZnFe_2O_4$ nano/thin films were related to extrinsic or intrinsic defects [150–153]. Extrinsic defects are created by doping with nonmagnetic elements [10]. Sometimes, the undoped nonmagnetic materials also display ferromagnetic properties and are linked to the local disorder or lattice vacancies [154]. This phenomenon is known as defect-induced magnetism (DIM) [150–154]. The potential use of introducing defects in a few magnetic oxides is to make them ferromagnetic. As such, there is no way to control the defects. In ferrites, the cation distribution plays the most important role in controlling magnetic properties. Cation distribution can be modified by applying heat treatment, employing different preparation and fabrication methods, or reducing the size of the materials below a certain permissible or critical size [86, 113, 126, 149].

The disorder distribution of Zn^{2+} and Fe^{3+} ions in $ZnFe_2O_4$ leads to unexpected magnetic ordering. In $ZnFe_2O_4$ thin films, low oxygen pressure creates large magnetic effects at room temperature and they usually show ferrimagnetic behavior [125, 155, 156]. The cations are swapped with the influence of oxygen displacement in $ZnFe_2O_4$ spinel ferrites [156,157].

X-ray magnetic circular dichroism (XMCD) and DFT calculations were employed to prove the abnormal magnetic growth in $ZnFe_2O_4$ films [147]. The unusual magnetic properties were not only due to cation inversion, but also due to the existence of oxygen ions. XMCD measurements revealed that Fe^{3+} ions occupy (A) and [B] sites, but the large magnetic contribution is only due to [B] sites. The highest magnetization was observed for

$ZnFe_2O_4$ thin films grown under low oxygen pressure conditions. This is due to an anion defect caused by incomplete oxidation of cation on the $ZnFe_2O_4$ deposited layer [70, 149, 158, 159].

$ZnFe_2O_4$ thin films grown on Mg (001) substrates using molecular beam epitaxy under Ar and Ar/O_2 atmosphere showed excellent electrical and magnetic properties [160]. These properties were mainly observed due to tuning the Zn^{2+} ions, and by controlling the oxygen partial pressure and temperature during the growth of thin films. The preliminary investigation showed that the Fe^{3+} ions' magnetic moment was antiparallel in the (A) and [B] sites. It is interesting to observe that with Zn^{2+} substitution, the Fe^{3+} ions migrate from the [B] site to the (A) site under oxygen partial pressure, thereby creating Zn^{2+} and Fe^{3+} vacancies on [B] sites.

Ultrasound velocity measurements were performed on a high-purity $ZnFe_2O_4$ single crystal fabricated by the flux method [161]. The electrostatic anomalies due to the magnetostructural phase transition were absent in $ZnFe_2O_4$. The elastic anomalies were observed at a particular temperature where the magnetic susceptibility separates from the Curie-Weiss law.

Guo et al. [141] reported the paramagnetic nature in annealed $ZnFe_2O_4$ thin films. Due to annealing, the ions redistribute in $ZnFe_2O_4$ thin films, causing the reversal effect to make $ZnFe_2O_4$ as a normal spinel ferrite. The ferrimagnetic nature of these thin films was due to dominant J_{AB} interactions, even at a higher annealing temperature. At 10 K, these thin films showed hysteresis loops with large remanence and coercivity values.

Powder neutron diffraction measurements and DFT calculations using EINES simulation with a CASTEP code were performed on $ZnFe_2O_4$ [162]. Sun et al. [162] modified the structure of $ZnFe_2O_4$ by introducing inversion parameters. They observed that the inversion between Fe^{3+} and Zn^{2+} ions does not deform the structure after carrying the full relaxation; the only change they observed is a slight increase in the unit cell volume. Their magnetic calculations showed that the magnetic moment with inversion values originates from oxygen (O) at 2p orbital and Fe^{3+} ions at 3d orbital. Usually, Zn^{2+} ions contribute partially for the net magnetic moment.

The cation inversion in $ZnFe_2O_4$ materials leads to A-O-A and A-O-B superexchange interactions between Fe^{3+} ions [163]. The strength of the superexchange interactions depends on the distance between the Fe^{3+}-O-Fe^{3+} ions. Due to inversion, the A-O-B superexchange interactions are stronger than A-O-A interactions. This kind of unusual interaction does not actually exist in a normal spinel ferrite such as $ZnFe_2O_4$ because all the Fe^{3+} ions reside in

[B] sites and the B-O-B antiferromagnetic coupling comes into play [51, 164]. When the inversion parameter increases in $ZnFe_2O_4$, a considerable number of Fe^{3+} ions transfer from [B] to (A) sites, thereby changing the structure locally. Due to this, Fe^{3+} ion concentration increases in (A) sites, thereby gradually increasing the inversion ratio. Therefore, $ZnFe_2O_4$ behaves like a ferromagnetic material.

$ZnFe_2O_4$ thin films were grown on an Si (111) substrate using the PLD technique and irradiated with an ion beam [165]. Due to irradiation, there were no significant changes in the grain sizes. But the only difference observed was that the irradiated $ZnFe_2O_4$ ferrite thin films showed tiny bumps on the surface of the film. By thermal treatment during the irradiation process, the smoothness of the film surface was modified. Due to heavy irradiation, the magnetization was observed to increase when compared to pristine $ZnFe_2O_4$ thin films [166]. With irradiation, a significant rearrangement of ions takes place inside the $ZnFe_2O_4$ thin films due to cation distribution, surface modification, crystal structure, crystal growth, etc. $ZnFe_2O_4$ thin films deposited on $SrTiO_3$ substrates using the PLD technique showed ferrimagnetic ordering at room temperature [167]. The oxygen and argon annealed $ZnFe_2O_4$ thin films showed more reduced magnetization than the pristine thin films. Therefore, the annealing atmosphere and annealing conditions were also observed to play important roles in controlling the structural and magnetic properties of $ZnFe_2O_4$ thin films. As a general observation, thin films prepared under high oxygen partial pressure showed less magnetization, and thin films lacking oxygen vacancies showed considerably high magnetization [168].

4 LOW TEMPERATURE MAGNETIC PROPERTIES OF $ZnFe_2O_4$ NANOPARTICLES AND THIN FILMS

Fig. 9 shows the zero field-cooled (ZFC) and field-cooled (FC) magnetization curves of the $ZnFe_2O_4$ thin films fabricated on an Al_2O_3 substrate using the PLD technique [120, 121]. The ZFC-FC curves were measured with an applied magnetic field of 1000 Oe at temperatures between 5 and 300 K. The blocking temperature T_B for $ZnFe_2O_4$ thin films was around 50 K. In comparison with other studies, the observed low magnetization values for $ZnFe_2O_4$ thin films indicate that the films fabricated by different techniques may have different magnetic properties. The ZFC-FC magnetization analysis for $ZnFe_2O_4$ thin films fabricated using sputtering was observed due to the interaction among clusters. The discrepancy between ZFC-FC curves

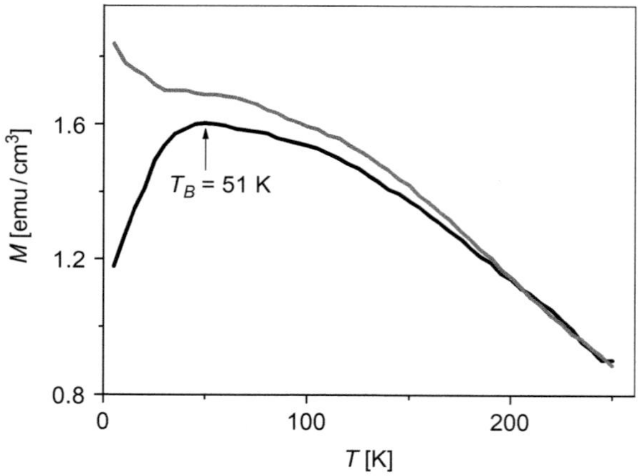

Fig. 9 Temperature-dependent magnetization (ZFC-FC) curves for $ZnFe_2O_4$ thin films measured under an applied field of 1000 Oe (*Red line*: FC; *black line*: ZFC).

below the T_{irr} for $ZnFe_2O_4$ thin films represents glassy behavior that is paramagnetic in nature and the absence of the long-range ferrimagnetic ordering [127].

The magnetic dynamics of nanoparticles can be described by Neel-Brown [86] (Eq. 6)

$$\tau = \tau_0 \exp\left(\frac{E_A}{k_B T}\right) \tag{6}$$

where τ is the superparamagnetic relaxation time of magnetization, E_A is the anisotropy energy barrier height, k_B is the Boltzmann constant, τ_0 is a time relaxation constant ($\sim 10^{-9}$s), and T is temperature. If the measurement/observation time is shorter than τ, the system cannot reach equilibrium during the time window of the single point measurement. Therefore, at low temperatures, the splitting between ZFC and FC $M(T)$ curves is expected. Thus, the outcome of the magnetic measurement depends on the measuring time related to each experimental technique. The blocking temperature T_B is defined as the temperature for which the measuring time is equal to the relaxation time. Above T_B, ZFC and FC curves overlap. For SQUID measurements, $\tau \approx 100$ s and $E_A \approx 25 k_B T_B$, providing an experimental estimate of the anisotropy energy barrier.

The ZFC-FC measurements for $ZnFe_2O_4$ having 31 nm particle sizes are shown in Fig. 10. These measurements were performed with different fields of 10, 100, and 1000 Oe to see the effect of the field on the T_B. It can be

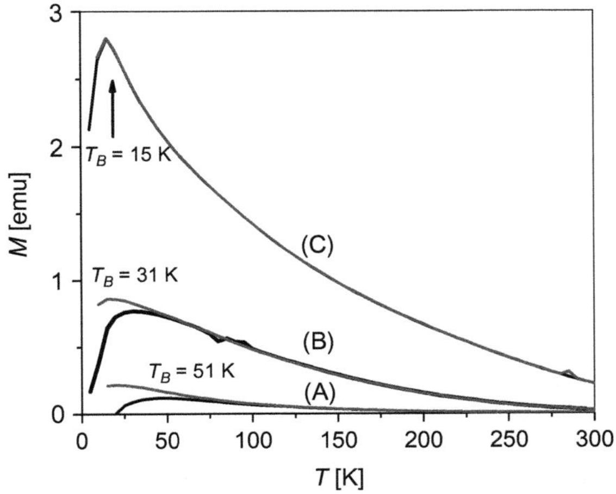

Fig. 10 Temperature-dependent magnetization (ZFC-FC) curves for $ZnFe_2O_4$ nanoparticles measured under the applied fields (A) 10 Oe, (B) 100 Oe, and (C) 1000 Oe of 100 Oe (*Red line*: FC; *black line*: ZFC).

observed from Fig. 10 that, with the increasing applied magnetic field, the blocking temperatures decreased, as expected for a superparamagnet [165]. The blocking temperature for the lowest applied field of 10 Oe is observed to be 51 K, for 100 Oe it is 31 K, and for 1000 Oe it is 15 K. In theory, the blocking temperature of a substance should decrease with increasing applied field. It eventually disappears when the field reaches a critical value of the coercive field (H_C) because the higher field is expected to lower the barriers between the two easy axis orientations, which makes a thermal transition more favorable.

Table 1 lists the values of the measured blocking temperature, the deduced anisotropy barrier height (U), and the effective magnetic anisotropy constant (K_{eff}). The effective magnetic anisotropy constant is calculated by considering that the energy anisotropy barrier of a single particle is proportional to the particle volume. Using the volume $V_{eff}=\Pi D^3/6$ of

Table 1 Applied magnetic field (H), blocking temperature (T_B), barrier height (U), and effective magnetic anisotropy constant (K_{eff})

T_B (K)	U (K)	K_{eff} (J/m^3)	H (Oe)
10	51	1275	1.1×10^3
100	31	775	0.68×10^3
1000	15	375	0.33×10^3

spherical particles, $K_{eff}=U/V_{eff}$ is estimated and the values are given in Table 1. The magnetocrystalline anisotropy constant of $ZnFe_2O_4$ nanocrystals is expected to be close to that of bulk γ-Fe_2O_3 ($\approx 5 \times 10^3 J/m^3$) [84], which is ferrimagnetic and contains only Fe^{3+} in the lattice. The calculated K_{eff} value determined for $ZnFe_2O_4$ nanocrystals with the lowest field (10 Oe) measured initially. It is found to be ($1.1 \times 10^3 J/m^3$) smaller than γ-Fe_2O_3. This drastic decrease may be attributed to the method of synthesis. From the other side, the Stoner-Wohlforth model [169] gives the energy needed for the reversal of single domain particle magnetization, according to $K_{SW}=M_S H_C$. For a sample annealed at 500°C, the K_{SW} was observed to be $4.3 \times 10^3 J/m^3$. The discrepancy between K_{eff} and K_{SW} might originate from nonhomogenous sizes, nonsingle axial anisotropy, and surface contributions. However, the estimated values are satisfactorily close to give an insight into the anisotropy density energy of the synthesized material.

$ZnFe_2O_4$ nanoparticles of 6.6 and 14.8 nm were synthesized by hydrolysis in a polyrol medium. The ZFC-FC curves were measured with different applied fields of 50, 500, and 1000 Oe. A cusp was observed for a 6.6-nm particle at 89 K and for a 14.8-nm particle at 20 K. Therefore, the nanoparticles were considered to be of a single domain. The blocking temperatures for these samples were calculated in accordance with other researchers [88, 98, 170]. The blocking temperature was observed to decrease with increasing applied magnetic field as a typical feature for superparamagnetic particles. The blocking temperature was observed to be less defined for a smaller particle size of 6.6 nm than for a bigger one of 14.8 nm. Dipole-dipole interactions were observed near T_B due to a lot of agglomeration in smaller particles.

ZFC-FC measurements were carried out on $ZnFe_2O_4$ nanoparticles prepared using the citrate method with the applied field of 50 Oe [94, 97]. A sharp cusp around the Néel temperature was observed. But, in the coprecipitation sample, the peak appeared at 30 K. This kind of ZFC behavior at a low temperature was due to the superparamagnetic nature of the samples. This peak in ZFC is designated as the blocking temperature T_B. The broadness of this peak depends on the particle size distribution [171].

Fig. 11 shows the ZFC-FC curves for $ZnFe_2O_4$ 8 nm nanoparticles synthesized using the oxalic acid route and measured with an applied field of 100 Oe [45]. $ZnFe_2O_4$ nanoparticles showed ferromagnetic properties at room temperature. The observed ferromagnetism at much higher temperatures is due to confinement effects that exist in nanoparticles. It should be noted that the T_B is as high as 99 K for $ZnFe_2O_4$. This indicates a strong

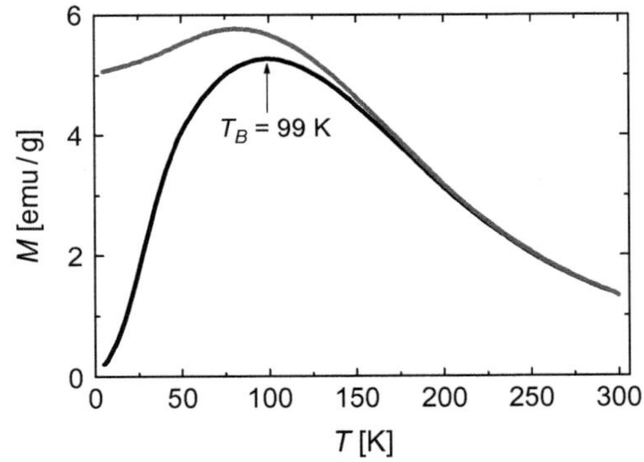

Fig. 11 Temperature-dependent magnetization (ZFC-FC) curves for $ZnFe_2O_4$ nanoparticles measured under the applied field of 100 Oe (*Red line*: FC; *black line*: ZFC).

anisotropy in $ZnFe_2O_4$. A narrow ZFC curve indicates due to the narrow particle size distribution in the synthesized samples [45, 172]. $ZnFe_2O_4$ nanoparticles prepared by the multistep homogeneous method showed the blocking temperature of 6 K for a 3-nm particle. The ZFC-FC curves appeared to be broad with increasing particle size. This indicates the presence of broad relaxation distribution for the metastable states. The observed T_B for these samples is higher than the Néel temperature [98].

$ZnFe_2O_4$ thin films were grown on an Si (111) substrate using rf magnetron sputtering at room temperature. These films were annealed at 200°C and 600°C. The ZFC-FC curves were separated throughout the measurement temperature range due to the small particle size on the film surface. Thin films annealed at 600°C showed a maxima of around 50 K, corresponding to the T_B. The appearance of T_B in these samples is due to a higher particle size on the film surface [97, 141, 155].

ZFC-FC analyses were carried out on $ZnFe_2O_4$ nanoparticles synthesized by mechano-activation on nano- and microsized particles [173]. These samples showed the T_B around 118 K. This confirms that the system goes from a blocked to an unblocked state around maxima. In the $ZnFe_2O_4$ nanoparticles, due to defects, the ferrimagnetic state was observed instead of the paramagnetic state. The T_C for such nanoparticles was higher than 300 K.

ZFC-FC measurements were performed with the applied field of 0.05 and 0.1 T on $ZnFe_2O_4$ incorporated in an amorphous matrix by the melt-quench method [126]. These curves showed broad peaks around 34 K, signifying the typical superparamagnetic behavior. The T_B was observed to be affected by the change in the particle size. The smaller particles showed higher T_B and the higher particle size showed lower T_B values. These T_B values were observed to be far higher than the bulk $ZnFe_2O_4$. The T_B values for $ZnFe_2O_4$ with the same particle size when prepared by other synthesis routes showed entirely different values [60, 105, 174, 175].

$ZnFe_2O_4$ synthesized by the nitrate route was irradiated to study its structural and magnetic properties [176]. The irradiated sample showed a 200% increase in the magnetization. The sample annealed at 300°C showed the T_B around 270 and 70 K under different applied fields. The T_B for unirradiated samples was observed to be 20 K and was unaffected with the applied magnetic fields. Irradiated samples showed divergence in ZFC-FC curves and in the unirradiated sample, they did not. This kind of behavior was due to the presence of defects or higher magnetic ordering in the samples. From the EPR analysis, it was observed that, with irradiation, the peaks get broader due to induced magnetic interactions and the increased strength of the magnetic moment [11, 176].

$ZnFe_2O_4$ sputtered thin films annealed at different temperatures showed interesting magnetic properties [66]. ZFC-FC measurements were performed with the applied field of 50 Oe. For the bulk $ZnFe_2O_4$, a sharp cusp was observed at 11 K, which is close to the antiferromagnetic Néel temperature of 10 K. From the neutron diffraction analysis on a $ZnFe_2O_4$ single crystal, it was observed that the long-range magnetic ordering in the normal $ZnFe_2O_4$ is not achieved, even at a very low temperature of 1.5 K. For as-prepared $ZnFe_2O_4$ thin films, the ZFC-FC curves typically represent the spin freezing behavior at a temperature higher than room temperature. The discrepancy between the ZFC-FC curves and the high spin-freezing temperature was due to the metastable phase and higher cation distribution. With the increase in the annealing temperature, the spin-freezing temperature was observed to shift toward lower temperatures.

Shim et al. [70] synthesized $ZnFe_2O_4$ nanoparticles using ball milling and measured ZFC-FC curves with applied magnetic fields of 500 Oe. These samples showed a net magnetic moment below temperature 460 K. In the ZFC-FC curves, a broad peak was observed around 150 K. This represents the magnetic moment of the synthesized $ZnFe_2O_4$ nanoparticles due to the occurrence of T_B. Surprisingly, the *M-H* curves for these samples were observed only at 5 K, and at room temperature they were superparamagnetic.

$ZnFe_2O_4$ thin films were deposited on $SrTiO_3$ substrates using the PLD technique [167]. These films showed higher coercivity of 1140 Oe at 5 K. From the ZFC-FC analysis, a broad maxima was observed around 105 K, indicating the wide grain size distribution. The variations in the magnetic properties were observed to be dependent on the annealing temperature.

$ZnFe_2O_4$ thin films were fabricated on Si (111) substrates using the PLD technique. These films were irradiated with 90 keV neon ions to see the structural and magnetic changes in the films [174]. ZFC-FC curves were observed to overlap above T_B, which might be due to the presence of small particle sizes in the samples. The maximum in ZFC curves was observed at $T_B \approx 216\,K$. After this temperature, ZFC decreased due to the transition from the superparamagnetic phase to the ferromagnetic phase. This transition activates the anisotropy, which pushes the magnetization toward easy axis. Therefore, the magnetic moment will orient randomly. A shift in the T_B was observed from 216 to 226 K for the irradiated samples [86]. The change in the magnetization may be due to a partial inversion in the spinel phase [177]. To know the effect of irradiation on the thin films, it can be assumed that the Fe^{3+} ions displace to (A) sites. During irradiation, a large number of ions and atoms are supposed to undergo displacement from the [B] to (A) sites [178]. Therefore, the irradiation showed small impact on shifting T_B compared to unirradiated films. It is confirmed that, due to irradiation, large amount of Fe^{3+} ions would dislocate, thereby showing enhanced magnetization.

5 CONCLUDING REMARKS

There are numerous publications that have focused on the unusual magnetic properties of $ZnFe_2O_4$ in nanomaterials. Before the existence of nanomaterials, $ZnFe_2O_4$ was always proved to be antiferromagnetic with a Néel temperature around 10 K. $ZnFe_2O_4$ in a nanostructured form started showing a net magnetic moment at room temperature due to the random distribution of Zn^{2+} and Fe^{3+} ions over (A) and [B] sites. The inversion in $ZnFe_2O_4$ is achieved due to different synthesis, preparation, and fabrication techniques. The inversion parameter contributes in changing the normal spinel ferrite to a mixed spinel ferrite. It is observed that the inversion parameter has great impact on achieving the ferromagnetic properties in $ZnFe_2O_4$ nano and thin films. They play an important role in shifting the T_B. Due to this, $ZnFe_2O_4$ would undergo to a metastable state to show ferromagnetic or ferrimagnetic behavior. The inversion can be tuned with particle size or annealing temperature. The dominant (A-B) interactions make $ZnFe_2O_4$ behave like ferromagnetic material. The (B-B) interactions induce antiferromagnetic interactions in [B] sites. The (A-B) interactions induce the ferrimagnetic coupling between the (A) and [B] sites. But still, the spin structure in the nanocrystalline $ZnFe_2O_4$ was observed to be controversial until now. $ZnFe_2O_4$ thin films behave ferromagnetically due to defects. The defects can be reduced in a normal spinel with annealing temperatures. By selecting a suitable synthesis technique and annealing temperatures, one can control the particle size in order to achieve the ferromagnetism in $ZnFe_2O_4$ nanoparticles or thin films. By inducing defects in $ZnFe_2O_4$ nanoparticles or thin films during the preparation or fabrication process, we can easily tune the magnetic properties. From the theoretical concept of understanding the magnetic properties of $ZnFe_2O_4$, it has become one of the most challenging tasks that is under consideration, and still there is no clear evidence for such magnetic behavior in nonmagnetic $ZnFe_2O_4$.

REFERENCES

[1] R. Singh, Unexpected magnetism in nanomaterials, J. Magn. Magn. Mater. 346 (2013) 58.
[2] H.-J. Freund, G. Pacchioni, Oxide ultra-thin films on metals: new materials for the design of supported metal catalysts, Chem. Soc. Rev. 37 (2008) 2224–2242.
[3] J.F. Quinn, A.P.R. Johnston, G.K. Such, A.N. Zelikin, F. Caruso, Next generation, sequentially assembled ultrathin films: beyond electrostatics, Chem. Soc. Rev. 36 (2007) 707–718.
[4] F. Baletto, R. Ferrando, Structural properties of nanoclusters: Energetic, thermodynamic, and kinetic effects, Rev. Mod. Phys. 77 (2005) 371–423.
[5] S.D. Bader, Opportunities in nanomagnetism, Rev. Mod. Phys. 78 (2006) 1–15.
[6] J. Martín, J. Nogués, K. Liu, J. Vicent, I.K. Schuller, Ordered magnetic nanostructures: fabrication and properties, J. Magn. Magn. Mater. 256 (2003) 449–501.
[7] S.P. Gubin, Y.A. Koksharov, G.B. Khomutov, G.Y. Yurkov, Magnetic nanoparticles: preparation, structure and properties, Russ. Chem. Rev. 74 (2005) 489.
[8] R. Singh, P. Kroll, Structural, electronic, and magnetic properties of 13-, 55-, and 147-atom clusters of Fe, Co, and Ni: a spin-polarized density functional study, Phys. Rev. B 78 (2008) 245404.
[9] N.H. Hong, N. Poirot, J. Sakai, Ferromagnetism observed in pristine SnO_2 thin films, Phys. Rev. B 77 (2008) 033205.
[10] T. Dietl, H. Ohno, F. Matsukura, J. Cibert, D. Ferrand, Zener model description of ferromagnetism in zinc-blende magnetic semiconductors, Science 287 (2000) 12019.
[11] N.H. Hong, Ferromagnetism in transition-metal-doped semiconducting oxide thin films, J. Magn. Magn. Mater. 303 (2006) 338.
[12] M. Venkatesan, C.B. Fitzgerald, J.M.D. Coey, Unexpected magnetism in a dielectric oxide, Nature (London) 430 (2004) 630.
[13] N.H. Hong, J. Sakai, N. Poirot, V. Brizé, Room-temperature ferromagnetism observed in undoped semiconducting and insulating oxide thin films, Phys. Rev. B 73 (2006) 132404.
[14] S.D. Yoon, Y. Chen, A. Yang, T.L. Goodrich, X. Zuo, D.A. Arena, K. Ziemer, C. Vittoria, V.G. Harris, Oxygen-defect-induced magnetism to 880K in semiconducting anatase TiO_{2-d} films, J. Phys. Condens. Matter 18 (2006) L355.

[15] N.H. Hong, J. Sakai, V. Brizé, Observation of ferromagnetism at room temperature in ZnO thin films, J. Phys. Condens. Matter 19 (2007)036219.

[16] J. Shen, Z. Gai, J. Kirschner, Growth and magnetism of metallic thin films and multilayers by pulsed-laser deposition, Surf. Sci. Rep. 52 (2004) 163–218.

[17] C.A.F. Vaz, J.A.C. Bland, G. Lauhoff, Magnetism in ultrathin film structures, Rep. Prog. Phys. 71 (2008) 056501.

[18] R. Singh, P. Kroll, Magnetism in strained pseudomorphic ultrathin films of fcc 3d-transition metals (Cr, Mn, Fe, Co and Ni) with lateral lattice parameters of bulk fcc-Cu(001), J. Magn. Magn. Mater. 321 (2009) 2827–2832.

[19] R. Singh, Magnetic coupling in pseudomorphic 2 ml overlayers and sandwich superlattice structures of Cr, Mn, Fe, Co and Ni on fcc Cu(001), Int. J. Mod. Phys. B 24 (2010) 405–412.

[20] I.M. Billas, A. Ch ̂atelain, W.A. de Heer, Magnetism of Fe, Co and Ni clusters in molecular beams, J. Magn. Magn. Mater. 168 (1997) 64–84.

[21] A.D. Zdetsis, Optical and electronic properties of small size semiconductor nanocrystals and nanoclusters, Rev. Adv. Mater. Sci. 11 (2006) 56–78.

[22] M.B. Knickelbein, Magnetic moments of bare and benzene-capped cobalt clusters, J. Chem. Phys. 125 (2006) 044308.

[23] L. Gamma, A.P. Diniz, A.C.F.M. Costa, S.M. Rezende, A. Azevedo, D.R. Corenjo, Magnetic properties of nanocrystalline NiZn ferrites doped with samarium, J. Magn. Magn. Mater. 384 (2006) 97.

[24] G.C. Lin, Q. Wei, J.X. Zhang, Direct measurement of the magnetocaloric effect in $La_{0.67}Ca_{0.33}MnO_3$, J. Magn. Magn. Mater. 300 (2006) 97.

[25] A.R. Dinesen, S. Linderoth, S. Morup, Direct and indirect measurement of the magnetocaloric effect in $La_{0.67}Ca_{0.33-}xSrxMnO_{3\ \pm\ \delta}$ (x = [0: 0.33]), J. Phys. Condens. Matter 17 (2005) 6257.

[26] Y. Xu, A.M. Meier, P. Das, M.R. Libischka, U. Hartmann, Perovskite manganites: potential materials for magnetic cooling at or near room temperature, Cryst. Eng. 5 (2002) 383.

[27] E.J. Choi, Y. Ahu, E.J. Hahn, Size dependence of the magnetic properties in superparamagnetic zinc-ferrite nanoparticles, J. Korean Phys. Soc. 53 (2008) 2090.

[28] A.T. Raghavender, K. Zadro, D. Pajic, Z. Skoko, N. Biliškov, Effect of grain size on the Néel temperature of nanocrystalline nickel ferrite, Mater. Lett. 64 (2010) 1144.

[29] A.T. Raghavender, N. Biliškov, Ž. Skoko, XRD and IR analysis of nanocrystalline Ni–Zn ferrite synthesized by the sol–gel method, Mater. Lett. 65 (2011) 677.

[30] A.T. Raghavender, D. Pajic, K. Zadro, T. Milekovic, P. Venkateshwar Rao, K. M. Jadhav, D. Ravinder, Synthesis and magnetic properties of $NiFe_{2-x}Al_xO_4$ nanoparticles, J. Magn. Magn. Mater. 316 (2007) 1.

[31] S.D. Shenoy, P.A. Joy, M.R. Anantharaman, Effect of mechanical milling on the structural, magnetic and dielectric properties of coprecipitated ultrafine zinc ferrite, J. Magn. Magn. Mater. 269 (2004) 217.

[32] M.G. Scott, Amorphous Metallic Alloys, Butterworths, London, 1983 151 p.

[33] H. Yang, X. Zhang, C. Huang, W. Yang, G. Qiu, Synthesis of $ZnFe_2O_4$ nanocrystallites by mechanochemical reaction, J. Phys. Chem. Solids 65 (2004) 1329.

[34] T. Sato, K. Haneda, T. Iijima, M. Seki, Morphology and magnetic properties of ultrafine $ZnFe_2O_4$ particles, J. Appl. Phys. A 50 (1990) 13.

[35] P. Uniyal, K.L. Yadav, Synthesis and study of multiferroic properties of $ZnFe_2O_4$-$BiFeO_3$ nanocomposites, J. Alloys Compd. 492 (2010) 406.

[36] A. Verma, T.C. Goel, R.G. Mendiratta, R.G. Gupta, High-resistivity nickel–zinc ferrites by the citrate precursor method, J. Magn. Magn. Mater. 192 (1999) 271.

[37] M. Zheng, X.C. Wu, B.S. Zou, Y.J. Wang, Magnetic properties of nanosized $MnFe_2O_4$ particles, J. Magn. Magn. Mater. 183 (1998) 152.

[38] Q. Chen, J.Z. Zhang, Size-dependent superparamagnetic properties of $MgFe_2O_4$ spinel ferrite nanocrystallites, Appl. Phys. Lett. 73 (1998) 3156.
[39] S. Kumar, V. Singh, S. Aggarwal, U.K. Mandal, R.K. Kotnala, Synthesis of nanocrystalline $Ni_{0.5}Zn_{0.5}Fe_2O_4$ ferrite and study of its magnetic behavior at different temperatures, Mater. Sci. Eng. B 166 (2010) 76.
[40] E. Hasmonay, J. Depeyrot, M.H. Sousa, F.A. Tourinho, J.C. Bacri, R. Perzynski, Y. J. Raikher, I. Rosenman, Magnetic and optical properties of ionic ferrofluids based on nickel ferrite nanoparticles, J. Appl. Phys. 88 (2000) 6628.
[41] C. Caizer, M. Stefanscu, Magnetic characterization of nanocrystalline Ni–Zn ferrite powder prepared by the glyoxylate precursor method, J. Phys. D 35 (2002) 3035.
[42] D.G. Wickham, Metal Iron (III) oxides, Inorg. Synth. 9 (1967) 152.
[43] M. Bremer, S. Fisher, H. Langbein, W. Topelmann, H. Scheler, Investigation on the formation of manganese-zinc ferrites by thermal decomposition of solid solution oxalates, Thermochim. Acta 209 (1992) 323.
[44] A.T. Raghavender, S.E. Shirsath, K. Vijaya Kumar, Synthesis and study of nanocrystalline Ni–Cu–Zn ferrites prepared by oxalate based precursor method, J. Alloys Compd. 509 (2011) 7004.
[45] N.H. Hong, A.T. Raghavender, O. Ciftja, M.-H. Phan, K. Stojak, H. Srikanth, Y. H. Zhang, Ferrite nanoparticles for future heart diagnostics, Appl. Phys. A Mater. Sci. Process. 112 (2013) 323.
[46] T. Sato, K. Haneda, M. Seki, T. Iijima, Proceedings of the International Symposium of Physics of Magnetic Materials, World scientific, Singapore, 1987. 210 p.
[47] T. Sato, K. Haneda, M. Seki, T. Iijima, Morphology and magnetic properties of ultrafine $ZnFe_2O_4$ particles, Appl. Phys. A Mater. Sci. Process. 50 (1990) 13.
[48] T. Sato, K. Haneda, T. Iijima, M. Seki, Proceedings of the International Conference on Ferrite, The Japan Society of Powder and Powder Metallurgy, Tokyo and Kyoto, 1992. 284 p.
[49] T. Kamiyama, K. Haneda, T. Sato, S. Ikeda, H. Asano, Cation distribution in $ZnFe_2O_4$ fine particles studied by neutron powder diffraction, Solid State Commun. 81 (1992) 563.
[50] E.K. Lotgering, The influence of Fe^{3+} ions at tetrahedral sites on the magnetic properties of $ZnFe_2O_4$, J. Phys. Chem. Solids 27 (1966) 139.
[51] R.N. Bhowmik, R. Ranganathan, R. Nagarajan, Coexistence of spin glass and superparamagnetism with ferrimagnetic order in polycrystalline spinel $Co_{0.2}Zn_{0.8}Fe_{1.95}Ho_{0.05}O_4$, J. Magn. Magn. Mater. 299 (2006) 327.
[52] R.N. Bhowmik, R. Ranganathan, Anomaly in cluster glass behaviour of $Co_{0.2}Zn_{0.8}Fe_2O_4$ spinel oxide, J. Magn. Magn. Mater. 248 (2002) 101.
[53] J. Geshev, L.G. Pereira, J.E. Schmidt, M. Mikhov, Dependence of the magnetization and remanence of single-domain particles on the second cubic anisotropy constant, J. Appl. Phys. 90 (2010) 6243.
[54] H.H. Hamedh, J.C. Ho, Magnetic properties of partially-inverted zinc ferrite aerogel powders, J. Appl. Phys. 81 (1997) 1851.
[55] Y. Ahn, E.J. Choi, S. Kim, D.H. An, K.U. Kang, B.G. Lee, K.S. Baek, H.N. Oak, Magnetization and Mossbauer study of nanosize $ZnFe_2O_4$ particles synthesized by using a microemulsion method, J. Korean Phys. Soc. 41 (2002) 123.
[56] C.N. Chinnasamy, A. Narayansamy, N. Ponpandian, K. Chattopadhyay, The influence of Fe^{3+} ions at tetrahedral sites on the magnetic properties of nanocrystalline $ZnFe_2O_4$, Mater. Sci. Eng. A 304–306 (2001) 983.
[57] J.P. Singh, R.S. Payal, R.C. Srivastava, H.M. Agrawal, P. Chand, A. Tripathi, R. P. Tripathi, Effect of thermal treatment on the magnetic properties of nanostructured zinc ferrite, J. Phys. Conf. Ser. 217 (2016) 012108.

[58] S.H. Yu, T. Fujino, M. Yoshimura, Hydrothermal synthesis of $ZnFe_2O_4$ ultrafine particles with high magnetization, J. Magn. Magn. Mater. 256 (2003) 420.

[59] F.S. Li, L. Wang, J.B. Wang, Q.G. Zhou, X.Z. Zhou, H.P. Kunkal, G. Williams, Site preference of Fe in nanoparticles of $ZnFe_2O_4$, J. Magn. Magn. Mater. 268 (2004) 332.

[60] R. Kumar, S.K. Sharma, A. Dogra, V.V. Shiva Kumar, S.N. Dolia, A. Gupta, M. Knobel, M. Sing, Magnetic study of nanocrystalline ferrites and the effect of swift heavy ion irradiation, Hyperfine Interact. 160 (2005) 143.

[61] R. Saez Pocha, M.J. Torralvo Fernandez, V. Balanco Gulierrez, R. Gomez, V. Marquina, M.L. Marquina, J.L. Perez Mazariego, R. Ridaura, Ferrites nanoparticles MFe_2O_4 (M = Ni and Zn): hydrothermal synthesis and magnetic properties, Bol. Soc. Esp. Ceram. 47 (2008) 133.

[62] N.M. Deraz, A. Alarifi, Synthesis and physicochemical properties of nanomagnetic zinc ferrite system, Int. J. Electrochem. Sci. 7 (2012) 3798.

[63] J.P. Singh, G. Dixit, R.C. Srivatsava, H. Kumar, H.M. Agrawal, P. Chand, R. Kumar, Observation of size dependent attributes on the magnetic resonance of irradiated zinc ferrite nanoparticles, Curr. Appl. Phys. 11 (2011) 532.

[64] M. Sulta, R. Singh, FMR studies on nanocrystalline zinc ferrite thin films, J. Phys. Conf. Ser. 200 (2010) 722090.

[65] M.K. Roy, B. Halder, H.C. Verma, Characteristic length scales of nanosize zinc ferrite, Nanotechnology 17 (2006) 232.

[66] S. Nakashima, K. Fujita, K. Tanaka, K. Hirao, T. Yamamoto, I. Tanaka, First-principles XANES simulations of spinel zinc ferrite with a disordered cation distribution, Phys. Rev. B 75 (2007) 174443.

[67] M. Hofmann, S.J. Campbell, H. Ehrhardt, R. Feyerharm, The magnetic behaviour of nanostructured zinc ferrite, J. Mater. Sci. 39 (2004) 5057.

[68] S.H. Lee, C. Broholm, W. Ratclife, G. Gasparovic, Q. Huang, T.H. Kim, S. W. Cheong, Emergent excitations in a geometrically frustrated magnet, Nat. Mater. 418 (2002) 856.

[69] C.N. Chinnasamy, A. Narayansamy, N. Ponpandian, K. Chattopadhyaya, H. Huerant, H. Greneche, Magnetic properties of nanostructured ferrimagnetic zinc ferrite, J. Phys. Condens. Matter 12 (2000) 7795.

[70] J.H. Shim, S. Lee, J.H. Park, S.J. Han, Y.H. Jeong, Y.W. Cho, Coexistence of ferrimagnetic and antiferromagnetic ordering in Fe-inverted zinc ferrite investigated by NMR, Phys. Rev. B 73 (2006) 064404.

[71] J.P. Singh, G. Dixit, R.C. Srivatsava, H.M. Agrawal, V.R. Reddy, A. Gupta, Observation of bulk like magnetic ordering below the blocking temperature in nanosized zinc ferrite, J. Magn. Magn. Mater. 324 (2012) 2553.

[72] J.C. Ho, H.H. Hamdeh, Y.S. Chen, S.H. Lin, Y.D. Yao, R.J. Willley, S.A. Oliver, Low-temperature calorimetric properties of zinc ferrite nanoparticles, Phys. Rev. B 52 (1995) 10122.

[73] G.F. Goya, H.R. Richenberg, M. Chen, W.B. Yelon, Magnetic irreversibility in ultrafine $ZnFe_2O_4$ particles, J. Appl. Phys. 87 (2000) 8005.

[74] H.H. Hamdeh, J.C. Ho, S.A. Oliver, R.J. Willey, J. Kramer, M. Daturi, G. Busca, Ferrimagnetic zinc ferrite fine powders, IEEE Trans. Magn. 31 (1995) 3808.

[75] M.R. Anantharaman, S. Jagathesan, K.A. Malini, S. Sindhu, A. Narayansamy, K. Seshan, R.H.H. Smits, H.H. Brongersma, On the magnetic properties of ultra-fine zinc ferrites, J. Magn. Magn. Mater. 189 (1998) 83.

[76] G.F. Goya, H.R. Rechenberg, Ionic disorder and Néel temperature in $ZnFe_2O_4$ nanoparticles, J. Magn. Magn. Mater. 196–197 (1993) 91.

[77] C.N. Chinnasamy, A. Narayansamy, N. Ponpandian, K. Chattopadhyay, K. Shinoda, B. Jeyadevan, K. Tohji, K. Nakatsuka, T. Furubayashi, I. Nakatani, Mixed spinel structure in nanocrystalline $NiFe_2O_4$, Phys. Rev. B 63 (2001) 184108.

[78] W. Schiessel, W. Potzel, H. Karzel, M. Steiner, G.M. Kalvius, A. Martin, A. Martin, M.K. Krause, I. Halevy, J. Gal, W. Schafer, G. Will, M. Hillberg, R. Wappling, Magnetic properties of the $ZnFe_2O_4$ spinel, Phys. Rev. B 53 (1996) 9143.
[79] K. Hanea, Recent advances in the magnetism of fine particles, Can. J. Phys. 65 (1987) 1233.
[80] A.H. Morrish, K. Haneda, X.Z. Zhou, Proceedings of NATO Advanced Study Institute on Nanophase Materials-Syntheses, Properties and Applications, Kluwer Academics, Netherlands, 1994. 515 p.
[81] M.T. Clerk, B.J. Evans, Enhanced magnetization and cation distributions in nanocrystalline $ZnFe_2O_4$: a conversion electron Mossbauer spectroscopic investigation, IEEE Trans. Magn. 33 (1997) 3745.
[82] X. Batle, A. Laborta, Finite-size effects in fine particles: magnetic and transport properties, J. Phys. D 35 (2002) R15.
[83] J. Sort, J. Nogues, S. Surinach, J.S. Munoz, M.D. Baro, E. Chappal, F. Dupout, G. Choutean, Coercivity and squareness enhancement in ball-milled hard magnetic–antiferromagnetic composites, Appl. Phys. Lett. 79 (2001) 1142.
[84] S. Ammar, N. Jouini, F. Fievet, Z. Beji, L. Smiri, P. Moline, M. Donot, J.-M. Greneche, Magnetic properties of zinc ferrite nanoparticles synthesized by hydrolysis in a polyol medium, J. Phys. Condens. Matter. 18 (2006) 9055.
[85] S.A. Oliver, V.G. Harris, H.H. Hamdeh, J.C. Ho, Large zinc cation occupancy of octahedral sites in mechanically activated zinc ferrite powders, Appl. Phys. Lett. 76 (2000) 2761.
[86] J.F. Hochepied, P. Bonville, M.P. Pilini, Nonstoichiometric zinc ferrite nanocrystals: syntheses and unusual magnetic properties, J. Phys. Chem. B 104 (2000) 905.
[87] M. Andrew-Verges, M. Martinez, E. Matijevic, Synthesis and characterization of zinc ferrite particles prepared by hydrothermal decomposition of zinc chelate solutions, J. Mater. Res. 8 (1993) 2916.
[88] G.F. Goya, H.R. Rechenberg, Magnetic properties of $ZnFe_2O_4$ synthesized by ball milling, J. Magn. Magn. Mater. 203 (1999) 141.
[89] V. Sepelak, U. Steinike, D.C. Vecker, S. Wissmann, K.D. Becker, Structural disorder in mechanosynthesized zinc ferrite, J. Solid State Chem. 135 (1998) 52.
[90] Y. Yafet, C. Kittel, Antiferromagnetic arrangements in ferrites, Phys. Rev. 87 (1952) 290.
[91] F.J. Burghart, W. Potzel, G.M. Kalvins, E. Schreier, G. Grusse, D. Noakes, R. W. Schafer, A. Marrin, M.K. Krauss, Magnetism of crystalline and nanostructured $ZnFe_2O_4$, Phys. B 289 (290) (2000) 286.
[92] G. Thirupathi, R. Singh, Magnetic properties of zinc ferrite nanoparticles, IEEE Trans. Magn. 48 (2012) 1.
[93] R.H. Kodama, Magnetic nanoparticles, J. Magn. Magn. Mater. 200 (1999) 359.
[94] A. Kundu, C. Upadhyay, H.C. Verma, Magnetic properties of a partially inverted zinc ferrite synthesized by a new coprecipitation technique using urea, Phys. Lett. A 311 (2003) 410.
[95] P.M. Prithviraj Swamy, S. Basavaraju, A. Laga Shetty, N.V. Srinivas Rao, R. Nijagunappa, A. Venkataraman, Synthesis and characterization of zinc ferrite nanoparticles obtained by self-propagating low-temperature combustion method, Bull. Mater. Sci. 34 (2011) 1325.
[96] K. Maaz, A. Mumtaz, S.K. Hasananin, C. Abdullah, Synthesis and magnetic properties of cobalt ferrite ($CoFe_2O_4$) nanoparticles prepared by wet chemical route, J. Magn. Magn. Mater. 308 (2007) 289.
[97] C. Upadhyay, H.C. Verma, V. Sathe, A.V. Pimpale, Effect of size and synthesis route on the magnetic properties of chemically prepared nanosize $ZnFe_2O_4$, J. Magn. Magn. Mater. 312 (2007) 271.

[98] L.D. Tung, V. Kolesnichenko, G. Caruntu, D. Caruntu, Y. Remond, V.O. Golub, C. J. O'Connor, L. Spinu, Annealing effects on the magnetic properties of nanocrystalline zinc ferrite, Phys. B 319 (2002) 116.

[99] R. Raeisi Shahraki, M. Ebrahimi, S.A. Seyyed Ebrahimi, S.M. Masoudpanah, Structural characterization and magnetic properties of superparamagnetic zinc ferrite nanoparticles synthesized by the coprecipitation method, J. Magn. Magn. Mater. 324 (2012) 3762.

[100] A.N. Birgani, M. Niyaifar, A. Hasanpour, Study of cation distribution of spinel zinc nano-ferrite by X-ray, J. Magn. Magn. Mater. 374 (2015) 179.

[101] L.-J. Cao, Q.-H. Zhou, G. Li, H.-X. Hong, Q.-H. Li, P. Lan, Y. Fang, Preparation and characterization of composite microspheres of nano zinc ferrite/poly (D,L-lactide-co-alanine), Trans. Nonferrous Metals Soc. China 22 (2012) 360.

[102] X. Huang, J. Zhang, S. Xiao, T. Song, G. Chen, Unique electromagnetic properties of the zinc ferrite nanofiber, Mater. Lett. 124 (2014) 126.

[103] M.G. Naseri, E.B. Saion, M. Hashim, A.A. Shaari, H.A. Ahangar, Synthesis and characterization of zinc ferrite nanoparticles by a thermal treatment method, Solid State Commun. 151 (2011) 1031.

[104] J.P. Singh, R.C. Srivathsava, H.M. Agrawal, R. Kumar, Magnetic behaviour of nanosized zinc ferrite under heavy ion irradiation, Nucl. Inst. Methods Phys. Res. B 268 (2010) 1422.

[105] J.P. Singh, G. Dixit, R.C. Srivatsava, H. Kumar, H.M. Agrawal, R. Kumar, Formation of latent tracks and their effects on the magnetic properties of nanosized zinc ferrite, J. Magn. Magn. Mater. 352 (2014) 36.

[106] J.P. Singh, G. Dixit, R.C. Srivatsava, H. Kumar, H.M. Agrawal, R. Kumar, Study of size dependent features of swift heavy ion irradiation in nanosized zinc ferrite, J. Magn. Magn. Mater. 324 (2012) 3306.

[107] S. Surinwong, A. Rujiwatra, Ultrasonic cavitation assisted solvothermal synthesis of superparamagnetic zinc ferrite nanoparticles, Particuology 11 (2013) 588.

[108] Y. Koseoglu, A. Baykal, A. Tuprak, F. Goznak, A.C. Basaran, A. Akthar, Synthesis and characterization of ZnFe2O4 magnetic nanoparticles via a PEG-assisted route, J. Alloys Compd. 462 (2008) 209.

[109] J.E.F. Westrum Jr., D.M. Grimes, Low temperature heat capacity and thermodynamic properties of zinc ferrite, J. Phys. Chem. Solids 3 (1957) 44.

[110] J.M. Hastings, L.M. Corliss, An antiferromagnetic transition in zinc ferrite, Phys. Rev. 102 (1956) 1460.

[111] S.A. Oliver, H.H. Hamdeh, J.C. Ho, Localized spin canting in partially inverted $ZnFe_2O_4$ fine powders, Phys. Rev. B 60 (1999) 3400.

[112] U. Konig, E.F. Bertuat, Y. Gross, M. Mitrikov, G. Chol, Models of the magnetic structure of zinc ferrite, Solid State Commun. 8 (1970) 759.

[113] S.J. Stewart, S.J.A. Figueera, J.M. Ramallo Lopez, S.G. Marchetti, J.F. Bengoa, R.J. Prado, F.G. Requejo, Cationic exchange in nanosized $ZnFe_2O_4$ spinel revealed by experimental and simulated near-edge absorption structure, Phys. Rev. B 75 (2007) 073408.

[114] M. Gharagozlou, R. Bayati, Low temperature processing and magnetic properties of zinc ferrite nanoparticles, Superlattice. Microstruct. 78 (2015) 190.

[115] A.T. Raghavender, N.H. Hong, Dependence of Néel temperature on the particle size of $MnFe_2O_4$, J. Magn. Magn. Mater. 323 (2011) 2145.

[116] K.V.M.P. Shafi, A. Gedanken, R. Prozorov, J. Balogh, Sonochemical preparation and size-dependent properties of nanostructured $CoFe_2O_4$ particles, Chem. Mater. 10 (1998) 3445.

[117] M. Rajendran, R.C. Pullar, A.K. Bhattacharya, D. Das, S.N. Chintalpudi, C. K. Majumdar, Magnetic properties of nanocrystalline $CoFe_2O_4$ powders prepared at room temperature: variation with crystallite size, J. Magn. Magn. Mater. 232 (2001) 71.

[118] X. Wang, L. Chen, Q. Fan, J. Fan, G. Xu, M. Yan, M.J. Hendersen, J. Courtois, K. Xiong, Lactoferrin-assisted synthesis of zinc ferrite nanocrystal: Its magnetic performance and photocatalytic activity, J. Alloys Compd. 652 (2015) 132.
[119] Q.-Y. Xiang, D. Wu, Y. Bai, K. Yan, W.-Q. Yao, L. Zhang, J. Zhang, J.-L. Cao, Atomic hydrogenation-induced paramagnetic-ferromagnetic transition in zinc ferrite, Ceram. Int. 42 (2016) 16882.
[120] A.T. Raghavender, Room temperature ferromagnetism in laser ablated Zn ferrite thin films, Mater. Lett. 65 (2011) 3636.
[121] A.T. Raghavender, N.H. Hong, E. Chikoidze, Y. Dumont, M. Kurisu, Effect of zinc doping on the structural and magnetic properties of nickel ferrite thin films fabricated by pulsed laser deposition technique, J. Magn. Magn. Mater. 378 (2015) 358.
[122] S. Ayyappan, S. Philip Raja, C. Venkateswaran, J. Philip, B. Raj, Ferromagnetism in vacuum annealed $ZnFe_2O_4$ nanoparticles, J. Appl. Phys. 96 (2010) 143106.
[123] G.F. Goya, H.R. Rechenberg, Ionic disorder and Néel temperature in $ZnFe_2O_4$ nanoparticles, J. Magn. Magn. Mater. 196–197 (1999) 191.
[124] S. Nakashima, K. Fujita, K. Tanaka, K. Hirao, T. Yamamoto, I. Tanaka, Thermal annealing effect on magnetism and cation distribution in disordered $ZnFe_2O_4$ thin films deposited on glass substrates, J. Magn. Magn. Mater. 310 (2007) 2543.
[125] M. Sultan, R. Singh, FMR studies on nanocrystalline zinc ferrite thin films, J. Appl. Phys. 105 (2009) 07A512.
[126] M. Bohra, S. Prasad, N. Kumar, D.S. Misra, S.C. Sahoo, N. Venkataramani, R. Krishnan, Large room temperature magnetization in nanocrystalline zinc ferrite thin films, Appl. Phys. Lett. 88 (2006) 262506.
[127] S. Nakashima, K. Fujita, K. Tanaka, K. Hirao, High magnetization and the high-temperature superparamagnetic transition with intercluster interaction in disordered zinc ferrite thin film, J. Phys. Condens. Matter. 17 (2005) 137.
[128] Y. Yamamoto, H. Tanaka, T. Kawai, The control of cluster-glass transition temperature in spinel-type ZnFe2O4-delta thin film, Jpn. J. Appl. Phys. 40 (2001) 545.
[129] K. Tanaka, M. Makita, H. Hirao, N. Soga, Effect of heat treatment on magnetic properties of ferrimagnetic zinc ferrite prepared by rapidly quenching method, J. Magn. Soc. Jpn. 20 (1998) 77.
[130] K. Tanaka, M. Makita, H. Hirao, N. Soga, Structure and high magnetization of rapidly quenched zinc ferrite, J. Phys. Chem. Solids 59 (1998) 1611.
[131] J.L. Dormann, N. Nogues, Magnetic structures in substituted ferrites, J. Phys. Condens. Matter 2 (1990) 1223.
[132] D. Fiorani, S. Viticoli, J.L. Dormann, J.L. Tholence, A.P. Murni, Spin-glass behavior in an antiferromagnetic frustrated spinel: $ZnCr_{1.6}Ga_{0.4}O_4$, Phys. Rev. 30 (1984) 2776.
[133] R.A. Brand, H. Georges-Gibert, J. Hubsch, J.A. Heller, Ferrimagnetic to spin glass transition in the mixed spinel $Mg_{1+t}Fe_{2-2t}Ti_tO_4$: a Mossbauer and DC susceptibility study, J. Phys. F Met. Phys. 15 (1985) 1987.
[134] R. Singh, S.C. Bhargava, Magnetization and Mossbauer studies of $Zn_{0.5}Co_{0.9}Fe_{1.2}Ti_{0.4}O_4$, J. Phys. Condens. Matter 7 (1995) 8183.
[135] S.C. Bhargava, A.H. Morrish, H. Kunkel, Spin-glass ordering in a spinel ferrite, $Mg(Al, Fe)_2O_4$, J. Phys. Condens. Matter 12 (2000) 101.
[136] J.A. Mydosh, Spin Glasses: An Experimental Introduction, Taylor and Francis, London, 1993.
[137] M. Itoh, I. Notori, S. Kubota, K. Motoya, Spin-glass behavior and magnetic phase diagram of $La_{1-x}Sr_xCoO_3$ ($0 \leq x \leq 0.5$) studied by magnetization measurements, J. Phys. Soc. Jpn. 63 (1994) 1486.
[138] S. Mukharjee, R. Ranganatha, P.S. Anil Kumar, P.A. Joy, Static and dynamic response of cluster glass in $La_{0.5}Sr_{0.5}CoO_3$, Phys. Rev. B 54 (1996) 9267.

[139] Y.-C. Liang, H.-Y. Hsia, Growth and crystallographic feature-dependent characterization of spinel zinc ferrite thin films by RF sputtering, Nanoscale Res. Lett. 8 (2013) 537.
[140] C.P. Luo, S.H. Liou, L. Gao, Y. Liu, D.J. Sellmyer, Nanostructured FePt: B_2O_3FePt:B_2O_3 thin films with perpendicular magnetic anisotropy, Appl. Phys. Lett. 77 (2000) 2225.
[141] D. Guo, J. Zhu, Y. Yang, X. Fan, G. Chai, W. Sui, Z. Zhang, D. Xue, High-frequency magnetic properties of Zn ferrite films deposited by magnetron sputtering, J. Appl. Phys. 107 (2010) 043903.
[142] N. Wakiya, K. Muraoka, T. Kadowaki, T. Kiguchi, N. Mizutani, H. Suzuki, K. Shinozaki, Preparation of ferromagnetic zinc-ferrite thin film by pulsed laser deposition in the magnetic field, J. Magn. Magn. Mater. 310 (2007) 2546.
[143] M. Sultan, R. Singh, Structural and optical properties of RF-sputtered $ZnFe_2O_4$ thin films, J. Phys. D 42 (2009) 115306.
[144] B. Jeyadevan, K. Tohji, K. Nakatsuka, Structure analysis of coprecipitated $ZnFe_2O_4$ by extended x-ray-absorption fine structure, Solid State Phys. 76 (1994) 6325.
[145] A.A. Timpopheev, A.M. Azevedo, N.A. Sobolev, K. Brachwitz, M. Lorentz, M. Ziese, P. Esquinazi, M. Grudmann, Magnetic anisotropy of epitaxial zinc ferrite thin films grown by pulsed laser deposition, Thin Solid Films 527 (2013) 273.
[146] S.A. Morrison, C.L. Cahill, E.E. Carpente, S. Calvin, R. Swaminathan, M. E. McHenry, V.G. Harris, J. Magnetic and structural properties of nickel zinc ferrite nanoparticles synthesized at room temperature, Appl. Phys. 95 (2004) 6392.
[147] S. Calvin, E.E. Carpenter, V.G. Harris, S.A. Morrison, Use of multiple-edge refinement of extended x-ray absorption fine structure to determine site occupancy in mixed ferrite nanoparticles, Appl. Phys. Lett. 81 (2002) 3828.
[148] A. Yang, Z. Chen, X. Zuo, D. Arena, J. Kirkland, C. Vittoria, V.G. Harris, Cation-disorder-enhanced magnetization in pulsed-laser-deposited $CuFe_2O_4$ films, Appl. Phys. Lett. 86 (2005) 252510.
[149] C.E. Rodriguez Torres, G.A. Pasquevich, P. Mendoza Zelis, F. Golmar, S.P. Heluan, S.K. Nayak, W.A. Adeagbo, W. Hergert, M. Hoffmann, E. Ernst, P. Esquinazi, S. J. Steware, Oxygen-vacancy-induced local ferromagnetism as a driving mechanism in enhancing the magnetic response of ferrites, Phys. Rev. B 89 (2014) 104411.
[150] O.V. Yazyev, Emergence of magnetism in graphene materials and nanostructures, Rep. Prog. Phys. 73 (2010) 056501.
[151] O. Volnianska, P. Boguslawaski, Magnetism of solids resulting from spin polarization of p orbitals, J. Phys. Condens. Matter 22 (2010) 073202.
[152] M. Stoneham, The strange magnetism of oxides and carbons, J. Phys. Condens. Matter 22 (2010) 074211.
[153] S. Calvin, E.E. Carpenter, B. Ravel, V.G. Harris, S.A. Morrison, Multiedge refinement of extended x-ray-absorption fine structure of manganese zinc ferrite nanoparticles, Phys. Rev. B 66 (2002) 224405.
[154] S.B. Ogale, Dilute doping, defects, and ferromagnetism in metal oxide systems, Adv. Mater. 22 (2010) 3125.
[155] K. Kamazawa, Y. Tsnoda, H. Kadowaki, K. Kohn, Magnetic neutron scattering measurements on a single crystal of frustrated $ZnFe_2O_4$, Phys. Rev. B 68 (2003) 024412.
[156] J. Philip, G. Gnanaprakash, G. Paneerselvan, M.P. Antony, T. Jayakumar, B. Raj, Effect of thermal annealing under vacuum on the crystal structure, size, and magnetic properties of $ZnFe_2O_4$ nanoparticles, J. Appl. Phys. 102 (2007) 054305.
[157] V. Stevanovic, M.D. Avezac, A. Zunger, Universal electrostatic origin of cation ordering in A_2BO_4 spinel oxides, J. Am. Chem. Soc. 133 (2011) 11649.
[158] W.A. Adegbo, G. Fischer, A. Ernst, W. Hergert, Magnetic effects of defect pair formation in ZnO, J. Phys. Condens. Matter 22 (2010) 436002.

[159] M. Khalid, M. Ziese, A. Setzer, P. Esquinaz, M. Lorenz, H. Hochmuth, M. Grudmann, D. Spemann, T. Butz, G. Braner, W. Anwand, W. Hergert, A. Ernst, Defect-induced magnetic order in pure ZnO films, Phys. Rev. B 80 (2009) 035331.
[160] D. Venkateshvaran, M. Althammer, A. Nielsen, G. Geprays, M.S. Ramachandra Rao, S.T.B. Goennenwein, M. Opel, R. Gross, Epitaxial $Zn_xFe_{3-x}O_4$ thin films: a spintronic material with tunable electrical and magnetic properties, Phys. Rev. B 79 (2009) 134405.
[161] T. Watanbe, S. Takita, K. Tomiyasu, K. Kamazawa, Acoustic study of dynamical molecular-spin state without magnetic phase transition in spin-frustrated $ZnFe_2O_4$, Phys. Rev. B 92 (2015) 174420.
[162] D. Sun, M.X. Wang, Z.H. Zhang, H.L. Tao, M. He, B. Song, Q. Li, Effects of inverse degree on electronic structure and electron energy-loss spectrum in zinc ferrites, Solid State Commun. 223 (2015) 12.
[163] R. Rameshbabu, R. Ramesh, S. Kanagesan, A. Karthigeyan, S. Ponnusamy, Structural, morphological and magnetic properties of hydrothermally synthesized $ZnFe_2O_4$ nanoparticles, J. Mater. Sci. Mater. Electron. 25 (2014) 2583.
[164] S.P. Gao, C.J. Pickard, M.C. Payne, J. Zhu, Y. Yuan, Theory of core-hole effects in 1s core-level spectroscopy of the first-row elements, Phys. Rev. B 77 (2008) 115122.
[165] M.P. Pileni, in: J.H. Fendler (Ed.), Nanoparticles and Nanostructured Films, Wiley-VCH, Weinheim, 1997. 71 p.
[166] S.N. Dolia, P.K. Sharma, M.S. Dhawan, S. Kumar, A.S. Prasad, A. Samariya, S.P. Pareek, R.K. Singhat, K. Asokan, Y.T. Xing, M. Alzamora, E. Saitovitach, Swift heavy ion irradiation induced modifications in magnetic and dielectric properties of Mn–Ca ferrite, Appl. Surf. Sci. 258 (2012) 4207.
[167] Y. Kumar, I. Lorite, M. Lorenz, P. Esquinazi, M. Grundmann, Effect of annealing on the magnetic properties of zinc ferrite thin films, Mater. Lett. 195 (2017) 89.
[168] Y.S. Chen, D. Spoddig, M. Ziese, Epitaxial thin film $ZnFe_2O_4$: a semi-transparent magnetic semiconductor with high Curie temperature, J. Phys. D Appl. Sci. 41 (2008) 205004.
[169] E.C. Stoner, E.P. Wohlfarth, A mechanism of magnetic hysteresis in heterogeneous alloys, Philos. Trans. R. Soc. Lond. A 240 (1948) 599.
[170] F. Grasset, N.D.L. Labhsetwar, D.C. Park, N. Saito, H. Haneda, O. Cador, T. Roisnel, S. Mornet, E. Dognet, J. Pontier, J. Etourneah, Synthesis and magnetic characterization of zinc ferrite nanoparticles with different environments: powder, colloidal solution, and zinc ferrite-silica core-shell nanoparticles, Langmuir 18 (2000) 8209.
[171] J.L. Dormann, D. Fiorani, M.E. Yamani, Field dependence of the blocking temperature in the superparamagnetic model: H23 coincidence, Phys. Lett. A 120 (1987) 95.
[172] B. Pandey, F.J. Litterst, E.M. Baggio-Saitovich, Preferential spin canting in nanosize zinc ferrite, J. Magn. Magn. Mater. 385 (2015) 412.
[173] K.L. Lopez-Maldonado, P. de la Presa, I. Betancourt, J.R. Farias Mancilla, J. A. Matutes Aquino, A. Hernando, J.T. Elizalde Galindo, Superparamagnetic response of zinc ferrite incrusted nanoparticles, J. Alloys Compd. 637 (2015) 443.
[174] E.V. Gofton, G. Bulai, O.F. Caltun, S. Cervera, S. Mace, M. Trassinelli, S. Steyde, D. Vernhet, Structural and magnetic properties of zinc ferrite thin films irradiated by 90 keV neon ions, Appl. Surf. Sci. 379 (2016) 171.
[175] B. Boucher, R. Buhl, R. di Bella, M. Perrin, Etude par des mesures de diffraction de neutrons et de magnetisme des proprietes cristallines et magnetiques de composes cubiques spine1les $C0_{3-x}Mn_x0_4$ ($0.6 \leq x \leq 1.2$), J. Phys. 379 (1970) 113.
[176] F. Sunder, C. Houpert, D. Groult, J.Y. Fan, A. Meltah, M. Toulemonde, Spontaneous magnetization induced in the spinel $ZnFe_2O_4$ by heavy ion irradiation in the electronic stopping power regime, Phys. Res. Sect. B 82 (1993) 91.

[177] M. Yokoyana, E. Ohta, T. Sato, T. Komaba, Size dependent magnetic properties of zinc ferrite fine particles, J. Phys. IV France 7 (1997) C1–C521.
[178] V.G. Vologin, V.D. Parkhomenko, S.F. Dubinin, Y.G. Chukalkin, B. N. Goshehitskii, S.K. Sidorov, V.V. Petrov, Exchange Fe^{3+} - Fe^{3+} interactions in ferrites with spinel structure, Phys. Status Solidi A 33 (1976) K83.

FURTHER READING

[179] S. Morup, F. Bodker, P.V. Hendriksen, S. Linderoth, Spin-glass-like ordering of the magnetic moments of interacting nanosized maghemite particles, Phys. Rev. B 52 (1995) 287.

CHAPTER 9

Modeling of Nanosheets Based on Transition-Metal Dichalcogenides for Spintronic Applications: Insights From First-Principles

Mohammed Benali Kanoun
Physics Department, College of Science, King Faisal University, Saudi Arabia

1 INTRODUCTION

The search for spintronic materials remains a continuous priority [1] due to promising new functionalities for future information technologies. The challenge is to identify or design devices that can be used to generate, detect, and manipulate spins and charges [2–5]. Diluted magnetic semiconductors (DMS) are of particular interest for applications in spintronics. They have attracted lots of attention due to their potential applications for spin-sensitive electronics [2–5]. Following the theoretical prediction [6], DMS are produced by doping semiconductors with transition-metal elements. A lot of work had been performed in typical systems based on III–V or II–VI semiconductors [7–17], particularly Mn-doped GaAs [11]. But this has a low Curie temperature [18] while wider band gap semiconductors such as ZnO and GaN were considered to be more promising for room temperature DMSs [12, 19, 20]. Experimental observations of room temperature DMSs are controversial, mainly due to their complicated bulk structures such as phase separation and cluster formation [11, 21, 22].

The recent discovery of two-dimensional (2D) materials [23] provides a way to overcome these difficulties as well as new research areas for materials science and technology, including spintronics. These 2D materials have revealed a large potential for new generations of low-dimensional transistors, photo-emitting devices, and spintronics devices, due to their distinct crystal structures and electronic properties [24]. Among the 2D materials, the transition-metal dichalcogenides (TMDs) with the chemical formula MX_2 (X=S, Se, Te; M=transition-metal) have attracted broad attention due

Nano-sized Multifunctional Materials
https://doi.org/10.1016/B978-0-12-813934-9.00009-8

to their potential for flexible and transparent electronics applications, owing to their sizeable band gaps ranging between 1 and 2 eV [25, 26]. Furthermore, 2D TMDs constitute an exciting family of transistors, including field-effect transistors (FETs) and heterostructure junctions [26–28]. Moreover, recent advances in synthesis processes for 2D TMDs have opened new opportunities toward novel materials based on spintronics [29, 30]. Recently, many theoretical and experimental investigations have been performed to explore the feasible ways to introduce magnetism to 2D TMDs, such as morphology fabricating [29, 31], external strain, [30] and impurity doping [32–42].

In this chapter, we provide an overall view of the structural and electronic properties of manganese-doped molybdenum dichalcogenides using first-principles calculations. We aim to study in order to manipulate the magnetic properties of MoS_2, $MoSe_2$, and $MoTe_2$ monolayers with the impurity Mn atom as the dopant. Our ambition is to contribute to a better understanding of the relationships between the magnetic properties and the electronic structures as a function of Mn-dopant concentration in the MoX_2 (X=S, Se, Te) monolayers in order to determine the presence of long-range ferromagnetism ordering.

2 THEORETICAL METHODS

Here, we present the first-principles calculations based on density functional theory (DFT) as implemented in the Vienna Ab initio simulation (VASP) code [43]. The interactions between electrons and ions were described by the projector-augmented wave (PAW) method [44]. In our calculations, the spin-polarized generalized gradient approximation (GGA) of the Perdew-Bruke-Ernzerhof (PBE) functional [45] was used for the exchange-correlation interaction between electrons. The energy cutoff of the plane wave expansion of the wave functions was fixed at 450 eV. The integration over the reciprocal Brillouin zone is performed within a Gamma-centered $4 \times 4 \times 1$ and $6 \times 6 \times 1$ Monkhorst-Pack k-meshes [46] sufficient for relaxation and electronic structure calculations, respectively. The onsite Coulomb interaction, U, and onsite exchange interaction, J, were applied to the 3d orbital of Mn atom of the Mn atom (4 and 1 eV, respectively) [42, 47, 48]. For geometry optimization, the convergence criterion of the energy is taken as 1.0×10^{-5} eV and the Hellmann-Feynman force between each atom is set to less than 0.01 eV/Å.

The layered structures of MoX_2 (X=S, Se, Te) have a sandwich structure that is composed of three atomic layers with one Mo layer between two chalcogens (S, Se, Te). MoX_2 layers have a space group symmetry of P-6m2 with the Mo atoms having a trigonal prismatic coordination with the chalcogen atoms, as shown in Fig. 1. A $4\times4\times1$ MoX_2 (X=S, Se, Te) monolayer supercell structure containing 32 S/Se/Te and 16 Mo atoms is built. A 15-Å vacuum slab is adopted along the c axis in order to avoid the interaction between two adjacent bilayer images due to the periodic boundary conditions.

3 ELECTRONIC STRUCTURE OF UNDOPED MONOLAYERS

MoS_2, $MoSe_2$, and $MoTe_2$ monolayers are usually stable in the 1H prototype structure with the optimized lattice constants $a=b=3.16$ Å, 3.28 Å, and 3.51 Å, respectively. To illustrate the electronic structure of the pristine monolayer MoS_2, $MoSe_2$,and $MoTe_2$, we calculated the total density of states (DOS), as displayed in Fig. 2. The analysis DOS profile with the band structure reported in other Refs. [25, 26, 38, 49–54] shows that the pristine monolayer MoX_2 is a nonmagnetic semiconductor. We found that the calculated band gaps of MoS_2, $MoSe_2$, and $MoTe_2$ are 1.85, 1.60, and 1.15 eV, respectively, agreeing well with the experimental values of 1.85 [25], 1.60 [50], and 1.15 eV [53], respectively, as well as other theoretical results [26, 49–52].

The point defects usually play an important role in inducing magnetism into low-dimensional materials. Therefore, a vacancy defect was considered in order to investigate the electronic and magnetic properties of undoped MoX_2 (X=S, Se, Te). These monolayers are studied by generating one

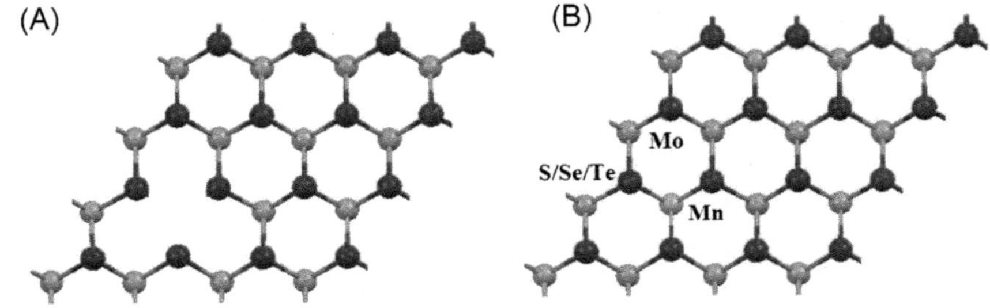

Fig. 1 (A, B) Top views of monolayers MoX_2 (X=S, Se, Te) with Mo vacancy and doped-Mn atom, respectively.

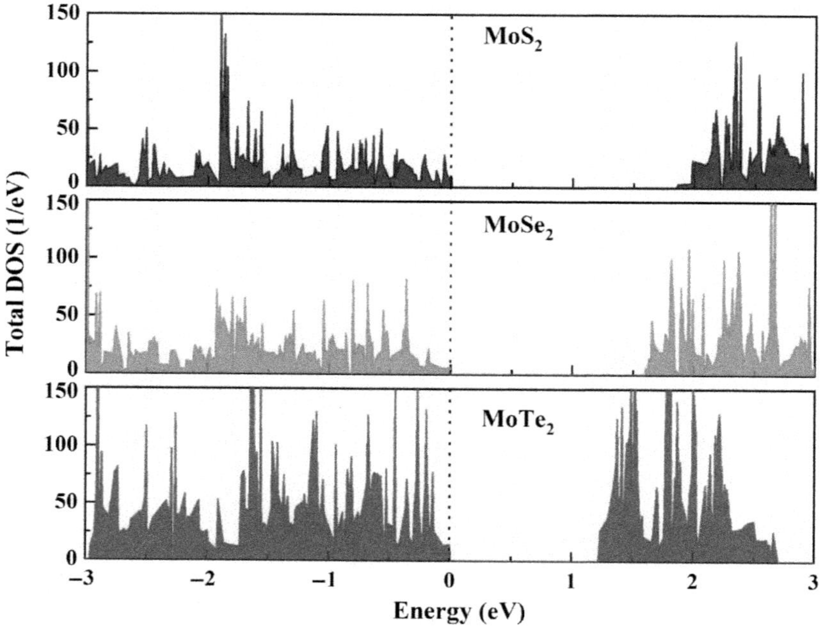

Fig. 2 Density of states of pristine MoS_2, $MoSe_2$, and $MoTe_2$ monolayers. The vertical dashed line denotes the Fermi level.

Mo vacancy in the supercell, leading to a Mo vacancy concentration of 6.25%, as shown in Fig. 3. For MoS_2, the symmetry of the spin-up and spin-down bands in Fig. 3 confirms the nonmagnetic nature, indicating that no split of the energy level near the Fermi level can be observed. An introduction of Mo vacancy in the monolayer MoS_2 does not destroy the structure symmetry C_{3v}, as illustrated in Fig. 3. This result is reflected by the appearance of unoccupied defect states in the band gap. Differently for undoped $MoSe_2$ and $MoTe_2$, Mo vacancy introduces magnetism by forming local moments that are reflected in the asymmetrical DOSs for majority and minority states below the Fermi level. Hence, the calculated total magnetic moments are 4.05 and 2.55 μ_B for $MoSe_2$ and $MoTe_2$, respectively. That result is mainly due to the relaxation of the neighboring Se/Te atoms. The main contribution to the magnetic moments occurs from the *p* states of the six nearest neighbor Se/Te sites surrounding the Mo vacancy, which has a local magnetic moment of about 0.16 μ_B. The d states of the six next neighboring Mo atoms have a little contribution to the magnetic moment (i.e., 0.10 and 0.06 μ_B for $MoSe_2$ and $MoTe_2$, respectively).

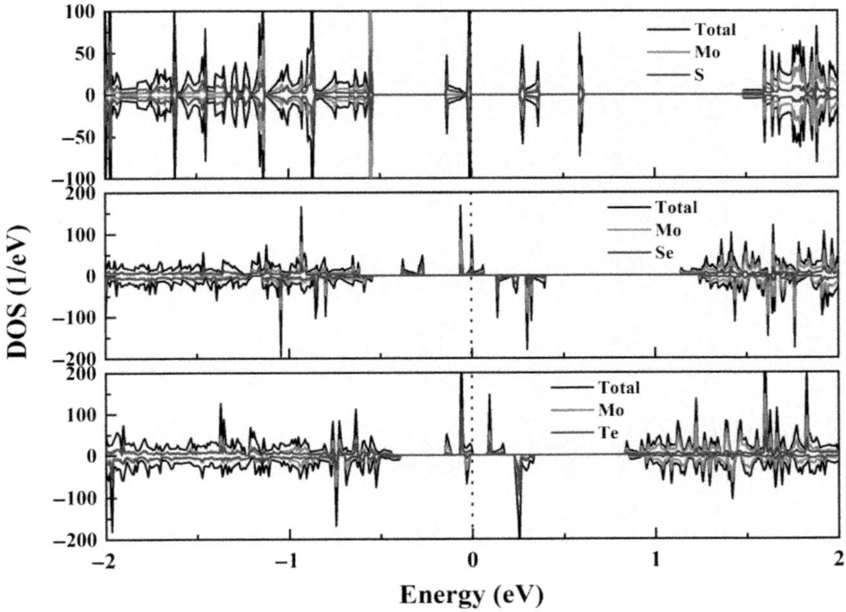

Fig. 3 Density of states of monolayers MoX_2 (X=S, Se, Te) with Mo vacancy.

4 MN DOPING-INDUCED MAGNETISM

In this section, we investigate the effect of manganese inclusion into MoX_2 (X=S, Se, Te) on the electronic structure and magnetic properties. To check the relative stability of the Mn dopant, we have calculated the binding energies, as gathered in Table 1. The binding energy is defined as $E_b = E_{supercell} - E_V - E_{TM}$, whereby, $E_{supercell}$, and E_V represent the total energy of the doped supercell and host system with Mo vacancy; E_{TM} is the energy of the isolated 3d transition-metal (TM) atom. According to our calculations reported in Table 1, the Mn atom can be easily embedded

Table 1 Binding energies and magnetic moments for 3d Mn doping of monolayers MoX_2 (X=S, Se, Te) and nearest neighbors S/Se/Te/Mo atoms

Mn doped	MoS_2	$MoSe_2$	$MoTe_2$
E_b (eV)	5.4	4.1	3.0
Total moment (μ_B)	1.0	1.0	1.0
Magnetic moment Mn dopant (μ_B)	2.34	2.73	3.40
Magnetic moment, Mo (μ_B)	−0.09	−0.11	−0.14
Magnetic moment, S/Se/Te (μ_B)	−0.06	−0.09	−0.12

with the Mo vacancy. Additionally, the binding energy decreases with the increasing atomic number of chalcogen atoms.

To explain the role of the Mn doping on the electronic structure and to acquire details on the roles of the individual atoms, we address the GGA+U density of states (DOSs). Figs. 4 and 5 display the total and partial DOSs of the Mn-doped MoX_2 (X=S, Se, Te) monolayers. The total DOSs showed that the substitution of Mn at the Mo site can induce both a spin polarized state and highly localized orbitals within the energy band gap region of the undoped compounds. It can be seen that Mn doping introduces in-gap states in the band gap region shift toward the Fermi energy to eventual occupied states. Indeed, the polarized charges mainly rise from the localized 3d states of the Mn atom with a weak contribution of the delocalized 4d states of the Mo atom and p states of chalcogen atoms. Furthermore, a spin majority impurity band appears around the Fermi level. This occupied band is mainly composed of the $d_{z^2-r^2}$ orbitals of the Mn spin majority channel, with substantial contributions of the Mo 4d and the chalcogenides p states, as shown in Fig. 5. In addition, the Mn states near the Fermi level split to an occupied

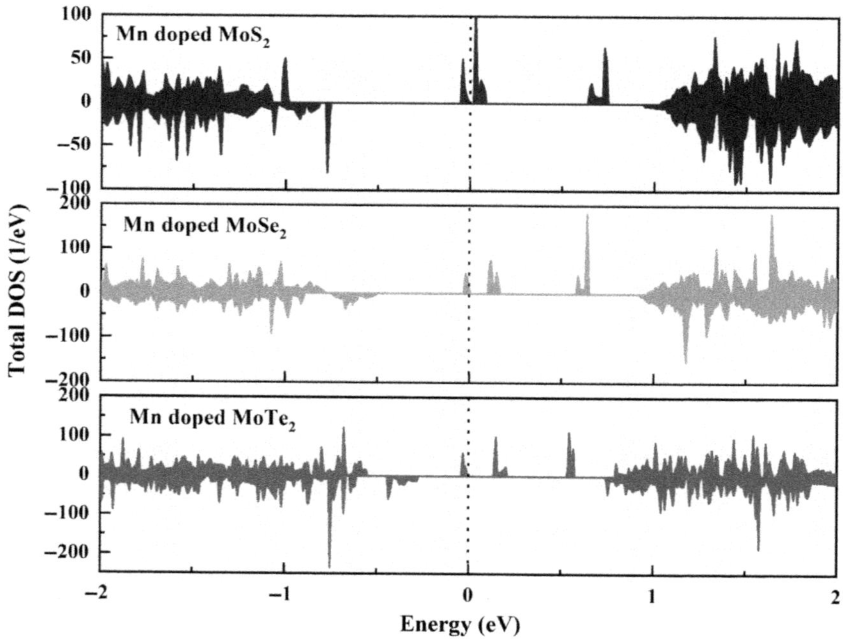

Fig. 4 Total density of states of the Mn-doped MoX_2 (X=S, Se, Te). Spin up and spin down correspond to positive and negative values, respectively.

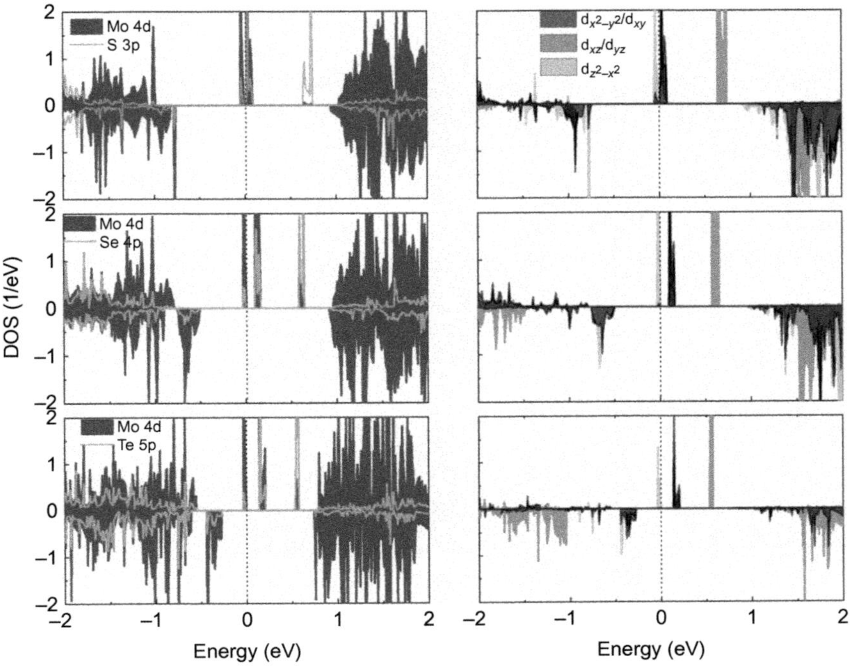

Fig. 5 Spin resolved atom-projected DOS showing the projected onto the five different *d* orbitals of the Mn atom (*right*) and the 4*d* and 5*p* states of its neighboring Mo and S/Se/Te atoms (*left*).

single state and an unoccupied twofold degenerate where the slitting distance between folds increases with the increasing atomic number of chalcogen atoms. An unoccupied spin majority impurity band that is composed of the $d_{xy}/d_{x^2-y}{}^2$ orbitals of the Mo lies just above the Fermi level. Moreover, we observed that the 3d electrons of the Mn atom induce a large local magnetic moment as of 2.34, 2.73, and 3.40 μ_B for MoS_2, $MoSe_2$, and $MoTe_2$, respectively. The additional contributions to the total magnetic moment appear to come from chalcogen atoms and Mo atoms to the Mn defect, as reported in Table 1. The spins induced on the six nearest neighboring S/Se/Te and Mo atoms are all antiparallel to the local spin on the dopant atom, which, in turn, produces a magnetic moment that is much smaller than the local magnetic moment of the Mn defect. This leads to a localized magnetic moment of 1.0 μ_B. As a result, the Mn doped system remains semiconducting without carrier doping. Furthermore, we have calculated the spin density iso-surface for all systems, as displayed in Fig. 6. It is obvious

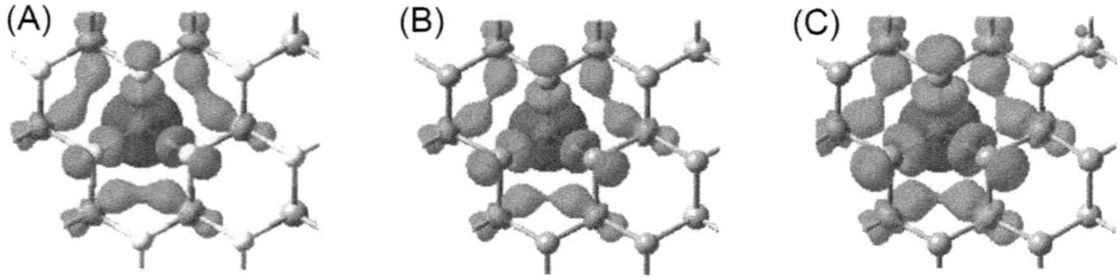

Fig. 6 Isosurface spin density plots of Mn-doped (A) MoS_2, (B) $MoSe_2$, and (C) $MoTe_2$ (the blue and orange isosurfaces represent positive and negative spin density $\pm 0.02\,e/Å^3$).

to observe after geometry optimization that the Mn dopant site can preserve trigonal prism symmetry. We also observed that the spins of the six nearest neighboring atoms are coupled antiferromagnetically with the Mn 3*d* states. This means that the induced spins on the nearby chalcogen atoms and Mo atoms are antiparallel to that of the impurity Mn atoms. Thus, the total magnetic moments of the considered systems are smaller than the local magnetic moments on the dopant. Moreover, spin-polarized *p* states of the chalcogen atoms can be shown to have similar features.

5 MAGNETIC COUPLING

At last, we have investigated the long-range ferromagnetism ordering MoX_2 (X=S, Se, Te) by substituting two Mn atoms at Mo sites $4 \times 4 \times 1$ monolayer supercell, giving rise to a doping level of 12.5%. To obtain the magnetic state, the total energy is computed for the spin alignments, showing the ferromagnetic (FM) or antiferromagnetic (AFM) ground state, where the energy difference $\Delta E = E_{FM} - E_{AFM}$ (per Mn atom) is negative. Three configurations with different Mn-Mn distances were considered, corresponding to doping Mn atoms at the first-, second-, and third-nearest neighboring sites. Fig. 7 displays our results corresponding to energy differences between the FM and AFM states for three considered configurations. The ferromagnetism was found to be the most energetically favorable. We can conclude that Mn dopants for all systems show strong ferromagnetism, indicating a long-range magnetic ordering that is in good agreement with the previous work of Mishra [40]. Moreover, the two substitutional Mn atoms were found to prefer a first nearest neighboring site. It is interesting to note that the Mn substitution in $MoSe_2$ exhibits a larger stability of FM ordering for first nearest neighbor interaction compared with those of Mn-doped MoS_2 and $MoTe_2$ monolayers.

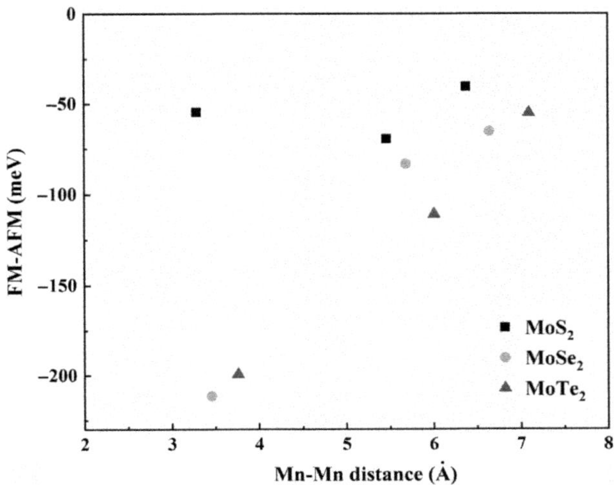

Fig. 7 Energy differences of the FM ordering over AFM ordering for Mn-doped MoX_2 (X=S, Se, Te) as a function of the distance between the two defects. A negative energy corresponds to FM ordering being more stable.

The magnetic interaction effects are important in establishing a long-range ferromagnetic ordering for diluted magnetic systems that play a crucial role in estimating the Curie temperature (T_C). However, due to the strong nearest-neighbor interaction, we can roughly estimate the Curie temperature using the relation $\frac{3}{2}k_BT_C = -\frac{\Delta E_{FM-AFM}}{n}$ based on the mean field approximation [54, 55], where k_B and n are the Boltzmann constant and an impurity concentration, respectively. The estimated Curie temperatures are 211 K, 818 K, and 770 K for MoS_2, $MoSe_2$, and $MoTe_2$, respectively.

6 CONCLUSION

In summary, we reported a systematic study about the electronic and magnetic properties of Mn atom-doped MoX_2 (X=S, Se, Te) monolayers. We demonstrated that the cation Mo vacancy in monolayers $MoSe_2$ and $MoTe_2$ can induce spin polarization while Mo vacancy in MoS_2 does not show a spin polarized state. However, Mn substitution in the Mo site produces a magnetic moment. The magnetic moment originates from the interaction between the electronic d states of the Mn dopant and the p states of the S/Se/Te atoms, indicating p–d hybridization. In addition, the induced spins on the nearby host atoms are antiparallel to that of the Mn dopants. The results show that the ferromagnetic interaction between the Mn impurities is found to be long range attributing to their FM ordering to exchange interactions between the localized Mn spins and the delocalized *p* spins of chalcogen atoms. Our calculations show that Mn doping is a good candidate to induce and manipulate the magnetism in 2D TMD for application in semiconductor spintronics.

REFERENCES

[1] T. Dietl, A ten-year perspective on dilute magnetic semiconductors and oxides, Nat. Mater. 9 (2010) 965–974.

[2] S. Wolf, D. Awschalom, R. Buhrman, J. Daughton, S. Von Molnar, M. Roukes, A. Y. Chtchelkanova, D. Treger, Spintronics: a spin-based electronics vision for the future, Science 294 (2001) 1488–1495.

[3] I. Zutic, J. Fabian, S. Das Sarma, Spintronics: fundamentals and applications, Rev. Mod. Phys. 76 (2004) 323–410.

[4] S. Maekawa, Concepts in Spin Electronics, vol. 13, Oxford University Press on Demand, Oxford, 2006.

[5] Y.P. Feng, L. Shen, M. Yang, A. Wang, M. Zeng, Q. Wu, S. Chintalapati, C.-R. Chang, Prospects of spintronics based on 2D materials, WIREs Comput. Mol. Sci. 7 (2017) e1313.

[6] T. Dietl, H. Ohno, F. Matsukura, J. Cibert, D. Ferrand, Zener model description of ferromagnetism in zincblende magnetic semiconductors, Science 287 (2000) 1019–1022.

[7] M.C. Jaworski, J. Yang, S. Mack, D.D. Awschalom, J.P. Heremans, R.C. Myers, Observation of the spin-Seebeck effect in a ferromagnetic semiconductor, Nat. Mater. 9 (2010) 898–903.

[8] C. Le Gall, L. Besombes, H. Boukari, R. Kolodka, J. Cibert, H. Mariette, Optical spin orientation of a single manganese atom in a semiconductor quantum dot using quasi-resonant photoexcitation, Phys. Rev. Lett. 102 (2009) 127402.

[9] H. Ohno, Making nonmagnetic semiconductors ferromagnetic, Science 14 (1998) 951–956.

[10] J.K.J. Furdyna, Diluted magnetic semiconductors, Appl. Phys. 64 (1988) R29–R64.

[11] K. Sato, L. Bergqvist, J. Kudrnovsky, P.H. Dederichs, O. Eriksson, I. Turek, B. Sanyal, H. Katayama-Yoshida, V.A. Dinh, First-principles theory of dilute magnetic semiconductors, Rev. Mod. Phys. 82 (2010) 1633.

[12] A. Titov, X. Biquard, D. Halley, S. Kuroda, E. Bellet-Amalric, H. Mariette, J. Cibert, A.E. Merad, M.B. Kanoun, E. Kulatov, Y.A. Uspenskii, X-ray absorption near-edge structure and valence state of Mn in (Ga,Mn)N, Phys. Rev. B 72 (2005) 115209.

[13] M.B. Kanoun, S. Goumri-Said, A.E. Merad, J. Cibert, First-principles investigation of electronic structure and magnetic properties in ferromagnetic $Ga_{1-x}Mn_xN$ and $Al_{1-x}Mn_xN$, J. Phys. D. Appl. Phys. 38 (2005) 1853.

[14] M.B. Kanoun, S. Goumri-Said, U. Schwingenschlogl, Ferromagnetism in Cr-doped passivated AlN nanowires, J. Mater. Chem. A 2 (2014) 9287–9290.

[15] M.B. Kanoun, Insight into the origin of magnetism in Iron-doped cadmium sulfide thin films from first principles calculations, Solid State Commun. 253 (2017) 10–13.

[16] M.B. Kanoun, S. Goumri-Said, U. Schwingenschlogl, A. Manchon, Magnetism in Sc-doped ZnO with zinc vacancies: a hybrid density functional and GGA + U approaches, Chem. Phys. Lett. 532 (2012) 96–99.

[17] M.B. Kanoun, S. Goumri-Said, A. Manchon, U. Schwingenschlogl, Ferromagnetism carried by highly delocalized hybrid states in Sc-doped ZnO thin films, Appl. Phys. Lett. 100 (2012) 222406.

[18] L. Chen, X. Yang, F. Yang, J. Zhao, J. Misuraca, P. Xiong, S. von Molnár, Enhancing the curie temperature of ferromagnetic semiconductor (Ga, Mn) as to 200 k via nano-structure engineering, Nano Lett. 11 (2011) 2584–2589.

[19] T. Fukumura, H. Toyosaki, Y. Yamada, Magnetic oxide semiconductors, Semicond. Sci. Technol. 20 (2005) S103.

[20] R. Janisch, P. Gopal, N.A. Spaldin, Transition metal doped TiO2 and ZnO present status of the field, J. Phys. Condens. Matter 17 (2005) R657.

[21] T. Dietl, H. Ohno, Dilute ferromagnetic semiconductors: physics and spintronic structures, Rev. Mod. Phys. 86 (2014) 187–251.

[22] A. Zunger, S. Lany, H. Raebiger, Trend: the quest for dilute ferromagnetism in semiconductors: guides and misguides by theory, Physics 3 (2010) 53.

[23] K.S. Novoselov, A.K. Geim, S.V. Morozov, D. Jiang, Y. Zhang, S.V. Dubonos, I. V. Grigorieva, A.A. Firsov, Electric field effect in atomically thin carbon films, Science 306 (2004) 666–669.

[24] H. Li, J. Wu, Z. Yin, H. Zhang, Preparation and applications of mechanically exfoliated single-layer and multilayer MoS2 and WSe2 nanosheets, Acc. Chem. Res. 47 (2014) 1067.

[25] K.F. Mak, C. Lee, J. Hone, J. Shan, T.F. Heinz, Atomically thin MoS_2: a new direct-gap semiconductor, Phys. Rev. Lett. 105 (13) (2010) 136805.

[26] Q.H. Wang, K. Kalantar-Zadeh, A. Kis, J.N. Coleman, M.S. Strano, Electronics and optoelectronics of two-dimensional transitional metal dichalcogenides, Nat. Nanotechnol. 7 (2012) 699–712.

[27] B. Radisavljevic, A. Radenovic, J. Brivio, V. Giacometti, A. Kis, Single-layer MoS_2 transistors, Nat. Nanotechnol. 6 (2011) 147–150.

[28] T. Cao, G. Wang, W. Han, H. Ye, C. Zhu, J. Shi, Q. Niu, P. Tan, E. Wang, B. Liu, J. Feng, Valley-selective circular dichroism of monolayer molybdenum disulphide, Nat. Commun. 3 (2012) 887.

[29] L. Cai, J.F. He, Q.H. Liu, T. Yao, L. Chen, W.S. Yan, F.C. Hu, Y. Jiang, Y.S. Zhao, T. D. Hu, Z.H. Sun, W. SQ, Vacancy-induced ferromagnetism of MoS_2 nanosheets, J. Am. Chem. Soc. 137 (2015) 2622–2627.

[30] H. Zheng, B. Yang, D. Wang, R. Han, X. Du, Y. Yan, Tuning magnetism of monolayer MoS_2 by doping vacancy and applying strain, Appl. Phys. Lett. 104 (2014) 132403.

[31] H. Pan, Y.W. Zhang, Edge-dependent structural, electronic and magnetic properties of MoS_2 nanoribbons, J. Mater. Chem. 22 (2012) 7280–7290.

[32] N. Feng, W.B. Mi, Y.C. Cheng, Z.B. Guo, U. Schwingenschlögl, H.L. Bai, Magnetism by interfacial hybridization and p-type doping of MoS2 in Fe4N/MoS2 superlattices: a first-principles study, ACS Appl. Mater. Interfaces 6 (2014) 4587–4594.

[33] Q.C. Sun, D. Mazumdar, L. Yadgarov, R. Rosentsveig, T. Tenne, J.L. Musfeld, Spectroscopic determination of phonon lifetimes in rhenium-doped MoS_2 nanoparticles, Nano Lett. 13 (2013) 2803–2808.

[34] Y.C. Cheng, Q.Y. Zhang, U. Schwingenschlögl, Valley polarization in magnetically doped single-layer transition-metal dichalcogenides, Phys. Rev. B 89 (2014) 155429.

[35] W.S. Yun, J.D. Lee, Unexpected strong magnetism of Cu doped single-layer MoS_2 and its origin, Phys. Chem. Chem. Phys. 16 (2014) 8990–8996.

[36] A. Ramasubramaniam, D. Naveh, Mn-doped monolayer MoS_2: an atomically thin dilute magnetic semiconductor, Phys. Rev. B 87 (2013) 195201.

[37] Y. Zhou, Q. Su, Z. Wang, H. Deng, X. Zu, Controlling magnetism of MoS_2 sheets by embedding transition-metal atoms and applying strain, Phys. Chem. Chem. Phys. 15 (2013) 18464–18470.

[38] Y. Ma, Y. Dai, M. Guo, C. Niu, J. Lu, B. Huang, Electronic and magnetic properties of perfect, vacancy-doped, and nonmetal adsorbed $MoSe_2$, $MoTe_2$ and WS_2 monolayers, Phys. Chem. Chem. Phys. 13 (2011) 15546–15553.
[39] X.-L. Fan, Y. An, W.-J. Guo, Ferromagnetism in transitional metal-doped MoS_2 monolayer, Nanoscale Res. Lett. 11 (2016) 154.
[40] R. Mishra, W. Zhou, S.J. Pennycook, S.T. Pantelides, J.-C. Idrobo, Long-range ferromagnetic ordering in manganese-doped two-dimensional dichalcogenides, Phys. Rev. B 88 (2013) 144409.
[41] A.-M. Hu, L.-l. Wang, W.-Z. Xiao, B. Meng, Electronic structures and magnetic properties in Cu-doped two-dimensional dichalcogenides, Phys. E. 73 (2015) 69–75.
[42] M.B. Kanoun, Tuning magnetic properties of two-dimensional $MoTe_2$ monolayer by doping 3d transition metals: insights from first principles calculations, J. Alloys Compd. 748 (2018) 938–942.
[43] G. Kresse, J. Furthmuller, Efficiency of ab-initio total energy calculations for metals and semiconductors using a plane-wave basis set, Comput. Mater. Sci. 6 (1996) 15–50.
[44] P.E. Blochl, Projector augmented-wave method, Phys. Rev. B 50 (1994) 17953.
[45] J.P. Perdew, K. Burke, M. Ernzerhof, Generalized Gradient Approximation Made Simple, Phys. Rev. Lett. 77 (1996) 3865.
[46] H.J. Monkhorst, J.D. Pack, Special points for Brillouin-zone integrations, Phys. Rev. B 13 (1976) 5188.
[47] S.L. Dudarev, G.A. Botton, S.Y. Savrasov, C.J. Humphreys, A.P. Sutton, Phys. Rev. B 57 (1998) 1505.
[48] N.H. Hong, M.B. Kanoun, S. Goumri-Said, J.H. Song, E. Chikoidze, Y. Dumont, A. Ruyter, M. Kurisu, The origin of magnetism in transition metal-doped ZrO2 thin films: experiment and theory, J. Phys. Condens. Matter 25 (2013) 436003.
[49] H. Shi, H. Pan, Y.-W. Zhang, B.I. Yakobson, Quasiparticle band structures and optical properties of strained monolayer MoS_2 and WS_2, Phys. Rev. B 87 (2013) 155304.
[50] S. Tongay, J. Zhou, C. Ataca, K. Lo, T.S. Matthews, J.B. Li, J.C. Grossman, J.Q. Wu, Thermally driven crossover from indirect toward direct bandgap in 2D semiconductors: $MoSe_2$ versus MoS_2, Nano Lett. 12 (2012) 5576–5580.
[51] A. Kumar, P.K. Ahluwalia, Magnetic relaxation measurements of exchange biased (Pt/Co), Eur. Phys. J. B 85 (2012) 185.
[52] J. Chang, L.F. Register, S.K. Banerjee, Ballistic performance comparison of monolayer transition metal dichalcogenide MX_2 ($M = Mo, W$; $X = S, Se, Te$) metal-oxide-semiconductor field effect transistors, J. Appl. Phys. 115 (2014) 084506.
[53] C. Ruppert, O.B. Aslan, T.F. Heinz, Optical properties and band gap of single- and few-layer $MoTe_2$ crystals, Nano Lett. 14 (2014) 6231–6236.
[54] J. Kudrnovsky, I. Turek, V. Drchal, F. Maca, P. Weinberger, P. Bruno, Exchange interactions in III-V and group-IV diluted magnetic semiconductors, Phys. Rev. B 69 (2004) 115208.
[55] F. Maca, J. Kudrnovsky, V. Drchal, G. Bouzerar, Magnetism without magnetic impurities in ZrO2 oxide, Appl. Phys. Lett. 92 (2008) 212503.

CHAPTER 10

Exploring Fundamental Properties of Organic Molecule Devices Based on Graphene: Density Functional Theory and Transport Theory

Souraya Goumri-Said
College of Science, Physics Department, Alfaisal University, Riyadh, Saudi Arabia

1 INTRODUCTION

For several years, the molecular electronics research field has been the subject of increasing attention. This research field consists of using molecules, most often to perform functions equivalent to those of transistors, diodes, switches, and other components used in microelectronics on silicon. The concept of molecular electronics can also be extended even when it is no longer about a single organic molecule such as organized monomolecular layers and more generally when we exploit the electronic properties of nanostructures for which at least one of the dimensions is in the scale of molecular dimensions. Besides this, molecular electronics devices clearly offer several advantages over conventional electronics on silicon, such as the miniaturization of components that theoretically can go down to the ultimate sizes of atoms and molecules. Furthermore, the associated gain of energy might be useful for lower-cost industrial production [1–3].

Organic molecules have also been used in organic photovoltaic cells. In parallel with the development related to the technological processes of solar cells, organic materials for third-generation solar cells are also potential materials of the future because of their lower cost and the potential simplifications of the device manufacturing process. Among these organic cells, there are hybrid solar cells and "all organic" solar cells or OPV (organic photovoltaics) cells based on conjugated molecules or polymers. There are two types of hybrid organic/inorganic solar cells: dye cells and organic/inorganic bulk heterojunction.

Nano-sized Multifunctional Materials
https://doi.org/10.1016/B978-0-12-813934-9.00010-4

Graphene is an organic material with promising properties due to the abundance of carbon. Graphene, as a two-dimensional (2D) carbon crystal, has attracted huge scientific interest due to its exceptional physical properties [4–7]. The most attractive behavior of graphene is the possibility of tailoring its physical properties by the presence of corrugation [8–12].

Recently, an MIT team modeled the use of these materials in photovoltaic cells. They discovered that by stacking a layer of carbon atoms–graphene–and a layer of molybdenum disulfide (MoS_2), the solar cell was obtained with an efficiency approaching 1%, which is infinitely small. In total, it generates 30 times more power per unit volume than the finest known solar cells (gallium arsenide, silicon, or indium selenium), is just one micron thick, and has a performance approaching 30%. Hence, the researchers have found that by stacking six layers—three layers of graphene and three layers of molybdenum disulfide—the efficiency could theoretically reach 10% for a thickness of only three nanometers.

For example, a formation of periodic ripples ranging from 2 nm to 5 μm was demonstrated to be strongly dependent on the nature of the substrate [13, 14]. The idea to see the buckling phenomena is very challenging from experimental and theoretical points of view. Very recently, progress on the resolution of aberration-corrected transmission electron microscopy (TEM) has allowed us to observe corrugation and buckling. In this chapter, we first review the fundamental properties of graphene and graphene nanoribons in armchair and zigzag structures. Then, we introduce the buckling phenomena. In fact, graphene buckling is very important to predict the strength and failure of graphene sheet-based composites. The control of optoelectronic properties form experiments in nanoscale is difficult, hence theoretical analysis of nanostructures is an efficient tool to perform prediction and guide future experiments.

2 THEORETICAL APPROACHES AND COMPUTATIONAL DETAILS: SEMIEMPIRICAL AND NEGF FORMALISM

In this chapter, we give the essential information about the computational used method. We are using the tight-binding method based on the Slater-Koster model and the extended Hückel (EH) model. It is well known in the Density Functional Theory (DFT) community that the Slater-Koster tight-binding model closely follows the DFTB formalism [15] whereas the extended Hückel model is based on the formalism described in Ref. [16].

In fact, the choice of this method for the present system (buckling graphene) is because of the lower computational cost of the empirical approach. The present material needs a large atomic-scale electron transport simulation model based on the Nonequilibrium Green's Function (NEGF) formalism [17]. Let us now give a brief summary of the EHT model and the NEGF simulation procedure.

The Extended Hückel Theory (EHT) grew out of the need to consider all valence electrons in a molecular orbital calculation. In the EHT method [18, 19], the wave function is considered as a product of a valence wave function and a core wave function. For the transport properties calculation, we used a typical device configuration, as illustrated in Fig. 1.

It is based on simulating device systems using the NEGF method. The system is divided into three regions: left, central, and right. We assume that the properties of the electrodes of the left and right zones are described by solving a bulk problem for the fully periodic electrode cell (i.e., the screening approximation). The current through the system will be sufficiently small that the electrode regions can be described by an equilibrium electron distribution, which fulfills the screening approximation. With periodic

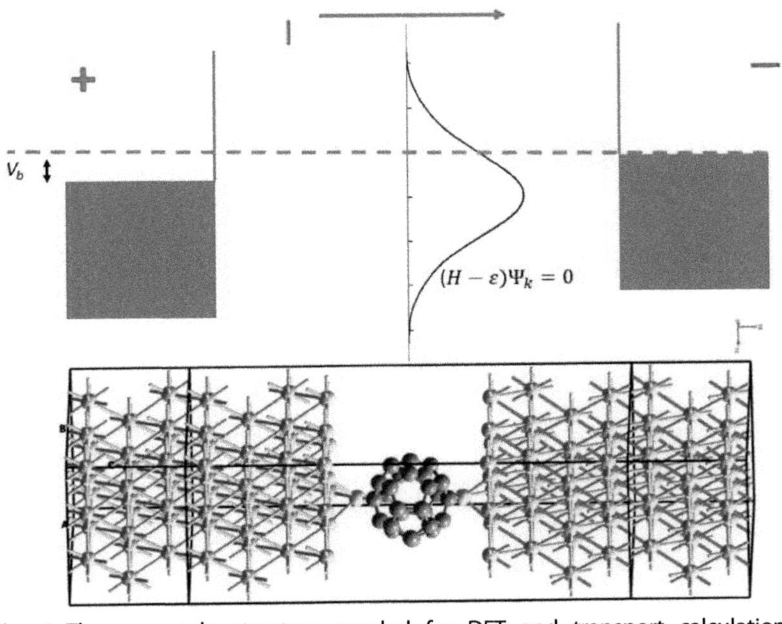

Fig. 1 The nanoscale structure needed for DFT and transport calculations is a semiinfinite device configuration. It consists of three main parts: left electrode, central region, and right electrode.

boundary conditions, the left and right electrodes are in equilibrium. Hence, DFT is used to compute the electronic structure. However, the calculations might be difficult as the electron distribution in the central region is in nonequilibrium. The most well-used assumption is to consider the system in a steady state where electron density is nonvariable in time in the central zone. The occupied states will give the electron density. The contribution of each electrode to the density is different and must be calculated independently. In fact, left and right density are computed by summing up the occupied scattering states as:

$$n^L(\mathbf{r}) = \sum_{\alpha} |\psi_\alpha(\mathbf{r})|^2 f\left(\frac{\varepsilon_\alpha - \mu_L}{kT}\right)$$
$$n^R(\mathbf{r}) = \sum_{\alpha} |\psi_\alpha(\mathbf{r})|^2 f\left(\frac{\varepsilon_\alpha - \mu_R}{kT}\right),$$

We use the NEGF method [20, 21] and divide the density matrix into left and right contributions as:

$$D = D^L + D^R.$$

And

$$D^{L,R} = \int \rho^{L,R}(\varepsilon) f\left(\frac{\varepsilon - \mu_{L,R}}{k_B T_{L,R}}\right) d\varepsilon$$
$$\rho^{L,R}(\varepsilon) = \frac{1}{2\pi} G(\varepsilon)\Gamma^{L,R}(\varepsilon)G^+(\varepsilon)$$

Where f is the Fermi function with an electron temperature T_L. G is the retarded Green's function, and $\Gamma^{L,R} = \frac{1}{i}\left(\Sigma^{L,R} - (\Sigma^{L,R})^\dagger\right)$ is the broadening functions of the left (right) electrodes. $\Sigma^{L,\ R}$ is A right/left density matrix contribution. Finally, for the central region, we calculate the retarded Green's function matrix, using the overlap (S) and Hamiltonian (H) matrices, respectively, of the entire system as:

$$G(\varepsilon) = \left[(\varepsilon + i\delta_+)S - H - \Sigma^L(\varepsilon) - \Sigma^R(\varepsilon)\right]^{-1}.$$

3 DISTORTED ARMCHAIR AND ZIGZAG NANORIBBONS GRAPHENE: TRANSPORT PROPERTIES

In general, graphene nanoribbons (GNRs) are essentially single layers of graphene that are cut from the infinite sheet in a suitable shape and particular pattern to give them certain electrical properties. Depending on how the unbonded edges are configured, they can either be in a zigzag or armchair configuration. In this section, we will calculate the band structure of ideal, infinite graphene nanoribbons of either the armchair or zigzag type. Then we will introduce defects on the edge of these GNRs in order to examine their effect on the transport properties. Also, we will study their influence on the transmission spectrum.

Now, it is important to recall some basic information on the band structure of an infinite 2D sheet of graphene. It is instructional for this particular system to compare the results of a DFT calculation with the semiempirical method based on the extended Hückel tight binding model. For this purpose, the Cerda graphite basis set is selected for carbons.

From our calculations, we may conclude that the Hückel method leads to a good accuracy for this system and will be used for the further systems such as GNRs, mainly to save time, as in Fig. 2.

We focus now on graphene nanoribbons (GNR). There are two primary ways to cut out such a ribbon, and these two structures are known as armchair or zigzag ribbons, a nomenclature that refers to the shape of the edges. We display in Fig. 3 the band structures of these ribbons.

The armchair ribbon is found to be semiconducting whereas the zigzag is metallic. Further studies have shown that the band gap is strongly dependent on the types of ribbons (armchair/zigzag) and different widths.

Now, we turn our attention to the metallic zigzag GNR. This system has no elastic scattering, leading to a perfect conductivity. In perfect device geometry, there is no scattering from the central region and so we can obtain the transmission spectrum from the properties of the electrode. We will also consider the system as a device by setting up electrodes from the GNR, as

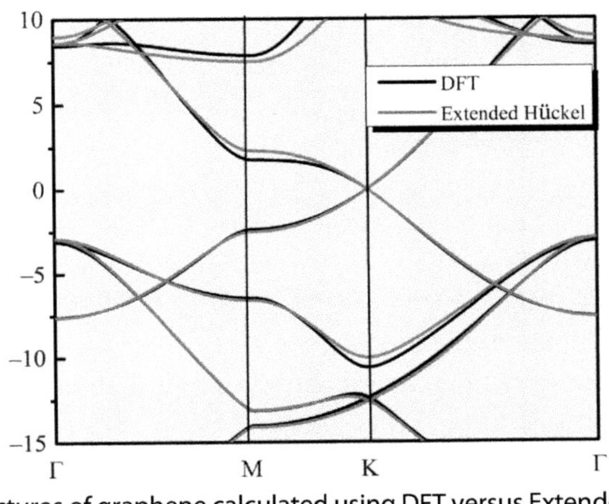

Fig. 2 Bandstructures of graphene calculated using DFT versus Extended Hückel model.

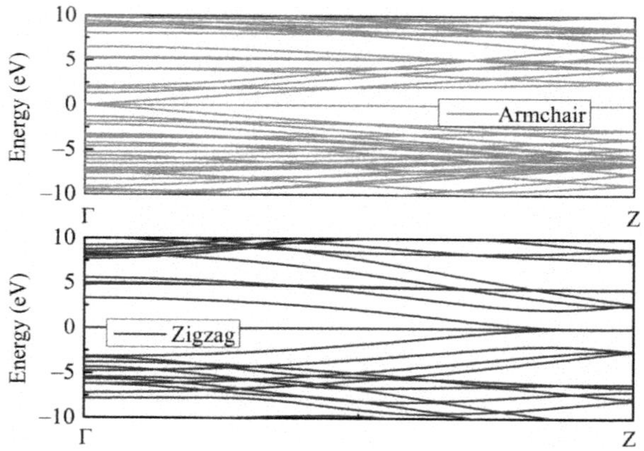

Fig. 3 Electronic band structures of armchair/zigzag graphene.

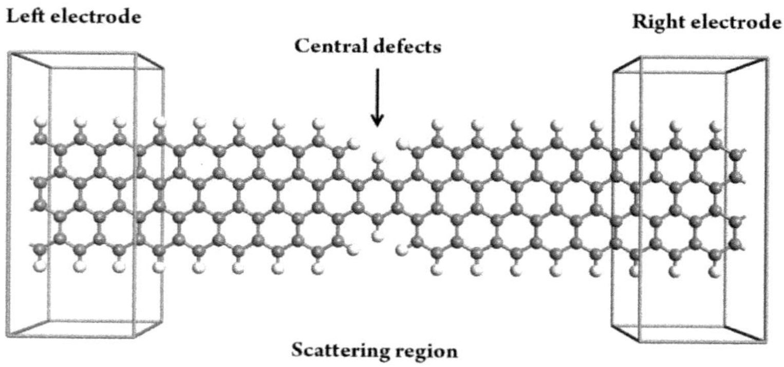

Fig. 4 Distorted armchair GNR device.

displayed in Fig. 4. This device is also distorted where we removed some carbon atoms from the center region of the ribbon. To avoid numerical problems and a surface effect, we passivated the dangling bonds with hydrogen.

Now we display the transmission spectrum of the ideal GNR without any defect and in bulk configuration together with the spectrum in Fig. 5 of the distorted GNR.

Comparing these results, one can see that the strong transmission at the Fermi level is also present in the distorted ribbon, but otherwise the transmission spectrum is strongly suppressed due to scattering by the distortion.

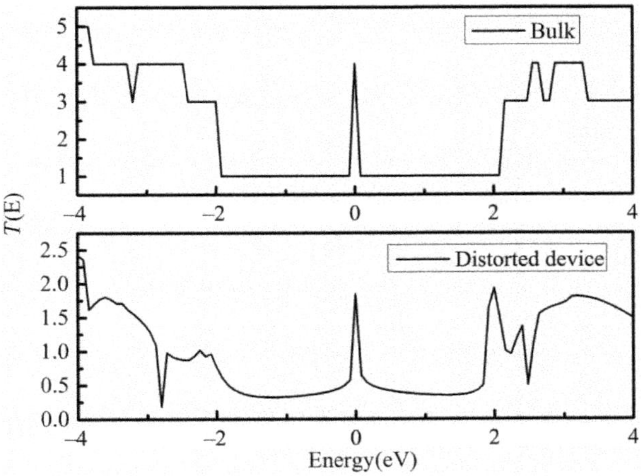

Fig. 5 Comparison of the transmission coefficient of a perfect ribbon and the distorted ribbon calculated nonself-consistent.

4 GRAPHENE BUCKLING

The buckling is characterized by a sudden failure of a structural member subjected to high compressive stress. A thin sheet tends to buckle under compression. From our literature review, we noticed that the electronic properties of graphene with and without buckling, critical to flexible devices, have been rarely discussed.

In the present section, we have considered GNR in the armchair phase. In order to be able to modify the structure of the graphene ribbons and to consider the buckling effect, we use the supercell approximation of $1 \times 1 \times 12$, leading to a final length of 49.44 Å in z-direction.

The buckled graphene is obtained on the supercell following the z-direction (Fig. 6). In order to create a buckling in this sheet, we are considering a function $f(y,z)$ added to the x coordinate of each atom, and essentially $f(y,z)=A^*\sin(y)^*\sin(z)$, where A is the buckling amplitude., and $\sin(y)$ and $\sin(z)$ are taking into account the periodicity in the y and z directions, respectively. Periodicity also induces a wavelength of the buckling in both directions. These wavelengths are named the width and length of buckling. It is important to notice that some adjustments are considered in order to have the possibility of an excluded zone on the edges in the z direction.

In our calculation, we have considered 2 and 1 periods following y and z directions, respectively. The Buckling amplitudes are 0.5–2.7 Å.

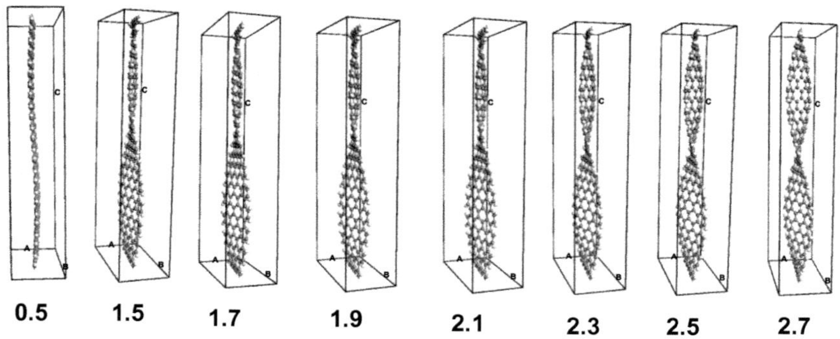

Fig. 6 Modeling the buckling in GNR.

We notice that from $A = 2.3$ Å, a breaking bond starts to appear but until 2.5 Å, the GNR structure still cohesive.

We start by studying the electronic properties of the buckled graphene systems and we compare their calculated bandstructures to the pure armchair. In Fig. 7, we present the calculated band structures for the considered buckling amplitudes. We may observe that an opening band gap is appearing when the amplitude increases from 0.5 to 2.3 Å. Then, the band gap is decreasing for 2.5 and 2.7 Å. It is important to notice that we didn't apply a larger amplitude than 2.7 Å in order to avoid a breaking of the structure. Our calculations confirm that armchair ribbons are semiconducting compared to the metallic zigzag ribbons. Compared to the armchair band structure, the buckled graphene BSs are showing a band folding but it doesn't affect the fundamental properties and we may easily extract the band gap.

In Fig. 8, we report the variation of the band gap versus the buckling amplitude. This curve shows an enhancement of the band gap with the increasing of the buckling amplitude.

5 CONCLUSION

Recently, graphene and graphene-based materials have been included in various fields of research. To improve the performances of devices based on graphene, original strategies should be made to efficiently change the physical properties. In this chapter, we have presented a theoretical approach to stimulate further investigations in organic electronic and organic photovoltaic fields for systems based on graphene. In summary, we reported the

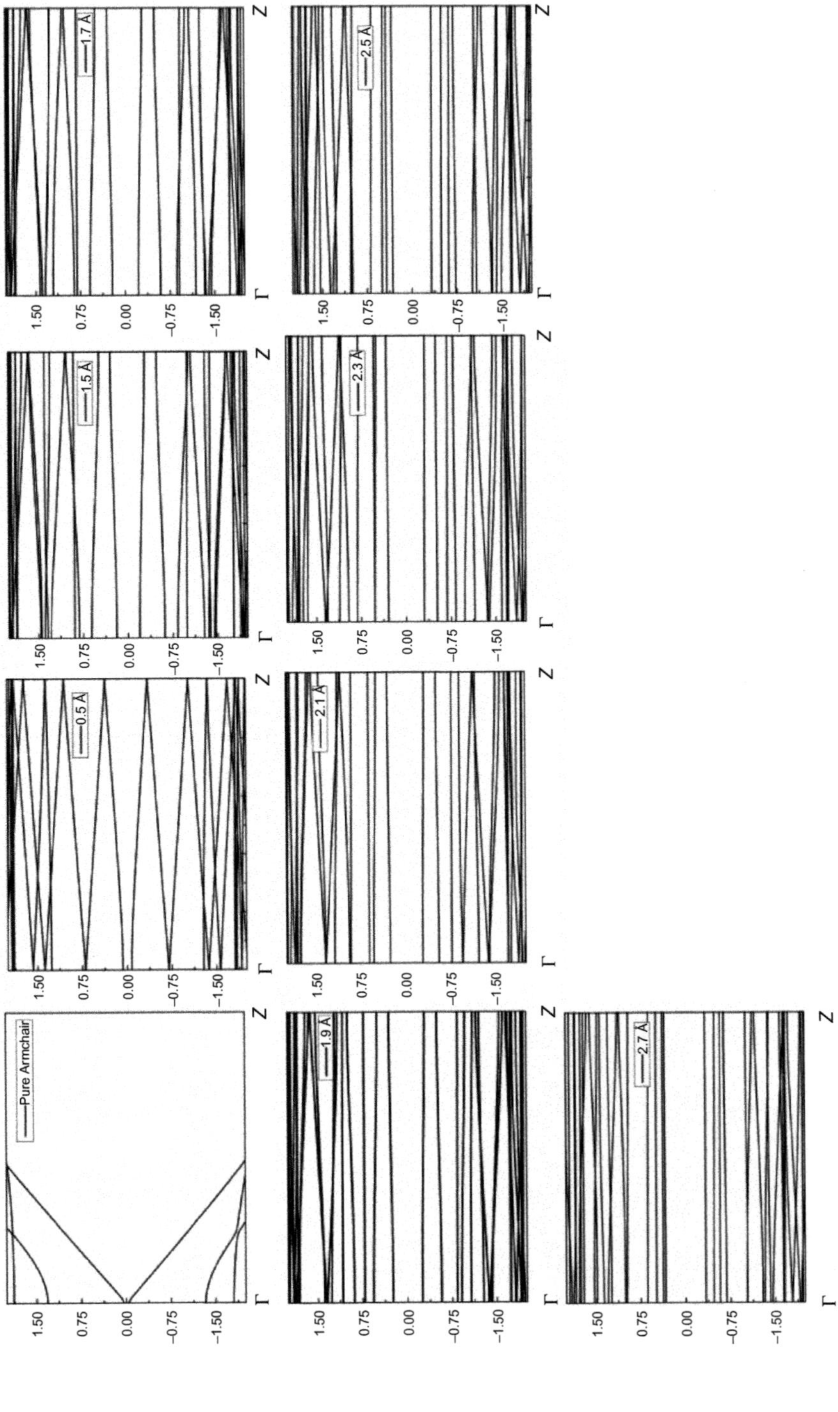

Fig. 7 Electronic band structures for different buckled amplitudes.

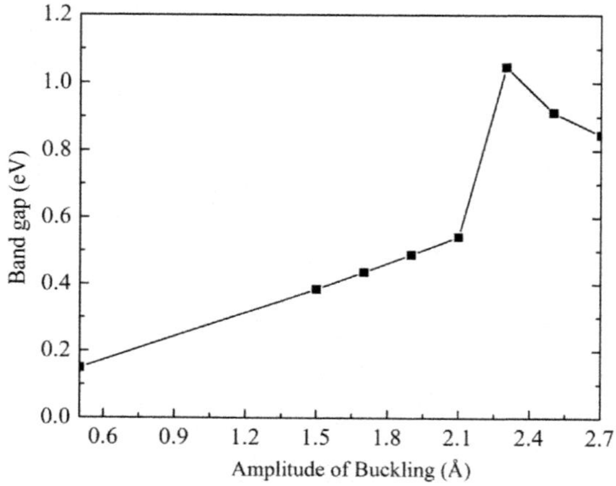

Fig. 8 An opening band gap is appearing by increasing the amplitude from 0.5 to 2.3 Å.

use of computational simulation approaches based on density functional theory and transport theory (Green's Function) to study electronic devices based on organic material: graphene. To understand the potential changes in the physical properties of graphene arising from the angstrom scale periodic buckling, we used DFT to calculate the electronic structures of the buckled graphene monolayers. The tunable physical properties of buckled graphene lead to promising applications in numerous fields, including flexible devices, photovoltaic, and energy storage. In order to improve the device performance, new strategies should be proposed to efficiently modify the physical properties of graphene-based materials.

ACKNOWLEDGMENTS

Financial support for this study is acknowledged in the form of an internal grant IRG 18418. QW ATK 2017 On-Demand! was used under Sabalcore's high-performance cloud computing service (http://www.sabalcore.com/).

REFERENCES

[1] H. Shirakawa, E.J. Louis, A.G. MacDiarmid, C.K. Chiang, A.J. Heeger, Synthesis of electrically conducting organic polymers: halogen derivatives of polyacetylene, (CH) x, J. Chem. Soc. Chem. Commun. (1977) 578–580, https://doi.org/10.1039/C39770000578.

[2] J.C. Cuevas, E. Scheer, Molecular Electronics: An Introduction to Theory and Experiment, second ed., World Scientific Publishing Company Pte Limited, 2010.

[3] (a) A. Mishra, M.K.R. Fischer, P. Bäuerle, Metal-free organic dyes for dye-sensitized solar cells: from structure: property relationships to design rules, Angew. Chem. Int. Ed.

48 (2009) 2474–2499. (b) A. Hagfeldt, G. Boschloo, L. Sun, L. Kloo, H. Pettersson, Dye-sensitized solar cells, Chem. Rev. 110 (2010) 6595–6663. (c) Y.-S. Yen, H.-H. Chou, Y.-C. Chen, C.-Y. Hsu, J.T. Lin, Recent developments in molecule-based organic materials for dye-sensitized solar cells, J. Mater. Chem. 22 (2012) 8734–8747.
[4] F. Schwierz, Graphene transistors, Nat. Nanotechnol. 5 (2010) 487–496.
[5] K. Kim, J.-Y. Choi, T. Kim, S.-H. Cho, H.-J. Chung, A role for graphene in silicon-based semiconductor devices, Nature 479 (2011) 338–344.
[6] K.S. Novoselov, V.I. Fal'ko, L. Colombo, P.R. Gellert, M.G. Schwab, K. Kim, A roadmap for graphene, Nature 490 (2012) 192–200.
[7] X. Jia, J. Campos-Delgado, M. Terrones, V. Meunier, M.S. Dresselhaus, Graphene edges: a review of their fabrication and characterization, Nanoscale 3 (2011) 86–95.
[8] L. Ma, J. Wang, F. Ding, Recent progress and challenges in graphene nanoribbon synthesis, ChemPhysChem 14 (2013) 47–54.
[9] P. Ruffieux, et al., Electronic structure of atomically precise graphene nanoribbons, ACS Nano 6 (2012) 6930–6935.
[10] J. Gao, J. Zhang, H. Liu, Q. Zhang, J. Zhao, Structures, mobilities, electronic and magnetic properties of point defects in silicene, Nanoscale 5 (2013) 9785.
[11] A.K. Geim, K.S. Novoselov, The rise of graphene, Nat. Mater. 6 (2007) 183.
[12] D.R. Dreyer, R.S. Ruoff, C.W. Bielawski, From conception to realization: an historial account of graphene and some perspectives for its future, Angew. Chem. Int. Ed. Eng. 49 (2010) 9336.
[13] H.-S. Shen, Y.-M. Xu, C.-L. Zhang, Graphene: why buckling occurs? Appl. Phys. Lett. 102 (131905) (2013), https://doi.org/10.1063/1.4799673.
[14] A. Fasolino, J.H. Los, M.I. Katsnelson, Intrinsic ripples in graphene, Nat. Mater. 6 (2007) 858.
[15] M. Elstner, D. Porezag, G. Jungnickel, J. Elsner, M. Haug, T. Frauenheim, S. Suhai, G. Seifert, Self-consistent-charge density-functional tight-binding method for simulations of complex materials properties, Phys. Rev. B 58 (1998) 07260.
[16] K. Stokbro, D.E. Petersen, S. Smidstrup, A. Blom, M. Ipsen, K. Kaasbjerg, Semiempirical model for nanoscale device simulations, Phys. Rev. B 82 (2010) 075420.
[17] H. Haug, A.-P. Jauho, Quantum Kinetics in Transport and Optics of Semiconductors, Springer Verlag, Berlin, 1996.
[18] R. Hoffmann, An extended Hückel theory. I. Hydrocarbons. J. Chem. Phys. 39 (6) (1963) 1397–1412, https://doi.org/10.1063/1.1734456.
[19] M. Wolfsberg, L.J. Helmholz, The spectra and electronic structure of the tetrahedral ions MnO_4^-, CrO_4^{--}, and ClO_4^-. J. Chem. Phys. 20 (5) (1952) 837, https://doi.org/10.1063/1.1700580.
[20] T.B. Martins, R.H. Miwa, A.J.R. da Silva, A. Fazzio, Electronic and transport properties of boron-doped graphene nanoribbons, Phys. Rev. Lett. 98 (2007) 196803.
[21] B. Huang, et al., Making a field effect transistor on a single graphene nanoribbon by selective doping, Appl. Phys. Lett. 91 (2007) 253122–253123.

CHAPTER 11

Nanobiomimicry at Quantum Scales: Synthetic Imperfections to Imitating Low-Dimensional Biomimetic Polymer Confinement of TiO_2 at Self-Assembled Heterogeneous Interfaces

Varsha Khare, Sanjiv Sonkaria
Institute of Advanced Machinery and Design, Seoul National University, Seoul, Republic of Korea

1 INTRODUCTION

Better accessibility to quantum materials with potentially useful applications (Fig. 1) is being met with the growing opinion that materials are increasingly tunable [1] through the capture of surface-directed crystalline states or triggering crystallization of metastable intermediates via amorphous states [2] operating through disorder-order transitions. The crystalline transition from amorphous calcium carbonate (ACC) occurs via the metastable phase [3] in many organisms [4] and the role of synthetic polymers namely polystyrene, can play an effective role in arresting the ion binding phase during biomineralization favoring the formation of amorphous assemblies [5]. The precrystallization phase and the extent of deviation from an ordered arrangement in both natural and synthetic processes brings into account the ability of polymorphism in which a solid material can exist in multiple lattice configurations coupled to enthalphic routes and energy barriers (Fig. 2). To account for diversity and complexity in structural assemblies, insightful mechanistic clues to crystal nucleation are being unveiled through better understanding of biomineralization. The biomimetic design of synthetic templates that hold the potential to synthetically mimic biohybrid crystalline behavior of inorganic materials has recently been inspired by a new concept described as "crystallization by particle attachment" or CPA [6]. CPA operates by

Nano-sized Multifunctional Materials
https://doi.org/10.1016/B978-0-12-813934-9.00011-6

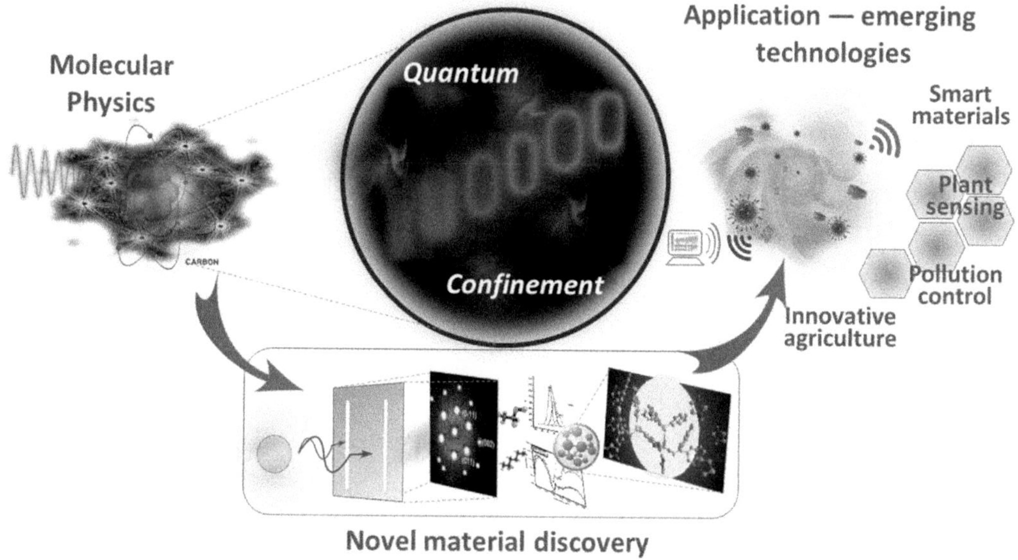

Fig. 1 Quantum materials. A multidisciplinary effort combining molecular physics and nanochemistry is the driving force to novel material discovery at the quantum scale, leading to innovative applications.

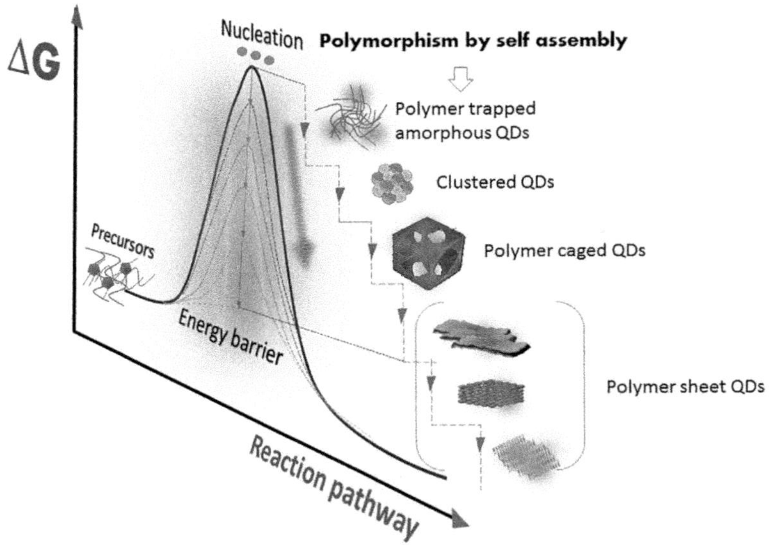

Fig. 2 Polymorphism of synthetic materials by self-assembly. Tunability by shape, size, and geometry of interchangeable nonequilibrium (metastable) structures governed by surface energy landscape intermediate between unstable and stable states.

taking advantage of free energy landscapes particularly at heterogeneous interfacial boundaries through forces that (1) liberate new chemical building blocks via surface catalysis overcoming energy barriers, (2) process the assembly of unit hybrid structures to larger assemblies driving new energy minimization routes (via bond associations/dissociations, stabilization of defect sites involving host chemical reactions), (3) increase stabilization via water bonding orientation, and (4) drive bulk to nano- and quantum-scale geometries through low-energy crystalline states or confinement of high-energy amorphous (nonequilibrium) metastable states. Hence, CPA represents a multiassembly process with implications in numerous biomimetic-related events shown as a cluster diagram (Fig. 3).

An interesting example of hierarchical assembly that takes advantage of the CPA regulating the biomineralization of higher-ordered structures at the nanoscale has been demonstrated by amelogenesis [7], a process coordinating enamel formation. Here, a multistep mineralization process leading to enamel crystals relies on oriented assembly via precluster formation that further undergoes morphological organization into bundles of chain-like structures directing the elongation of minerals [8]. The mechanisms defining preclustering in biologically mediated events are poorly understood.

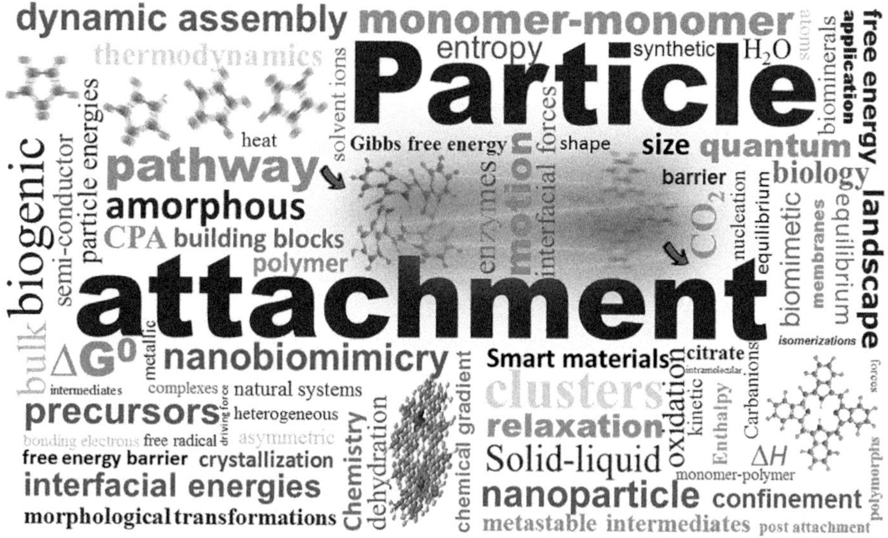

Fig. 3 Crystallization by particle attachment (CPA). A cluster-type representation of CPA signifying biomimetic-related events during particle assembly.

However, the biomimetic pathways of synthetic origin driving metal-organic assemblies are showing greater relevance and overlap with supramolecular morphologies. An important area of overlap is transient structures of an amorphous nature stabilized, for example, by amelogenin [9] during calcium phosphate biomineralization. Such mechanisms driving space-confined nanospheres and chain formation could be adapted to synthetic approaches utilizing biomimetic pathways by tuning interfacial energies, particularly at heterogeneous boundaries. This philosophy drives the notion that dissimilarity in bulk protein architecture and organometallic polymers, although strikingly different, could be intrinsically similar to generate kinetic stabilities around more challenging morphological structures. Similar to biogenic amorphous states in which organisms can dynamically adapt crystalline morphological structures to suit environmental conditions, synthetic materials are tunable by stabilizing interchangeable nonequilibrium structures desirably in a stimuli-responsive manner (Fig. 2).

Some of the overlapping biomimetic parameters in self-assembly that likely prevail in biogenically controlled environments during crystallization and pre- and post-crystallization events are summarized in Table 1. Cooperativity among these parameters can steer the reaction dynamics of multiphase nucleation processes enriched by heterogeneous interfaces to preferred energy landscapes influencing shape, size, and morphological structure. This is evident from the products of biomineralization that often evolve from simple to complex morphological structures different to the less elaborate structures that likely result from conventional pathways. In reality, competing nucleation pathways may be as diverse as the energy landscapes through which they operate where overcoming such barriers could lead to unresolved precrystalline states—amorphicity. If crystalline and noncrystalline pathways occur concurrently [10], the directional size and shape effects should dominate underlying nucleation mechanisms both for crystalline- and precursor-based routes. Particle-based approaches prevalent in biomineralization [11], however, from a synthetic point of view could enhance the availability of quantum-sized materials through higher-order geometrical shapes. Here, the time scale of nucleation rates becomes an important consideration in multiphase nucleation processes encompassing synergistic or independent nucleation centers. The solubility concerns of self-assembled building blocks such as monomers, precursors, and clustered groups could affect nucleation rates. For example, colloidal environments governed by the availability of precursor groups fueling polymer growth could have interesting contributory effects on spontaneous inhibitory nucleation or

Table 1 Factors governing energy landscapes affecting morphological synthetic properties in template driven self-assembly

Biomimetic characteristics of self-assembled growth from core bulk materials (without additives)					
Chemical environment		**Particle attachment**		**Nucleation**	**Routes to crystallinity**
Template Nonsacrificial or sacrificial Molecular: – Structure – Chem. composition, e.g, colloidal – Space (pores, crevices, restricted volumes) – Structure directing – Pore curvature **Charge** Magnitude distribution **Orientation reaction dynamics** Particle motion Solute-solvent Transport Diffusivity – Rotational – Transitional Vibration Collision Aggregation	Intrinsic / Extrinsic Factors	***Particle formation*** Monomers (primary/secondary) Preclusters (multiion complexation) Interaction: Chemical functionality – Adsorption – Desorption Interactions: Covalency Electrostatic (Columbic) Hydrogen Lewis/Acid base Van Der Waals Brownian Water associations Shape Size Morphology Crystal (ordered) Amorphous – Disordered – Disordered/Ordered (multiphase) Products: Catalytic Noncatalytic	Directional / Nondirectional	***Nucleation*** Directed Nondirected ***Energy barrier accessibility*** Surface energy Temperature – Glass transition (T_g) – Or $\geq/\gg T_g$ Supersaturation Free energy landscape (Enthalphic pathways) – Kinetic – Thermodynamic – Entropic ***Phase*** Singular (homogenous interface) ***Multiphase*** Multinucleation (Heterogeneous interface: multiinterfacial boundaries), e.g., polymer-metal interface – Cooperative – Independent Crystal Growth: – Critical nuclei size – Expansion – Shrinkage – Directional/nondirectional – Strain/stress confinement (nano/quantum state)	Amorphicity/metastability Stability and degree of structural resemblance to crystalline structure Static: Particle aggregation and dehydration Nonstatic: Stimuli responsive trigger, e.g., pH, solvent, photon, thermal

Interfacial Forces – Weak/strong – Attractive/repulsive Surface energy Solid-liquid Liquid-gas **Surface dynamics** Ion motilities Ion distribution Porosity (bulk/micro/nano) **Surface** Metal defects sites **Surface catalysis** Primary and secondary chemical building blocks	Polymerization of organic monomers Geometrical confinement Cluster size Out-growth/in-growth Curvature Stress/strain Defects Size and motion: – Bulk – Nano – Quantum ***Alignment and attachment*** Lattice plane complementation Equilibrium – Structured Nonequilibrium – Unstructured Polymorphs	Postnucleation ***Surface ligands*** Inhibitory Noninhibitory Bridges ***Complexation*** Aligned Misaligned ***Time-scale of nucleation growth rate (multiphase self-assembly)*** Cooperative – Synergistic growth with same rate – Synergistic growth at different rates – Inhibitory – Simultaneous inhibitory growth of one nuclei (e.g., polymer (matrix) nucleation confinement of semiconductor metal growth) Independent Simultaneous independent nuclei growth: – Interfacial boundaries – Noninterfacial boundaries	

its redirection to amorphous growth while regulating crystal size [12] by postattachment in the solution phase [13]. This scenario becomes more interesting when low polydispersity results from heterogeneous nucleation of polymer organometallics in which the spectrum of metastable disordered amorphous structures lies in a relatively narrow disorder-to-order transition range. Such materials are demonstrably desirable mimics of bioamorphous materials, likely representing a new class of biomimetic synthetics.

The foreseeable challenge in the context of this discussion is focused on using CPA to direct the biomimetic assembly of novel architectures that have the ability to nucleate around nucleation sites of TiO_2 undergoing bulk to subnanometer transition. The ability to impose shape and size restriction preferentially exhibiting quantum growth-scale characteristics mimicking bioorganisms would serve as a cornerstone to obtain technologically important materials. The simultaneous control of nucleation growth patterns governing chemical functionality, particle assembly and motion, orientation through alignment and attachment, directional control of shape and size of nucleating materials occur through enthalphic compensation by driving multiparticle complexation as depicted in Fig. 4.

Surface charge and ion distribution of mineral-specific proteins have proved to be important determinants in regulating the physical and chemical nature of inorganic materials and their ability to form polymorphs. It is known, for example, that differences in the phase solubility of soluble anionic proteins and insoluble matrix proteins from the mollusk shell enable mechanistic cooperativity between morphological crystal growth and mineralized plate separation [14], respectively. This observation highlights the possible role of cationic/anionic chemical environments as artificial mimics of spatial orientation and as structure-directing agents. In view of the special interest in employing template-driven reactions toward the synthesis of subnanometer particles, the adsorption of template constituents with molecular properties (e.g., shape, size charge, functional groups) at catalytic semiconductor interfaces could generate interesting outcomes. For example, polymer immobilization of metals to the nanoscale [15] could be an effective strategy to drive nucleation processes through spatial networks involving polymer frameworks to achieve metal growth confinement. Decades on, conventional methods such as thermal decomposition precursors in organic compositions and chemical reduction of metals in polymers of bulk metals in polymer solutions have yielded little success for quantumization of bulk TiO_2. To date, very few successes have yielded quantum-size TiO_2 and some of these methods are summarized in Fig. 5. This enforces the opinion

Fig. 4 Particle assembly characteristics during CPA. The need for desirable material characteristics such as size, shape, and geometry requires simultaneous control over aspects of particle motion of monomers, surface charge, orientation, clustering of monomers to form precursor molecules, alignment, attachment, chemical functionality, and controlled accessibility to enthalphic (kinetic, thermodynamic and entropic) routes.

that interfacial energies at heterogeneous surfaces are better facilitated by structural matching of surfaces, shape, and size with precursors to allow a high degree of congruency in particle attachment and phase transformation. Such events become superior control mechanisms for photonic semiconductor materials that can be manipulated by sacrificial and nonsacrificial templates.

A more promising strategy to better mirror the characteristics of nanobiomimicry at heterogeneous interfaces might require combining catalytic surfaces of semiconductors with ordered polyionic surfaces in the form of a soft, biologically inspired monomer-precursor template. Such monomers that can undergo catalytic polymerization from a nucleating phase and can adapt around nucleating semiconductor species by self-assembly through amorphous, semicrystalline, and crystalline phases have much exploratory potential in the era of new materials discovery. This work speculates on introducing the dynamics of polymer architecture from nucleation

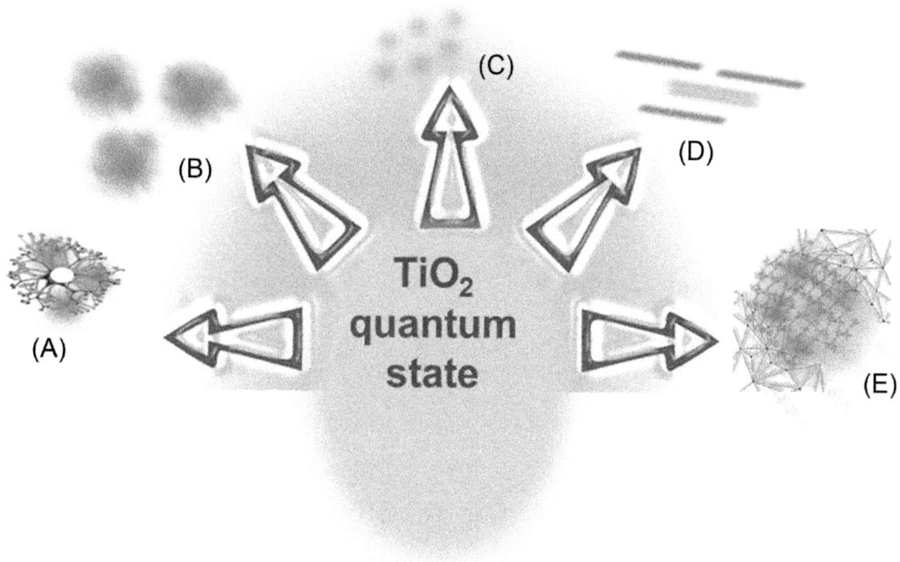

Fig. 5 Quantum-scale morphologies of TiO_2. A few successful attempts in the quantumization of TiO_2 have been reported to form (A) dendrimer templated quantum dots, (B) mesoporous spherical aggregates, (C) template-free quantum dots, (D) one-dimensional nanowires, and (E) polymer-caged TiO_2.

(and not by addition of preassemblies) in which hybridization at multiinterfaces can essentially mimic biomimicry. Further, the nature of self-assembled hybridized polymers in transitioning from bulk to nano pore-size geometries affecting its mechanical properties at the nano scale [16] (e.g., relaxation of interfacial bond stress and strain) can introduce new bending properties and orientations as biologically smart responsive structures correlated to pore size and shape. However, the departure of nucleation from the polymer crystalline state toward a glassy amorphous phase is tunable through the existence of cooperative length scales if metastable structures are close to equilibrium allowing disorder-order transitions. The metastable amorphicity of structures is behaviorally truly biomimetic, providing multifunctionality and tunability at the semiconductor interface.

In contrast to the self-assembly of polystyrene chains from the congruent precursor that readily undergoes crystallization, we demonstrate the assembly of a supramolecular polystyrene-porphyrin TiO_2 caged amorphous complex designated PS-P-TiO_2 (PS=polystyrene, P=porphyrin) that self-assembles without the addition of vinyl precursor molecules. Fig. 6 shows a simplified route to the quantum confinement of PS-P-TiO_2 via phosphonium IL (ionic liquid) chemistry, resulting in the catalytic evolution of styrene and porphyrin monomers at the bulk IL-TiO_2 interface.

2 TEMPLATING PROPERTIES OF A TETRA-ALKYL PHOSPHONIUM IONIC LIQUID-[P_{66614}][$N(CN)_2$]

2.1 Dynamics of Nanobiomimicry Through Synthetic Templating Induced by a Cationic-Anionic Ionic Liquid Environment

The importance of guided constraints imposed by a templated structure [17] by restricting the spatial environment for self-assembly to occur is gaining recognition as a tunable strategy toward biomimetic applications. Biological species occur both as charged biomolecules and chemical entities that are directionally important in steering specific nucleation events during assembly processes and regulating key surface interactions in numerous biological processes such as catalysis and folding. Applying physical-, chemical-, and shape-driven geometrical constraints by mimicking a mobile but ordered heterogeneous population of charged molecules composed of bulk cationic and anionic molecules or as pendant groups in more elaborate structures (e.g., alkyl chains with a terminal multivalent cationic element) could be an important biomimetic strategy (Fig. 7). Such diverse environments can

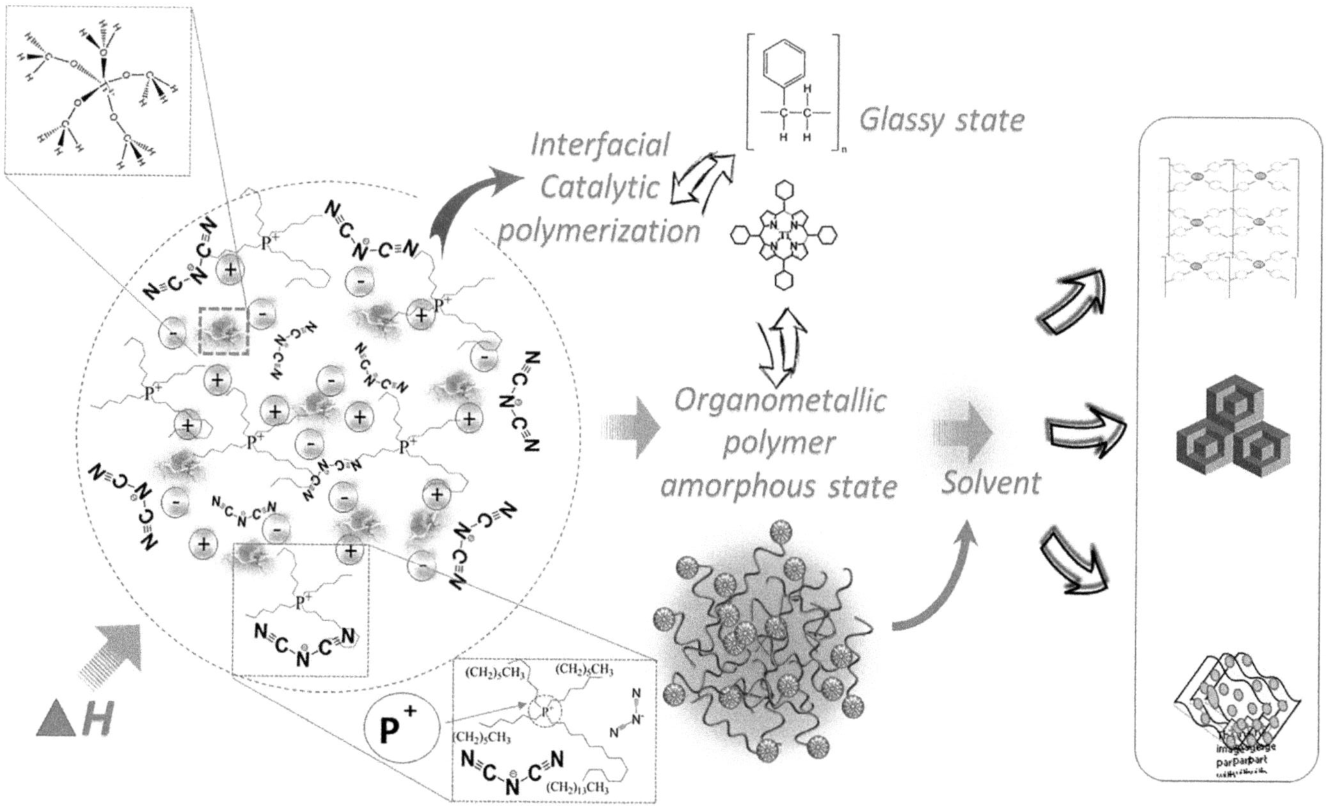

Fig. 6 Assembly of quantum-confined TiO_2 in the polystyrene-porphyrin [PS-P-TiO_2] assembly complex at the phosphonium dicyanmide ionic liquid-TiO_2 interface. Ionic liquid-induced confinement of TiO_2 via the growth arrest of the semiconductor oxide through ring confinement in the porphyrin and polystyrene matrix. Shape-dependent morphologies induced in the solvent from the amorphous state of the PS-P-TiO_2 complex provide control and access to semicrystalline and crystalline nanocage and sheet structures.

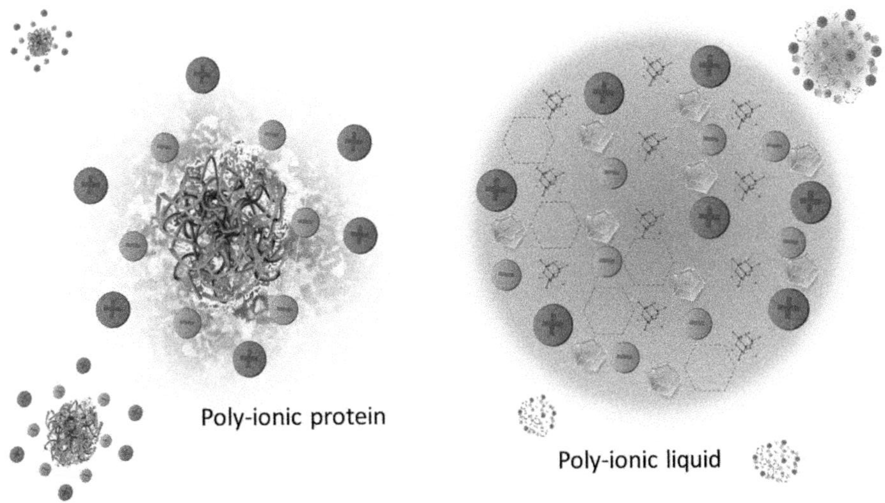

Fig. 7 Charge distribution of poly ionic protein and poly ionic liquid surfaces. The organization of cationic-anionic charges in ionic liquids as potential biomimetic templates mimicking the surface effects of anionic, cationic, or mixed cationic-anionic poly ionic proteins.

be modeled using simple or complex structured ILs by positioning contrasting bulk materials in compact spaces orientated by charge. The templating properties of ILs can be controlled by thermal activation of the reaction medium to generate specific behavioral effects observed in biological systems using alternative pathways.

The use of group-selective structural templates with TiO_2 bulk precursors [2] provides selective differential chemical binding targeted against multiple reaction centers in a mixed cationic-anionic environment, enhancing a natural selection of surface directed metal-polymer ligands. Such templates exhibit multifunctionalities that include (1) amphiphile features [18] of alternating cationic-anionic charge that cater for growth-inhibitory characteristics through metal surface catalyzed self-assembled passivating ligands at TiO_2 nucleating sites inhibiting growth, (2) strong interfacial interaction between thermally induced cationic defect vacancy sites on reduced titanium oxide with reactive nitrogen anionic templates, (3) regulating towards size control of crystal growth through controlled addition of organic/inorganic additives, and (4) metal surface catalyzed heterogeneous polymerization at the metal-organic interface driving nanocage quantumization of TiO_2 metal oxide cores. This process translates to an ensemble of multiple events that shares a multitude of parallel characteristics with biological processes pertaining to (a) biomineralization, (b) protein or multisubunit protein folding, (c) polymorphism in proteins, (d) dynamics of protein-water complexation hydration, (e) amorphous and glass transition protein states (T_g) demonstrating an exceptionally extraordinary biomimetic route to bioconvergence engineering of quantum size- and shape-dependent semiconductor materials, and (f) water dynamics. First, rapid supersaturation ensures a biomineralization-like route to quantum confinement of TiO_2 through a detectable amorphous phase signifying a time scale of events too short to allow adequate ordering of chemical bonds at nucleating surfaces. Second, competing kinetic and thermodynamic events during the quantum arrest of TiO_2 are essentially nonequilibrium in nature. This is shown by multiple kinetically trapped metastable amorphous assemblies bearing a strong tendency to exhibit polymorphism of synthetically compartmentalized TiO_2. These form amorphous networks and follow a stimuli-responsive shape dependency transitioning between the amorphous phase to quantum dots, nanospheres, nanorods, nanocages, and nanosheets (Fig. 8).

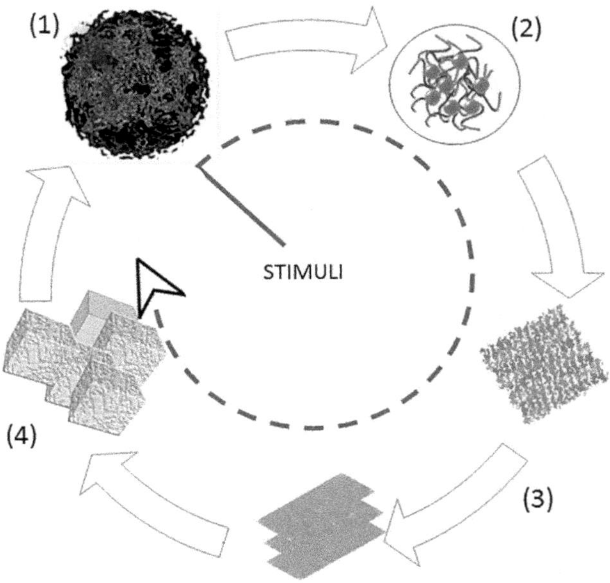

Fig. 8 Crystallization via an amorphous route. Stimuli-response dependent pathway to crystallization (e.g., solvent) initiated from the (1) amorphous state triggering the interconversion of polymorphic structures from (2) polymer-embedded quantum dots to (3) sheet and (4) cage-like morphologies.

2.2 Intrinsically Driven Pathway Selectivity at the TiO_2-IL Interface by Self-Assembly

Adopting a strategic approach shown in Fig. 9 allows self-assembly processes to survey and select for favorably driven kinetic and thermodynamic routes from a potential wealth of chemically divergent or convergent reactive nucleation sites formed at the TiO_2 semiconductor metal oxide layer and polymer interface. Here, the extent of structural accessibility to form polymer nitrogen (N) and phosphonium (P) coordinated complexes of titanium oxide has been explored [19].

N and P doping of bulk surfaces by high temperature pyrolysis is highly reactive to critically positioned oxygen defects that are a source of kinetically derived nanoarchitectures with strongly influencing electrocatalytic properties crucial for energy applications. The entry of nitrogen within these trapping sites at the TiO_2 surface, however, via a comparatively low thermal route altering the enthalphic competition between thermodynamic and kinetic pathways can arguably induce a metastable state, potentially steering

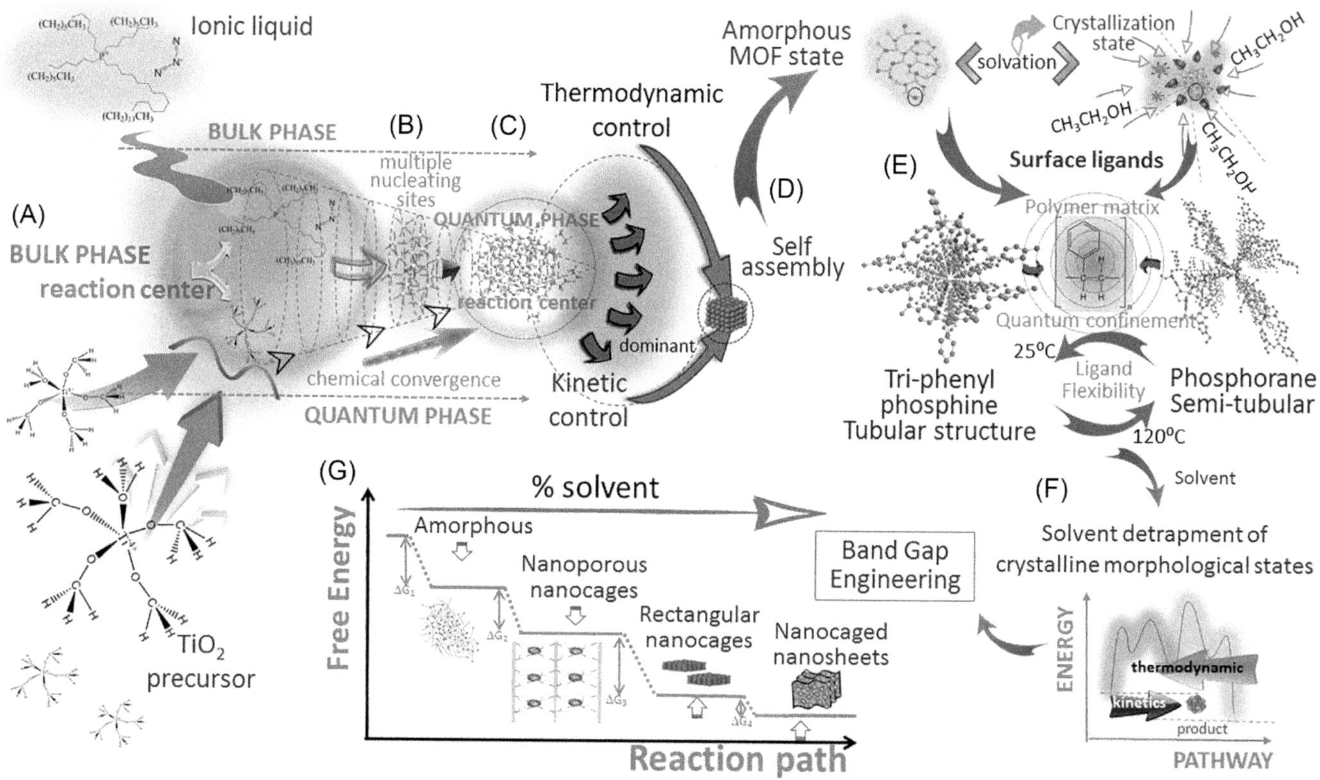

Fig. 9 Step-wise schematic representation of the key events leading to the formation of TiO2 quantum dot nanocages embedded in a polystyrene-porphyrin matrix (PS-P-TiO2) formed via independently converging crystallization pathways. (A) Bulk reaction between the TiO_2 precursor (titanium isopropoxide) and phosphonium ionic liquid polymer (tetradecyl phosphonium dicyanamide). (B) Interacting at the metal oxide-polymer interface through multiple nucleating sites where the interactions are (C) under kinetic and thermodynamic control proceeded from a quantum reaction center phase leading to an enthalpically favored pathway resulting from the chemical convergence of the bulk and quantum phase. (D) The self-assembly of the amorphous (disordered) polystyrene-TiO_2 MOF complex undergoes solvent-assisted crystallization (ordered) and (E) demonstrates temperature-dependent structural ligand flexibility coordinated at the interface of the metal oxide and polystyrene matrix. (F) Solvent detrapment of crystalline morphological states of the MOF complex releasing the (G) anisotropic solvent-driven dependency of hidden morphological structures opening up the potential of band gap engineering of photoresponsive materials.

precursor chemistry at the intersection of TiO_2, phosphonium, carbon, and nitrogen from the reaction environment. Using this strategy applied to an isopropoxide titanium bulk material, we propose that a metastable pyrrolic-N metal complexed supramolecular state attained through thermal capping unlocks an alternative pathway of selectivities to form stable intermediary states that compositionally comprise a series of trapped thermodynamic and kinetic states within a noncrystallographic plane. In recognition of the dynamic surface effects introduced by N doping on titanium surfaces and its potential to considerably enhance the photoresponsive properties at the semiconductor interface, the approach combines nitrogen-rich dicyanamide precursor chemistry. This is combined with the growth inhibitory characteristics of phosphonium derived ligand passivation using the reactive cationic-anionic properties of the alkyl phosphonium dicyanamide IL, $[P666_{14}][N(CN)_2]$ as the reaction medium. Fig. 10 shows that phosphonium and nitrogen play a central role in the assembly of precursor and stabilized states, leading to the quantum arrest of TiO_2.

2.3 Amorphicity of Nucleating Polymer Growth Around the TiO_2-Ionic Liquid Interface

Amorphous materials are high-energy trapped intermediates that may lose their transitory short-lived phase to become metastable in nature to exist in an excited state compared to the more stable ground state. In all probability, this is a rare occurrence but some organisms have incorporated such mechanisms in the biology of biomineralization and such routes to crystallization via transitory metastable phases are energetically unlikely to proceed from forbidden transitions. It is, however, evident from this observation that rerouting kinetic pathway selectivity via programmable surface interactions is indeed possible and this phenomena, reputably confined to biological systems, may arguably be applicable to nonbiological or synthetic processes. An obvious correlation between biological and nonbiological systems converges upon overlapping or diverging glassy behavioral characteristics common to material assembly from nucleation to fabricating supramolecular crystals and other underlying patterns that might help bridge a better understanding of material controllability. An interesting and probably one of the least understood phenomena of glassy materials and their subclasses is the extent to which aperiodicity of materials affects their deviation from crystal-like behavior. In other words, can materials be characteristically more crystal-like under the right conditions and yet exhibit strongly disordered diffraction patterns that cannot be distinguished from highly irregular

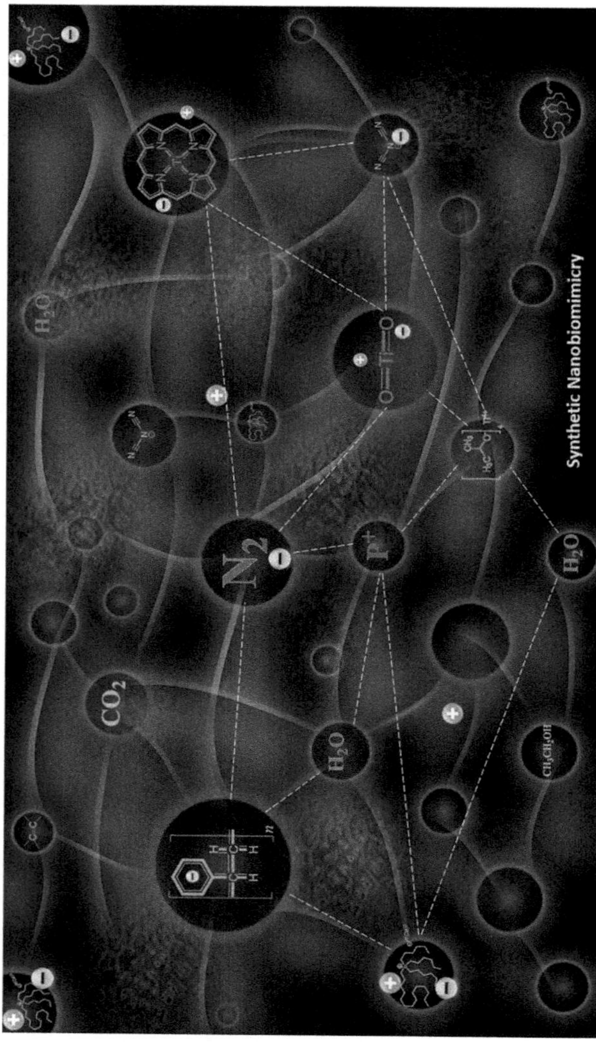

Fig. 10 Interconnection between cationic phosphonium and anionic nitrogen chemistry in precursor chemistry of PS-P-TiO_2.

material-types? The answer to this is undoubtedly more complex for materials with heterogeneous interfaces (e.g., colloidal metal-polymers). The suggestion that crystal structures can reversibly be formed from amorphous states or *vice versa* [20] as for some biomolecules might indicate that their configuration in the dysfunctional state could disassemble in a highly ordered concerted way, reducing energy barriers to crystallization. The preparatory steps to crystallization in the amorphous phase proceed to an energy minimum in which kinetically steered surface interactions are unperturbed and play a dominant role in predetermining the structural configurations that fall short of crystallization. This scenario exhibits elements of protein folding governed by a given peptide sequence in the unfolded state, suggesting that kinetic pathways coupled to amorphous systems with long relaxation periods may also have an evolutionary origin and are selective antecedently ahead of folding events on the landscape funnel time scale. This implies that if nonbonding interactions in the amorphous state are sufficiently tunable by the input of energy to allow for controlled structural rearrangements that permit bonding, it is possible to induce stimuli-responsive behavior to acquire polymorphic forms that are otherwise inaccessible from the most stable arrangements formed via more conventional routes. Hence, surface-driven energy landscape trajectories of protein assemblies that are consequential in directing thermodynamic and kinetic fates of biological structures could be equally important in generating amorphous trapped energy materials by regulating these transitions through temperature, which is intimately coupled to enthalphic effects. Like biology, materials chemistry can cater to synthetic materials that can freely adapt to shape and size contours through available configurational space of nonbonded or weak associations via energetically undemanding disorder-order transitions by chemical templating of bulk materials. This, to some extent, dispels the notion that disorder in synthetic structures is merely a consequence of surface defects and stabilization through passivation is random. However, templating plays an important role in ordering even a seemingly discontinuous disordered amorphous phase through mechanisms that are far from understood. An interesting development here is the role of growth polymers at the metal interface not only affecting routes to crystallization but arresting the growth phase of a semiconductor metal. This line of thought leads to much deliberation in unearthing the relationship between self-catalyzed polymer growth of semiconductor metals undergoing nucleation, the pore size (e.g., micro- to nanoscale) and shape of polymers that might be constrained, distorted, or exhibit irregular growth patterns, and how such characteristics affect glass transition properties (e.g., temperatures close to T_gs of polymers). For example, stress-strain at polymer metal interfaces can significantly reduce polymer pore dimension while contributing to growth restrictions at the metal surface (Fig. 11).

Long multichain polymer precursor molecules designed as nonmetallic polyatomic cations such as phosphorous covalently bonded by polymer alkyl chains of varying length spatially restricted by the presence of negatively charged dicyanamide anions (trihexyl tetradecyl phosphonium) have been used as templated IL-based reaction centers to growth arrest precrystalline ordered states of TiO_2. This approach represents a more structurally elaborate investigation of the contribution of chain alkylated phosphine oxide chemistry toward the assembly of quantum-sized colloidal semiconductor particles [21] using a spatial charge oriented system. Thus, the emergence of subnanometer metal nucleating species composed of TiO_2 that act as catalytic centers driving radial multithread-like nucleation of heterogeneous nanopolymers at the metal oxide surface requires deeper exploration. The time scale on which orientated polymer growth occurs between bulk precursor states and their temperature-dependent dissociation to monomer building blocks is connected to a number of molecular variables that govern (1) space and orientation between reacting precursor chains and the effect of monomer distribution, (2) chemical functionality of intermediate short and long polymer formation, (3) cooperativity in polymer assembly that depends on the multiplicity of precursor chains and chain length, (4) a signature pattern of polymer nucleation and growth multiplicity and its interdependency on temperature that suppresses metal oxide growth without extinguishing nucleation, a process of events that may require accessibility to new kinetic routes, and (5) favoring quantum metal oxide growth suppression around the critical nucleus size [22].

Copolymer blending to subnanometer dimensions by synthetic mixing has not proved advantageous to nanosizing during polymerization to the same effect as self-assembly. Cationic centered precursor alkyl chains and their catalytic activation at the TiO_2 surface into open-chain or partially ringed reactive monomers restricted within a cationic-anionic templated environment provide a diverse assembly of orientated cross-linked structures comprising both the main chain and bridging polymers from surface-adsorbed precursor molecules. We rationalized these template properties to initiate metal-polymer associations on the TiO_2 bulk precursor surface using an alkyl-rich phosphonium-centered cation counterbalanced with a dicyanamide anions alkyl IL precursor with essential templating properties

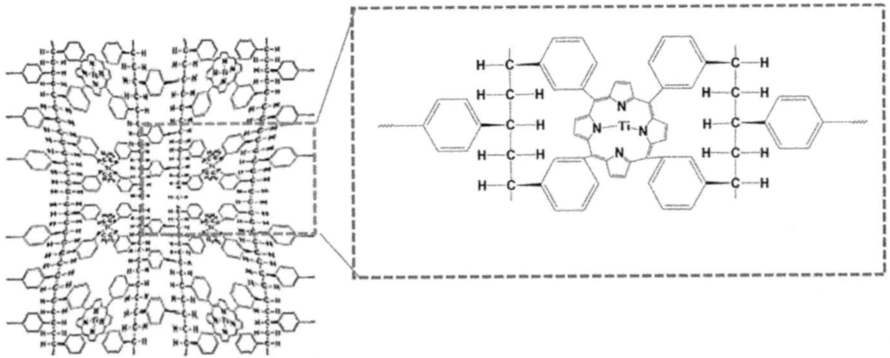

Fig. 11 Stress-strain effects at the polymer-TiO2 interface. Geometrical stress and strain at heterogeneous interfaces affect growth properties and material characteristics, significantly affecting polymer-metal outgrowth due to pore dimensions of polymers from bulk to nanoscale.

to first (a) orient monomer constituents formed as products of metal-assisted catalysis at the TiO_2 surface from the alkyl precursor chain, and their (b) subsequent assembly into short- and long-chain polymers from a rich supply of monomers extending from the metal surface via self-assembly. This approach enables compatible interfacial and bonding associations to occur freely at the metal-polymer interface that may otherwise be hindered and restricted by supplying preassembled polymer constructs. This rationale may have important consequences to cease enthalphic pathways that show greater dependence on temperature, chain length properties, and pendant group functionality of growing polymers and their intercalation. Better accessibility to drive free energy changes that maximize polymer chain conformations curbed by bulk templated groups favoring homogenous charge and shape specific effects for harmonizing surface effects of monomer assembly and orientation is highly desirable. This scenario might be particularly relevant in seeding the primary growth of multiple polymer-driven nucleation events rapidly advancing a systematic assembly of multiple polymer chains around metal surfaces, profoundly limiting crystal growth. The accumulation of chains at the interface of the organic-metal growth front could potentially be a barrier to postnucleation dominated by mass effects rather than polymer chain size and their interconnectivity at moderate temperatures.

Bulk TiO_2 precursor chemistry thermally driven to nano- or subnanometer scale nucleation under reducing conditions in the presence of a phosphonium cationic multialkyl chain linked polymer that is highly reactive to the catalytic metal oxide surface can appreciably alter nucleation kinetics at the soft-hard interface. This can be considerably complex for several reasons: (1) soft interaction of charged alkyl polymer chains at selective sites on the TiO_2 surface can be catalytically cleaved to form monomers that could participate to form more elaborate structurally controlled polymer assemblies around nucleation sites, and (2) highly stable dicyanamide anions present in the ionic liquid composition could play a crucial templating role in directionally orienting the polymerization of monomers and decisively shaping supramolecular polymer assemblies. Polymer comixtures have been shown to cooperatively impart maximum stability on the folding of biomolecules under conditions of decreased polymer dimension together with shape dynamics favorable to the caged structures.

Organic polymer ligands synthetically assembled within a metal organic framework construct that undergo stereoselective conformational changes through UV [23] and photo excitation [24] are a recently reported phenomena. Here, we propose tunability by anisotropic encapsulation of PS-P-TiO2 in the amorphous state in which the tacticity of polystyrene from the irregular *cis-trans* isotactic conformation to a regular *cis-trans* syndiotactic arrangement of styrene groups can be controllably initiated from the disordered atactic configuration via a solvent concentration-dependent process. Solvent polarity and its effect on hydrogen bonded networks at weak interaction sites [25] and the mobility of water molecules could affect stereospecific chain orientation. This shows that, while we observe a sizable order of change in the T_g value for the vinyl derivative polymer, the intrinsic property of polymer tacticity of polystyrene is retained in the bifunctional copolymer.

3 POLYMER-METAL COOPERATIVE NUCLEATION GROWTH CONFINEMENT OF RUTILE PHASE TiO_2: QUANTUM-SCALE CATALYTIC BULK PRECURSOR CHEMISTRY AT THE AMORPHOUS POLYMER-METAL OXIDE INTERFACE

3.1 A Building-Blocking Route to TiO_2 Quantum Confinement in PS-P-TiO_2 Complexation

We observed that the tetraalkylphosphonium IL reaction medium sufficiently generated chemical precursors that initiated the complexation of a phosphonic-centered TiO_2 polymer embedded in an extended polystyrene matrix from the interaction with titanium isopropoxide. An effective chemical barrier to the growth restriction of metal oxide nanoparticles has been reported for phosphonic acids [26] introduced independently into a suspension of preformed quantum-sized materials. This phenomenon is in agreement with our observations but fundamentally differs from our approach, suggesting that quantum confinement of TiO_2 is achievable at the onset of nucleation as a single "one-pot" reaction. Heterogeneous nucleation that occurs by metal center activation of surface-anchored substrates that are catalytically processed to synergistically release multiple reaction products via aromatization and dearomatization [27] events has recently been demonstrated. Here, the reactive TiO_2 isopropoxide precursor Ti[(OCH $(CH_3)_2)_4$ in the presence of a sacrificial cationic-ionic tryhexyltetradecyl phosphonium dicyanamide $[P_{6,6,6,14}][N(CN)_2]$ reaction medium generates lateral interactions dominated by interdependent reactions driving (a) tetrapyrrole ring confinement of TiO_2 embedded in (b) polystyrene via the cationic polymerization of styrene monomers and interlocked as supramolecular metalorganic matrix by a (c) surface-to-surface phosphonium-centered metal-ligand cooperative process. The convergence of complex events that involves the simultaneous evolution of TiO_2 QDs in a complex polystyrene matrix in a single reaction, to our knowledge, has not thus far been reported as a specific or general case for metal oxides [28]. An interesting comparison to the bulk reaction involving the ligation of a diphenylphosphine-polystyrene resin to the metal surface of nickel bromide forming an efficient catalytic complex driving polymerization at the organometallic interface [29] suggests a high degree of chemical programmability that connects the dynamic parallel between bulk and quantum pathways. A sufficiently well-matched surface energy interaction between the continuous phase of the organic environment and the metal oxide surface of the multiphasic reaction mediated by the phosphonic alkyl layer can potentially favor the stabilization of higher surface area particles between the nano- and quantum-sized domain.

The participation of titanium isopropoxide in olefinic chemistry has not been reported as a general phenomenon as for metal oxides of the same class [30] until recently in a study demonstrating its involvement as a cocatalyst in cyclic ring closure metathesis [31] favored by Lewis acid interactions via metal coordination. The substrate selectivity of Lewis acid binding sites at the precursor surface toward olefin terminal residues of phosphonium alkyl chains is consistent with bond vibrations and binding energies related with olefin metathesis. In this context, we rationalize that (1) terminal olefin structures formed at the end of tetraalkyl chains of the phosphonium cation can be important reaction centers for chain elongation catalyzed by metal oxide surfaces, (2) metal catalyzed cyclization or ring opening of polymerized products can influence nucleation affecting size, shape, and structure as a function of metal-ligand coordination chemistry and (3) phosphorane/phosphine ligands characteristically exhibit spectator properties and can affect the growth pattern of the metal oxide via stable metal coordination. Modulation of electron transitions of semiconducting materials presents interesting possibilities to alter the quantum physics of wide-band gap semiconductors. Fig. 12 depicts the evolution of particle clusters that originate from monomers possibly released as catalytic products from the interaction of the sacrificial template and the TiO_2 surface under a temperature constraint.

In this study, the ruthenium complex $RuH_2(PPh_3)_4$ structurally resembling the metal-bound triphenylphosphine ligand —$PN(CH_3)_2[(PPh_3)_3]$ at RT or O-$P[(PPh_3)_3]$ at 120°C (and residual traces with the configuration —$P[(PPh_3)_3]$ with Ti (Fig. 13) in the 4^+ valance state can coordinately

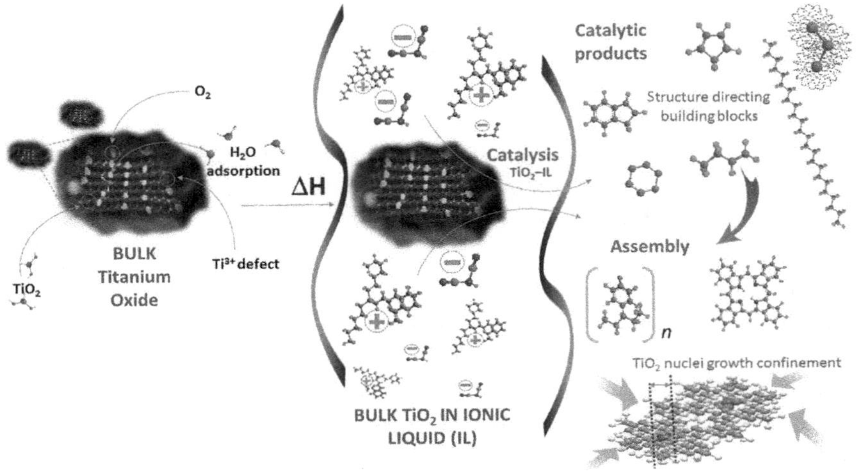

Fig. 12 Process of particle growth, attachment, and growth confinement at the TiO_2-polymer interface. Monomers generated as catalytic products at the bulk TiO_2 surface from a sacrificial template (substrate) participate in disassembly-assembly events and particle clustering, creating opportunities for polymer assembly around nucleating TiO_2 sites. Polymer growth, surface ligands, and complex bridge structures are important determinants in interfacial growth patterns toward growth confinement.

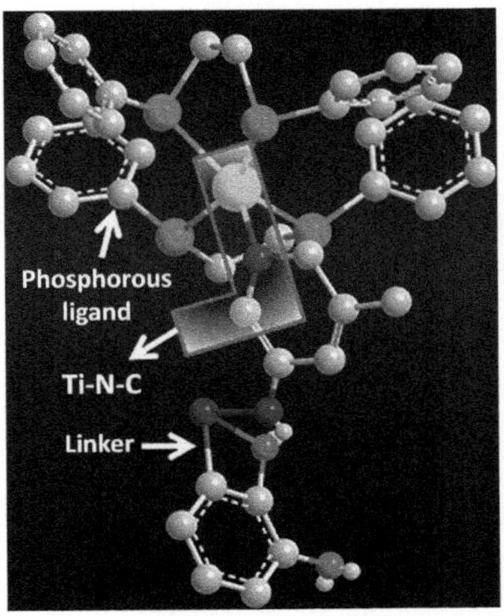

Fig. 13 Stoichiometry of the phosphorous ligand to TiO_2 via Ti—N—C bond formation and linker bridges.

interact with titanium oxide generated olefins at the heterogeneous catalytic metal oxide surface, frequently yielding ethylene as a low olefin hydrocarbon content product. In this specific instance, the synthetically driven complexation of the reaction of ruthenium intermediate $RuH_2(PPh_3)_4$ with ethylene forming $Ru(C_2H_4)[(PPh_3)_3$ can be rearranged in the presence of styrene with the central metal atom coordinated to the phosphonic ligand, allowing the styrene moieties to bind the metal with double-bond coordination.

The extent of the polystyrene matrix is distinguished from repetitive units of the styrene monomer $(\text{—}CH_2CH(C_6H_5)\text{—})_n$, revealing the association of the vinyl groups linked to the phenyl ring moiety of the C1s binding energy of a strong carbon-carbon backbone from photoelectron spectroscopy (XPS) analysis. Polymer hydrophobicity with intermediatory binding characteristics shared between adsorptive surface sites and the templating medium has been shown to favor an extended polymer conformation and its exposure along interfacial sites and its shell-like coverage subdue growth nucleation at accelerated rates [32], as evidenced here. As inferred, the shape and size of vinyl-based polymer compositions might conceivably maximize growth suppression, which might be hypothesized to occur by the passivation of crystalline regions via copolymer growth at noncompeting sites strongly favorable to growth suppression but not necessarily a hindrance to the crystallization of entrapped metal centers. A probable schematic visual representation of this scenario is shown in Fig. 14. Here in Fig. 14A,

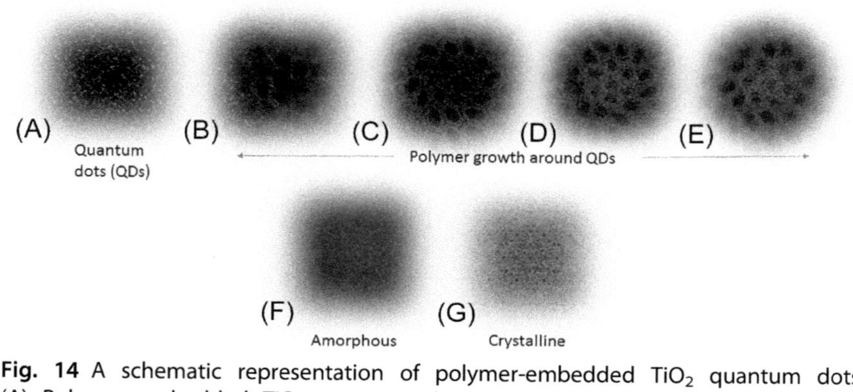

Fig. 14 A schematic representation of polymer-embedded TiO_2 quantum dots. (A) Polymer embedded TiO_2 quantum dots encapsulated by (B)–(E) network of nucleating polymer chains, (F) kinetically stabilized in a polymer amorphous state, (G) Metastable amorphous state tunable to crystalline states via an external stimuli.

quantum dots are embedded within the self-assembled polymer and the growth repression of nucleating quantum dots is strongly influenced by the nucleation of progressive polymer networks growth shown in Fig. 14B–E. If the nucleation rate of polymer growth occurs on a time scale too rapid to permit periodic ordering of polymers, a state of nonequilibrium results, captured in an (f) amorphous configuration entrapping TiO_2 as a subnanometer crystal. For the polymer-TiO_2 semiconductor crystal, this has biomimetic implications that reflect a stimuli-responsive tunable polymer shell to the (g) crystal state. In amorphous PS-P-TiO_2, polystyrene tacticity is solvent-responsive in which the chain configuration is solvent-dependent, transitioning from a noncrystalline (atactic) to a semicrystalline (isotactic) geometry and finally to the crystalline (syndiotactic) [2] state (Fig. 15). Such growth characteristics occupy central importance in the growth manipulation of subnanometer-sized particles in a supramolecular polymer matrix where crystallinity at the metal-polymer interface is

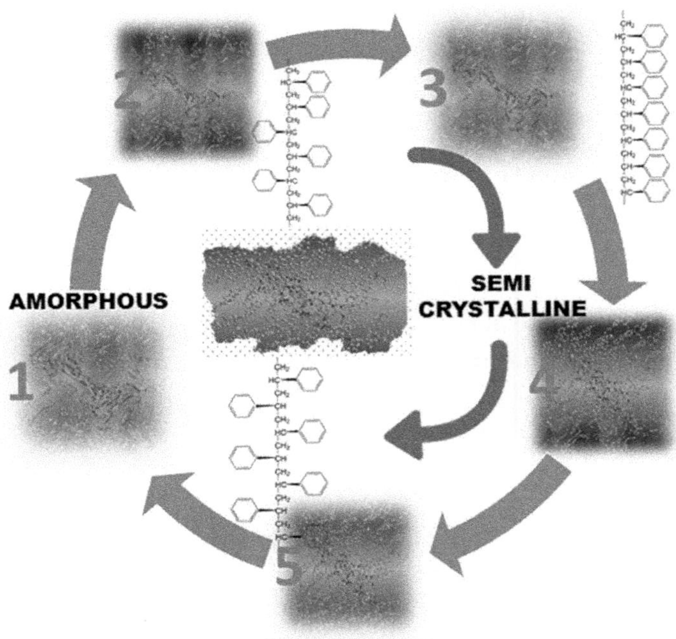

Fig. 15 Disorder-order polystyrene tacticity at the polymer-metal oxide interface. Stimuli-controlled polymer tacticity of polystyrene chains tunable through the amorphous to semicrystalline and crystalline phases.

crowded by the effects of distortion in the atomic alignment of polymer networks, leading to amorphous growth that originates from bond relaxation.

3.2 The Dual Role of Bulk and Nanoconfined Water Molecules

Differences in the behavioral properties of molecular water are distinguished by the thermogravimetric analysis (TGA) temperature decomposition pattern in ethanol and can be differentiated by their binding preferences that may be predominately lattice or coordinated [33]. The difference in the thermogravimetric pathway in reactions performed in the absence of ethanol is attributed to variations in the structural order, bond assembly, and phase space volume occupied by the polymer (polystyrene) and phosphated titania, demonstrating key differences in the coordination chemistry associated with polystyrene and TiO_2. The burnout temperature of polystyrene (i.e., approaching the glass transition state T_g) in the nanocomposite shows a 32% weight loss in the decomposition phase component, signifying differences in the structural ordering of water inside the polymer. This implies that the interaction parameters in ethanol between the polymer and the IL surface-modified TiO_2 may be dependent on altered repulsive and attractive intermolecular forces of the polymer chains governed by the interaction potential, which is more prominent at lower temperatures far from the T_g. The solution characteristics of polystyrene in ethanol show a different chemical configuration in the absence of solvent doping. Ethanol-induced diffusion of polymer chains increases the cooperativity of cross-linking activity, which proceeds in the presence of phosphoric acid that is undetectable [19]. The dynamic equilibrium of "loosely" and "tightly" bound water molecules particularly in reference to polymer confined spaces on the nanodimensional scale and its relationship to the respective band vibration of bonds between 3000 and 3500 cm^{-1} (O–H stretch) and the "blue shift" to1652 cm^{-1} points to an altered hydrated state. This establishes not only a reorientation in the binding modes of the water molecules but also leads to a possible reordering of neighboring groups affecting the equilibrium of surface states among other different functional groups. The presence of water was established by thermogravimetric analysis and fourier-transform infrared spectroscopy (FTIR), signifying the functional involvement of water in terms of its binding orientation. Based on this evidence, a schematic representation of freely bound, intermediate, and surface-bound water in nanocage assemblies is exemplified in Fig. 16. This scenario becomes apparent if the phase transition ordering is interrupted above the T_g of the polymer

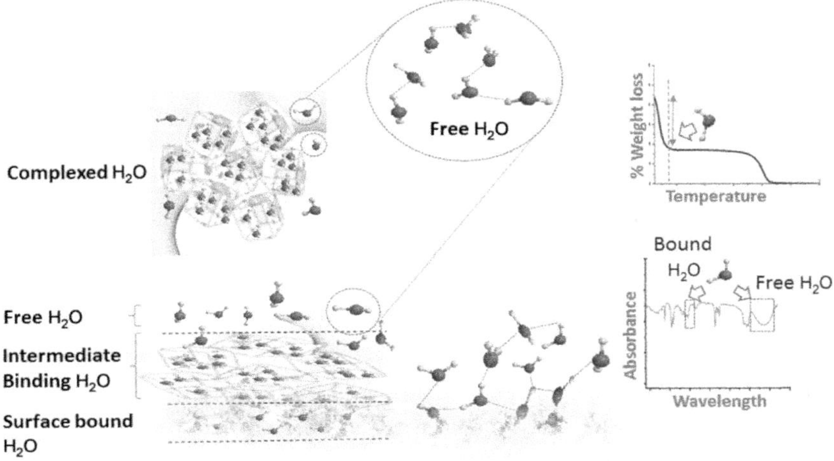

Fig. 16 Complexation of water in PS-P-TiO_2 nanocages. A schematic showing the structural presence of water in PS-P-TiO_2 nanocages complexed as (1) freely bound water, (2) intermediate bound water (between free and surface trapped), or (3) surface bound water.

and redirected to allow for interfacial restructuring through the interpositioning of water molecules assisted by the solvent. One example of this is the increase in the intensity of the conjugated carbonyl band component visible along the regional stretch between 1670–1820 cm^{-1} at 120°C but not at RT, suggesting that the stabilization of water molecules is less critical in the noncrystalline phase demonstrated by TGA. The stretching of the carbonyl bond around 1720 cm^{-1} (in the band range 1650–1800 cm^{-1}) in the vicinity of the H—O—H bending frequency and the blue shift vibrational state of water at 1652 cm^{-1} is representative of a temperature-dependent change and can be correlated to increased hydration with greater flexibility in the rotational freedom of water to the more rigid confinement at the low frequency end at 3550 cm^{-1}. Evidence of the interaction of the hydrocarbon chain of P_{66614} with carbonyl groups in the TiO_2 complex is supported by the FTIR study of the hydration of mono-oleoylglycerol [34], demonstrating structurally the important role of water in forming hydrogen-bonded linkages within the polystyrene matrix.

The noticeable association of stretching of the carbonyl bands with spontaneous increases in both O—H stretching and H—O—H bending vibrations at 120°C in ethanol and the possible existence of energy barriers preventing complete water dissociation potentially indicate that the geometrical interaction of water in the crystallographic planes is arranged differently within the organometallic matrix of noncrystallographic planes. Indeed, the energy burden in releasing chemisorbed water from polymer surfaces is significantly less than expelling polymer-embedded water, suggesting differences in the bonding order of water and the degree of hydration [35].

The interacting water molecules may also exhibit significantly different binding modes such as bending conformations to its more conventional —OH stretching adaptation synonymous with hydrogen bonding activities. Overall, the complete detachment of water from the complex at 450°C supports the idea that water is more loosely bound compared to bulk water or undergoes surface-adsorbed associations. This provides hints that water-based interactions might be directed toward thermosetting events at higher temperatures offset by products of reactions from the catalytic process at the metal-polymer interface. We envisage that loosely bound water molecules internalized in the polymer matrix provide essential spatial ordering of TiO_2.

We conclude that loosely attached lattice water molecules solvent-stabilized by ethanol are critical to the dynamic positioning of styrene monomers in crystalline polystyrene forming extensive hydrogen cavity oriented bonded networks maintaining phosphine-bridged TiO_2 ligands embedded in the polymer matrix different to coordinated water molecular arrangements at RT that are thermally resistant to decomposition up to 300°C. The dependence of interlocked lattice water on the ethanol content shown in the FTIR spectra is correlated to an increase in the absorption peak at $1652\,cm^{-1}$ and the increased state in hydration is accompanied by a reduced peak at $2142\,cm^{-1}$ substantiating a significant change on surface-exposed carboxylate groups.

The close packing density of TiO_2 polystyrene-coated spheres observed in the amorphous phase might form as a function of decreasing radial size aided by three-dimensional confinement of spheres, as reported in the specific case of a hexadecylamine hydrocarbon-coated ZnS nanoparticle system [36]. This further raises geometrical interest in favoring the structural ordering of high surface area exposed semiconducting materials that to some extent exhibit shape-dependent resistance (e.g., spheres versus rods) to polymer-generated repulsive forces in the aqueous surfactant, increasing the volume fraction of light harvesting complexes in devices. The morphological changes observed with increasing ethanol doping continue to be dominated by the kinetic growth of TiO_2 quantum dots above T_g during thermal nucleation before effective growth inhibition occurs within the embedded polystyrene matrix, cooperatively enforced by the inhibitory growth capping tendency of phosphine ligands.

3.3 Solvent-Induced Crystallization of Polymer-Confined TiO_2 at the Quantum Scale: Stimuli-Responsive Induced Structural Polymorphism Shaped by Polymer Tacticity and Surface Ligand Anisotropy

The stimuli responsive properties of the polymeric shell in PS-P-TiO_2 evidenced by morphological changes in the structural properties in ethanol in a concentration-dependent manner correlate well with the dynamic nature of living systems, notably cytoskeleton structures that transition between different states of polymerization [37] with unidirectional functionality. The availability of monomer building blocks originating from substrate (template) disassembly, monomer assembly into polymers and nitrogen-rich ring structures, and aromatic surface ligands is a critical step for TiO_2 confinement supramolecular scale crystal order clusterization to proceed (Fig. 17). We assert that pathway selection by self-assembly occurs via an amorphous route that allows biomimicry of (1) quantum-scale sizing of

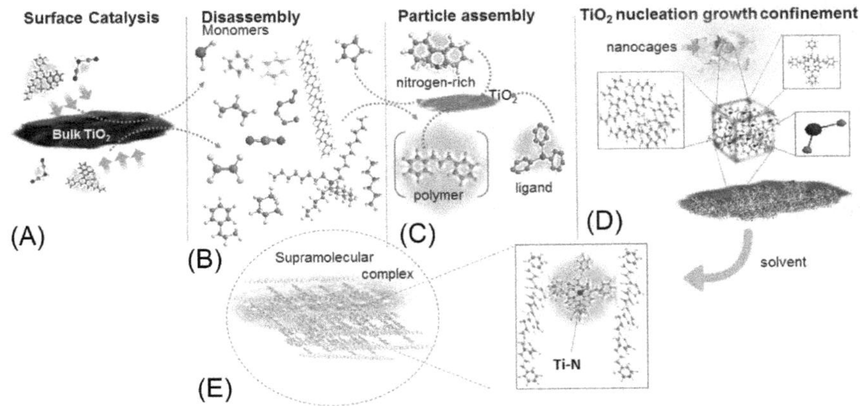

Fig. 17 Schematic illustration of a possible mechanism in the quantum confinement of TiO_2. (A) Bulk phase interaction of a nitrogen-rich environment at the metal oxide TiO_2 surface composed of orderly spaced charge separated molecules. (B) Catalytically driven dissociation of interacting molecules into monomers at the TiO_2 bulk surface followed by (C) particle assembly of functional monomers into larger hierarchical structures. (D) The growth confinement of TiO_2 nucleation via Ti–N bond activation and their amorphous entrapment in the nanocage cavity of self-assembled porphyrin molecules via particle attachment. (E) Solvent-driven supramolecular organometallic assembly forming a crystal complex.

semiconductors, and (2) morphological control via stimuli response as nature intends.

The high degree of solvent induced anisotropy proceeds as function of surface free energy and shows a dependency on solvent concentration in line with the suggestion that molecular motion of polymer chains and trapped water states at temperatures below T_g are dominated by the glassy state perturbation. This signifies the transition of multiple solvent-displaced substates corresponding to a minimum of metastable configurations of sufficiently perturbed mobility owing to the high colloidal viscosity preventing rapid structural metamorphosis of states across kinetic barriers that are otherwise enthalpically accessible. The visible range of crystalline polymorphism by high-resolution microscopy resulting from the controlled addition of solvent to the amorphous phase indicates that the equilibrium fallout characteristic of a glassy disordered state shows reversible annealing of misaligned polymer networks, demonstrating the high state of fragility and low resistance to restructure following solvation. The degree of fragility may indeed be correlated to polymer compression at interfacial sites as a function of crosslinking density. The notable higher-order glass transition in the polymer-TiO_2 complex is indicative of increased stiffness that may originate from restrictions around chain conformations and reduced segmental mobility [38]. Protein function is inherently hierarchical in a nature that strongly converges on structural competence and a regulator of biomolecular activity is often linked to the glass transition state below which many fail to function. This transition to the glass phase mirrors a rigid inflexible state that occurs around 200–220 K for proteins but is broader for synthetic polymers in the range of 140–350 K. These values can be substantially higher for polymer blends from the uncomplexed homogeneous states, altering the fragility index of the matrix.

The extent of the hydrophobic-hydration interactions that conform to a continuum of dynamic events captured in a sequential series of amorphous, semicrystalline, and crystalline phase interrupted intermediates is shown to be chemically immobilized in an ethanol solvent-driven process. The crystalline rearrangement of styrene rings around the phosphine-capped TiO_2 is likely stabilized by lattice water molecules forming an extensive hydrogen bonded network between successive polystyrene monomers, reducing the steric hindrance assisting the self-organization of the freely mobile rings to more stable configurations around the immobilized hydrocarbon backbone and leaving the more accessible ring structures to engage in forming nanocaged TiO_2 extended through dicyanamide bridging networks.

Our work shows that quantum dots (QDs) appear interaggregated in an amorphous PS network forming noncrystalline spherical morphologies at both ambient (RT) and elevated temperatures (120°C).

In contrast, a colloidal suspension of the amorphous material in ethanol revealed an anisotropic behavioral dependency on solvent concentration that progressed from the initiation of the crystallized state of a metal organic framework (MOF) in 10-fold excess of ethanol to a crystalline nanoporous state in 20-fold excess of solvent. The ease in the control of metal-polymer morphology was demonstrated further by varying solvent concentration, generating rectangular nanocages and layered sheet structures in 40-fold and 60-fold ethanol reaction mixtures. The assemblies imaged under different volumes of alcohol were characteristically similar to the sheet-like structures previously seen at 1:60, clearly emphasizing the importance of an ethanol-driven concentration effect in the self-assembly process at the organometallic polymer interface. This observation suggests that the multidirectional nature of the anisotropic self-assembly of a metal-centered homopolymer matrix system [39] might effectively adopt an alternative isotropic behavioral pattern while being equally susceptible to an anisotropic pathway, as reported here triggered by the solvent characteristics. Under isotropic conditioning, a governing critical solvent concentration may be thermodynamically influential in favoring a predominately solvent-assisted assembled structural framework restricting the mobility of the phosphate-doped metal oxide centers that may be further reinforced by the addition of increasing amounts of solvent. These events define a free energy complexation state (ΔG) that is not necessarily offset by the maximum enthalpy (ΔH) and entropy (ΔS) contributions, but is only sufficient to generate and stabilize a series of morphologically trapped structural states. This potentially caters to a series of metastable structures [40] that are thermodynamically distinguished by a succession of energy barriers relaxed through the formation of intermediary states in which the sum total of the minima defines the overall Gibbs free energy of the lowest ground state. The magnitude of the amorphous state is mirrored by the number of defect orientations that are proportional to size of the supramolecular structure and indicative of the number of metastable states exposed by the interactions with solvent molecules. The equilibrium state as a prerequisite to self-assembly suggests that the change in free energy from the redistribution of water molecules at local sites surrounding the amorphous structure might occur as a result of a polarity change introduced by the solvation of ethanol in a concentration-dependent manner. This scenario could significantly alter the rotational and mobile phase of the metastable state, significantly affecting the energy barriers permitting interchangeability between allowed morphological crystallized states. Hence, molecular associations that are essentially thermally generated forces at moderate temperatures as evidenced here that neither show crystalline nor reversible dissociative behavior but highly porous structurally adaptive conformations might offer unprecedented opportunities for solvent-directed tunability.

The importance of electronic and steric orientation in metal-mediated catalysis of PS was demonstrated by a nickel-centered MOF [29] that was shown to require the assistance of ethanol binding at the metal center, preferentially favoring syndiotacticity. As a structural regulator in PS, tacticity likely plays a prominent role in bond selectivity, permitting cross-linking molecules and bridging structures of PS layers and allowing the construction of a supramolecular matrix. This also signifies that the interfacial assembly is critical to the crystalline phase of the PS matrix. This observation is analogous to the copolymer self-assembly of a PS derivative driven by a combined thermal and solvent vapor annealing process [41] initiated above the glass transition temperature (T_g) of the reactants. The nature of a polymer-like PS with the intrinsic ability to supply an internal network of long-range intermolecular bonds that greatly exceeds the radial size of TiO_2 QDs in harmony with repulsive forces can take advantage of thermodynamic scaling effects to provide new material properties. In this way, the manipulation of a kinetically trapped but structurally flexible polymer metal oxide material that is a product of spontaneous chemical ordering via a self-assembly process can be rerouted to form crystallized structures as energy-harnessing materials. If such materials are transitory in nature, the intermolecular forces to which they are subject should be sufficiently strong to tolerate short-term aging but sufficiently weak to chemical alteration on an instantaneous time scale. We found this to be true for the TiO_2 polymer hybrid materials synthesized here.

The lack of polymer crystallinity that defines the amorphous phase likely supports the assembly of an organometallic exhibiting porous coordination polymer (PCP) type characteristics of the rutile phase. An important consideration in the stability of TiO_2 dispersions with polyvinyl derivatives and the colloidal stability in the mixed interface in ethanol and water as solvent and by-product is the lower critical solution temperature (LCST) and the thermoresponsive behavior of the medium. An important observation here, however, is that a precocrystallized state is an inherent feature of the amorphous phase and the paths to crystallization may be dictated by the dominant synthetic route steered by the thermal gradient. The simultaneous cocrystallization of TiO_2 with PS in the assembly of MOFs is a novel phenomenon that has not thus far been reported elsewhere. An alternative route to achieving ligand diversity might occur through ligand flexibility via a responsive stimulus such as temperature in addition to chemisorption of impurity atoms at the metal oxide surface.

There is a growing need to understand the basis of ligand assembly at the semiconductor surface structure on the geometrical shape and order of binding selectivity underlining the phase transformation from the amorphous to the crystalline state in the PS matrix using techniques to study the effects of temperature and colloidal metal stability in ethanol. The morphological diversity of nanostructures triggered in the presence of the precursors trioctylamine, oleic acid, and alkanes has previously shown to be dominated by directional crystal splitting [42]. Here, the amorphous growth of densely populated spherical TiO_2 QDs shows little tendency to diversify structurally. This observation serves as a defining feature of unprohibited stoichiometric access to the potential wealth of nucleating sites in the alkylated phosphonium IL medium in this study. In view of this finding, attention is directed toward the coassembly of PS in influencing the morphological growth pattern, particularly in correlation to the effect of a temperature gradient applied to the samples obtained at RT and 120°C. Phosphine ligand chemistry is implicitly favored as commonly formed catalytic products at the polymer-metal oxide nuclei interface for a broad selection of transition metal oxides. The possibility to trigger alternative routes for the assembly of more novel phosphonium ligand architectures from the interaction with freely accessible open hydrocarbon chain conformations in [P66614][N(CN)2] instead revealed chemical convergence between the bulk [29] and quantum phase that proceeded via the chemical ligation of olefinic (ethylene) products at the metal surface.

As discussed in Section 3.1, nitrogen complexation through N-incorporated precursors at the bulk TiO_2 surface is recognized not only to passivate defect vacancy sites but is also known to reduce the nucleation growth rate that might operate to the same effect as phosphine ligand chemistry. Hence, it is not surprising to find that the catalytic chemistry of the TiO_2 is favorable to the formation of higher-order monomer nitrogen precursor groups facilitated by self-assembly nitrogen-rich cavities at TiO_2 vacant sites. Interestingly, the use of tin pyridinetriphenyl porphyrin (SnPyTriPP) as a precursor molecule has previously been shown to accumulate as the key structural component in the self-assembly of micrometer-sized, hollow-faced nanocubes from nanosheets by a change in the reaction environment under ethanol at 90°C and further cooled without stirring [43]. The striking similarity in the shape geometry of nanoscale cages observed in the present investigation demonstrates the use of solvent-specific polarizability and charge distribution effects within defined thermal boundaries to control the scalability of nanostructures of N-metalated porphyrins. Further, it is plausible to note that the flexibility around polystyrene chains can provide the necessary architectural constraint to accommodate porphyrin rings stabilized by pendant phenyl ring moieties of polystyrene. Solvent-induced tacticity of polystyrene transcending the amorphous to semicrystalline to a crystalline morphology correlates well with the structural rearrangement necessary to allow an ordered polystyrene-porphyrin-TiO_2 organometallic assembly from the disordered state.

4 CONCLUDING COMMENTS

Synthetic processing of supramolecular structures exhibiting biomechanical rigidity and flexibility confined in a dynamic biomimetic framework transitioning from the amorphous phase to size- and shape-tunable crystallizable states of the functional polymer is exciting for driving technological applications. This achievement becomes a more important directive when such materials are coupled to biomimicry at the quantum scale. However, accessibility to hard-to-achieve atomic configurations with new catalytic, optical, and electronic functionalities at the metal polymer interface are from understood. Although the intrinsic nature of TiO_2 is relevant to these findings, the investigation strongly supports the morphological tunability of amorphous materials via template ordering of chemical bonds and composition and important opportunities to understand structure-function relationships of noncrystalline metal-centered anisotropic polymeric materials. We envision that the dominant chemical bonding interactions, shape, and size constraints imposed by the phosphonate IL might be of broader applicability toward improved synthesis of nanometer and subnanometer materials.

Our efforts as an experimental group have focused on developing new synthetic approaches to acquire quantum TiO_2 by introducing new interfacial properties from nonconventional nucleation events that are otherwise inaccessible due to hidden energy barriers. This approach necessarily involves exploring the reaction interfaces of contrasting materials that present sufficiently divergent surfaces that occur at hard and soft interfaces. We show that the resulting interfacial associations take advantage of biomimetic routes via an amorphous phase that provides opportunities to tune material properties in terms of shape using an external stimulant while growth is restricted to the quantum scale through the introduction of new surface chemistry. This approach draws upon a number of parallels between biomineralization and chemically driven synthetic processes that lead to the compartmentalization of TiO_2. It has therefore been possible using a multi-elemental templating to restrict the nucleation growth of TiO_2 using what essentially is a biomimetic strategy. This occurs via an amorphous precursor phase forming nonequilibrium states and progressing to a crystallized phase that is dependent on a number factors, including (1) particle size effects, (2) prenucleation clustering and confinement, (3) precrystallization transient molecular events specific to the amorphous phase, and (4) heterogeneous to homogeneous crystallization. Ligand flexibility at the metal surface and its accessibility to the solvent environment assisted by the spatial crystalline geometrical polymer networks restricting the position and dimensional size of well-nucleated TiO_2 plays a critically important role in the (1) quantum confinement of TiO_2 controlling core particle growth morphology in segregated core-shell systems, (2) structural interdependence on the altered stability of the colloidal phase, and (3) enhancement of the photocatalytic performance of polymerized TiO_2 as a direct consequence of quantum confinement via complexion of host-guest interactions in the ligand polymer supramolecular structure.

In view of the soft matter perspective on protein crystal assembly [44] and our current understanding of protein architecture, the dependency of a three-dimensional biomolecular structure on physical and chemical parameters has provided much insight on synthetic material behavior correlating to mineralization processes and phase change characteristics. Valuable clues to novel synthetic material development at the quantum-nano scale derive from a mixed phase concurrence of amorphous and crystalline states of biogenic origin. This work has important implications in distinguishing the persistence of an H_2O rich bionanomineral phase that precedes a dehydration and crystallization phase change. The loss of the stability phase leading to the formation of an amorphous phase semiconductor (TiO_2) in the glass temperature range of surface-catalyzed polymers provides new routes to tunable geometries that are shape- and size-dependent. Importantly, a periodicity that defines disordered structures does not impede their availability as bioinspired materials with important technological implications.

Moreover, an amorphous supramolecular matrix approach offers an obvious advantage of disengaging the interdependency of size and band gap properties by intervention using unconventional approaches to fine-tune photocatalytic characteristics by altering the intrinsic nature of confined states. The generality of establishing solvent or mixed solvent effects for a stabilized amorphous state can predictably [45] offer a potential route to modify the limits of interaction of HOMO-LUMO orbitals along mobile peripheries of a predominantly tunable amorphous composition.

REFERENCES

[1] D.N. Basov, R.D. Averitt, D. Hsieh, Toward properties on demand in quantum materials, Nat. Mater. 16 (2017) 1077.
[2] S. Sonkaria, S.-H. Ahn, C.S. Lee, V. Khare, "On the Dot"—the timing of self-assembled growth to the quantum scale, Chem. Eur. J. 23 (34) (2017) 8104–8117.
[3] A.V. Radha, T.Z. Forbes, C.E. Killian, P.U.P.A. Gilbert, A. Navrotsky, Transformation and crystallization energetics of synthetic and biogenic amorphous calcium carbonate, Proc. Natl. Acad. Sci. 107 (38) (2010) 16438–16443.
[4] S. Weiner, Y. Levi-Kalisman, S. Raz, L. Addadi, Biologically formed amorphous calcium carbonate, Connect. Tissue Res. 44 (1) (2003) 214–218.
[5] P.J.M. Smeets, K.R. Cho, R.G.E. Kempen, N.A.J.M. Sommerdijk, J.J. De Yoreo, Calcium carbonate nucleation driven by ion binding in a biomimetic matrix revealed by in situ electron microscopy, Nat. Mater. 14 (2015) 394.
[6] J.J. De Yoreo, P.U.P.A. Gilbert, N.A.J.M. Sommerdijk, R.L. Penn, S. Whitelam, D. Joester, H. Zhang, J.D. Rimer, A. Navrotsky, J.F. Banfield, A.F. Wallace, F. M. Michel, F.C. Meldrum, H. Cölfen, P.M. Dove, Crystallization by particle attachment in synthetic, biogenic, and geologic environments, Science 349 (6247) (2015).
[7] R.A. Giacaman, V.A. Perez, C.A. Carrera, Chapter 5—Mineralization processes in hard tissues: teeth, in: C. Aparicio, M.-P. Ginebra (Eds.), Biomineralization and Biomaterials, Woodhead Publishing, Boston, 2016, , pp. 147–185.
[8] P.-A. Fang, J.F. Conway, H.C. Margolis, J.P. Simmer, E. Beniash, Hierarchical self-assembly of amelogenin and the regulation of biomineralization at the nanoscale, Proc. Natl. Acad. Sci. 108 (34) (2011) 14097–14102.
[9] (a) S.-Y. Kwak, F.B. Wiedemann-Bidlack, E. Beniash, Y. Yamakoshi, J.P. Simmer, A. Litman, H.C. Margolis, Role of 20-kDa amelogenin (P148) phosphorylation in calcium phosphate formation in vitro, J. Biol. Chem. 284 (28) (2009) 18972–18979. (b) X. Yang, L. Wang, Y. Qin, Z. Sun, Z.J. Henneman, J. Moradian-Oldak, G. H. Nancollas, How amelogenin orchestrates the organization of hierarchical elongated microstructures of apatite, J. Phys. Chem. B 114 (6) (2010) 2293–2300.
[10] M.H. Nielsen, S. Aloni, J.J. De Yoreo, In situ TEM imaging of $CaCO_3$ nucleation reveals coexistence of direct and indirect pathways, Science 345 (6201) (2014) 1158–1162.
[11] M. Li, L. Wang, W. Zhang, C.V. Putnis, A. Putnis, Direct observation of spiral growth, particle attachment, and morphology evolution of hydroxyapatite, Cryst. Growth Des. 16 (8) (2016) 4509–4518.
[12] M. Kumar, H. Luo, Y. Román-Leshkov, J.D. Rimer, SSZ-13 crystallization by particle attachment and deterministic pathways to crystal size control, J. Am. Chem. Soc. 137 (40) (2015) 13007–13017.
[13] M. Cargnello, T.R. Gordon, C.B. Murray, Solution-phase synthesis of titanium dioxide nanoparticles and nanocrystals, Chem. Rev. 114 (19) (2014) 9319–9345.

[14] A.M. Belcher, X.H. Wu, R.J. Christensen, P.K. Hansma, G.D. Stucky, D.E. Morse, Control of crystal phase switching and orientation by soluble mollusc-shell proteins, Nature 381 (1996) 56.
[15] A.D. Pomogailo, Polymer-immobilized nanoscale and cluster metal particles, Nanostruct. Mater. 12 (1) (1999) 291–294.
[16] A.J. Crosby, J.Y. Lee, Polymer nanocomposites: the "nano" effect on mechanical properties, Polym. Rev. 47 (2) (2007) 217–229.
[17] G.M. Whitesides, M. Boncheva, Beyond molecules: self-assembly of mesoscopic and macroscopic components, Proc. Natl. Acad. Sci. 99 (8) (2002) 4769–4774.
[18] a A.M. Seddon, H.M. Patel, S.L. Burkett, S. Mann, Chiral templating of silica–lipid lamellar mesophase with helical tubular architecture, Angew. Chem. Int. Ed. 41 (16) (2002). b J.S. Chappell, P. Yager, Formation of mineral microstructures with a high aspect ratio from phospholipid bilayer tubules, J. Mater. Sci. Lett. 11 (10) (1992) 633–636.
[19] D. Choi; S. Sonkaria; S. J. Fox; S. Poudel; S.-Y. Kim; S. Kang; S. Kim; C. Verma; S. H. Ahn; C. S. Lee; Khare, V., Quantum scale biomimcry by hierarchical particle attachment: an unusual complex amorphous precursor route to TiO_2 band confinement by interfacial polymer-chain nucleation 2018 (in submission) 2018.
[20] Y.-C. Wang, W. Zhang, L.-Y. Wang, Z. Zhuang, E. Ma, J. Li, Z.-W. Shan, In situ TEM study of deformation-induced crystalline-to-amorphous transition in silicon, NPG Asia Mater. 8 (2016) e291.
[21] F. Wang, R. Tang, J.L.F. Kao, S.D. Dingman, W.E. Buhro, Spectroscopic identification of tri-n-octylphosphine oxide (TOPO) impurities and elucidation of their roles in cadmium selenide quantum-wire growth, J. Am. Chem. Soc. 131 (13) (2009) 4983–4994.
[22] C. Grigoriadis, H. Duran, M. Steinhart, M. Kappl, H.-J. Butt, G. Floudas, Suppression of phase transitions in a confined rodlike liquid crystal, ACS Nano 5 (11) (2011) 9208–9215.
[23] I.-H. Park, A. Chanthapally, Z. Zhang, S.S. Lee, M.J. Zaworotko, J.J. Vittal, Metal–organic organopolymeric hybrid framework by reversible [2+2] cycloaddition reaction, Angew. Chem. Int. Ed. 53 (2) (2014) 414–419.
[24] I.-H. Park, R. Medishetty, H.-H. Lee, C.E. Mulijanto, H.S. Quah, S.S. Lee, J.J. Vittal, Formation of a syndiotactic organic polymer inside a MOF by a [2+2] photo-polymerization reaction, Angew. Chem. Int. Ed. 54 (25) (2015) 7313–7317.
[25] T. Uemura, Y. Ono, K. Kitagawa, S. Kitagawa, Radical polymerization of vinyl monomers in porous coordination polymers: nanochannel size effects on reactivity, molecular weight, and stereostructure, Macromolecules 41 (1) (2008) 87–94.
[26] E.M. Wong, P.G. Hoertz, C.J. Liang, B.-M. Shi, G.J. Meyer, P.C. Searson, Influence of organic capping ligands on the growth kinetics of ZnO nanoparticles, Langmuir 17 (26) (2001) 8362–8367.
[27] C. Gunanathan, D. Milstein, Metal-ligand cooperation by aromatization-dearomatization: a new paradigm in bond activation and "green" catalysis, Acc. Chem. Res. 44 (8) (2011) 588–602.
[28] M. Hood, M. Mari, R. Muñoz-Espí, Synthetic strategies in the preparation of polymer/inorganic hybrid nanoparticles, Materials 7 (5) (2014) 4057.
[29] E. Duquesne, J. Habimana, P. Degée, P. Dubois, Nickel-catalyzed supported ATRP of methyl methacrylate using cross-linked polystyrene triphenylphosphine as ligand, Macromolecules 38 (24) (2005) 9999–10006.
[30] S. Lwin, I.E. Wachs, Olefin metathesis by supported metal oxide catalysts, ACS Catal. 4 (8) (2014) 2505–2520.

[31] S. Muthusamy, D. Azhagan, Titanium isopropoxide promoted tandem self-cross and ring-closing metathesis approach for the synthesis of macrotetralides, Eur. J. Org. Chem. 2014 (2) (2014) 363–370.
[32] C.J. Schram, L.S. Taylor, S.P. Beaudoin, Influence of polymers on the crystal growth rate of felodipine: correlating adsorbed polymer surface coverage to solution crystal growth inhibition, Langmuir 31 (41) (2015) 11279–11287.
[33] Y. Gao, R. Broersen, W. Hageman, N. Yan, M.C. Mittelmeijer-Hazeleger, G. Rothenberg, S. Tanase, High proton conductivity in cyanide-bridged metal-organic frameworks: understanding the role of water, J. Mater. Chem. A 3 (44) (2015) 22347–22352.
[34] A. Nilsson, A. Holmgren, G. Lindblom, An FTIR study of the hydration and molecular ordering at phase transitions in the monooleoylglycerol/water system, Chem. Phys. Lipids 71 (2) (1994) 119–131.
[35] A. Michaelides, A. Alavi, D.A. King, Different surface chemistries of water on Ru {0001}: from monomer adsorption to partially dissociated bilayers, J. Am. Chem. Soc. 125 (9) (2003) 2746–2755.
[36] M. Akbulut, A.R.G. Alig, Y. Min, N. Belman, M. Reynolds, Y. Golan, J. Israelachvili, Forces between surfaces across nanoparticle solutions: role of size, shape, and concentration, Langmuir 23 (7) (2007) 3961–3969.
[37] S. Sonkaria, G. Fuentes, C. Verma, R. Narang, V. Khare, A. Fischer, D. Faivre, Insight into the assembly properties and functional organisation of the magnetotactic bacterial actin-like homolog, MamK, PLoS One 7 (5) (2012) e34189.
[38] G. Choudalakis, A.D. Gotsis, Free volume and mass transport in polymer nanocomposites, Curr. Opin. Colloid Interface Sci. 17 (3) (2012) 132–140.
[39] P. Akcora, H. Liu, S.K. Kumar, J. Moll, Y. Li, B.C. Benicewicz, L.S. Schadler, D. Acehan, A.Z. Panagiotopoulos, V. Pryamitsyn, V. Ganesan, J. Ilavsky, P. Thiyagarajan, R.H. Colby, J.F. Douglas, Anisotropic self-assembly of spherical polymer-grafted nanoparticles, Nat. Mater. 8 (4) (2009) 354–359.
[40] V.V. Brazhkin, Metastable phases and 'metastable' phase diagrams, J. Phys. Condens. Matter 18 (42) (2006) 9643.
[41] S.S. Dinachali, W. Bai, K.-H. Tu, H.K. Choi, J. Zhang, M.E. Kreider, L.-C. Cheng, C. A. Ross, Thermo-solvent annealing of polystyrene-polydimethylsiloxane block copolymer thin films, ACS Macro Lett. 4 (5) (2015) 500–504.
[42] A.T. Kelly, I. Rusakova, T. Ould-Ely, C. Hofmann, A. Lüttge, K.H. Whitmire, Iron phosphide nanostructures produced from a single-source organometallic precursor: nanorods, bundles, crosses, and spherulites, Nano Lett. 7 (9) (2007) 2920–2925.
[43] Z. Wang, Z. Li, C.J. Medforth, J.A. Shelnutt, Self-assembly and self-metallization of porphyrin nanosheets, J. Am. Chem. Soc. 129 (9) (2007) 2440–2441.
[44] D. Fusco, P. Charbonneau, Soft matter perspective on protein crystal assembly, Colloids Surf. B: Biointerfaces 137 (2016) 22–31.
[45] M.T. Lusk, R.T. Collins, Z. Nourbakhsh, H. Akbarzadeh, Quantum confinement of nanocrystals within amorphous matrices, Phys. Rev. B 89 (7) (2014) 075433.

INDEX

Note: Page numbers followed by *f* indicate figures and *t* indicate tables.

E

F

G

H

I

U

V

W

CPI Antony Rowe
Chippenham, UK
2018-12-12 18:29